TechOne: Automotive Engine Repair

TechOne: Automotive Engine Repair

Elisabeth H. Dorries

Professor of Automotive Technology
Vermont Technical College
Randolph Center, VT

Jack Erjavec, Series Editor

Professor Emeritus,
Columbus State Community College
Columbus, OH

THOMSON
DELMAR LEARNING

Australia Canada Mexico Singapore Spain United Kingdom United States

TechOne: Automotive Engine Repair

Elisabeth H. Dorries

Vice President, Technology and Trades SBU:
Alar Elken

Editorial Director:
Sandy Clark

Senior Acquisitions Editor:
David Boelio

Developmental Editor:
Matthew Thouin

Marketing Director:
Dave Garza

Channel Manager:
William Lawrensen

Marketing Coordinator:
Mark Pierro

Production Director:
Mary Ellen Black

Production Editor:
Barbara L. Diaz

Art/Design Specialist:
Cheri Plasse

Technology Project Manager:
Kevin Smith

Technology Project Specialist:
Linda Verde

Editorial Assistant:
Andrea Domkowski

Printed in the United States of America
1 2 3 4 5 XX 05 04 03

For more information contact
Thomson Delmar Learning
Executive Woods
5 Maxwell Drive, PO Box 8007,
Clifton Park, NY 12065-8007
Or find us on the World Wide
Web at www.delmarlearning.com

Library of Congress Cataloging-in-Publication Data:

Dorries, Elisabeth H.
TechOne : automotive engine repair / Elisabeth H. Dorries
p. cm.
Includes index.
ISBN 1-4018-5941-0
1. Automobiles—Motors—Maintenance and repair. I. Title.
TL210.D645 2005
629.15'028'8—dc22 2004057974

NOTICE TO THE READER

Contents

Preface

THE SERIES

Welcome to Thomson Delmar Learning's *TechOne*, a state-of-the-art series designed to respond to today's automotive instructor's and students' needs. *TechOne* offers current, concise information on ASE and other specific subject areas, combining classroom theory, diagnosis, and repair into one easy-to-use volume.

You'll notice several differences from a traditional textbook. First, a large number of short chapters divide complex material into small chunks. Instructors can give tight, detailed reading assignments that students will find easy to digest. These short chapters can be taught in almost any order, allowing instructors to pick and choose the material that best reflects the depth, direction, and pace of their individual classes.

TechOne also features an art-intensive approach to suit today's visual learners—images drive the chapters. From drawings to photos, you will find more art to better understand the systems, parts, and procedures under discussion. Look also for helpful graphics that draw attention to key points in features like You Should Know and Interesting Fact.

Just as important, each *TechOne* starts off with a section on Safety and Communication, which stresses safe work practices, tool competence, and familiarity with workplace "soft skills," such as customer communication and the roles necessary to succeed as an automotive technician. From there learners are ready to tackle the technical material in successive sections, ultimately leading them to the real test—an ASE practice exam in the Appendix.

THE SUPPLEMENTS

TechOne comes with an **Instructor's Manual** that includes answers to all chapter-end review questions and a complete correlation of the text to NATEF standards. A **CD-ROM**, included with each Instructor's Manual, contains **PowerPoint Slides** for classroom presentations, a **Computerized Testbank** with hundreds of questions to aid in creating tests and quizzes, and an electronic version of the Instructor's Manual. Chapter-end review questions from the text have also been redesigned into adaptable **Electronic Worksheets**, so instructors can modify questions if desired to create in-class assignments or homework.

Flexibility is the key to *TechOne*. For those who would like to purchase jobsheets, Thomson Delmar Learning's NATEF Standards Job Sheets are a good match. Topics cover the eight ASE subject areas and include:

- Engine Repair
- Automatic Transmissions and Transaxles
- Manual Drivetrains and Axles
- Suspension and Steering
- Brakes
- Electrical and Electronic Systems
- Heating and Air Conditioning
- Engine Performance

  Plus,
- Advanced Engine Performance
- Fuels and Emissions

Visit **http://www.autoed.com** for a complete catalog.

OTHER TITLES IN THIS SERIES

TechOne is Thomson Delmar Learning's latest automotive series. We are excited to announce these future titles:

- Automatic Transmissions
- Suspension and Steering
- Heating and Air Conditioning
- Advanced Automotive Electronic Systems
- Advanced Engine Performance
- Automotive Fuels & Emissions

  Check with your sales representative for availability.

A NOTE TO THE STUDENT

There are now more computers on a car than aboard the first spacecraft, and even gifted backyard mechanics long ago turned their cars over to automotive professionals for diagnosis and repair. That's a statement about the nation's need for the knowledge and skills you'll develop as you continue your studies. Whether you eventually choose a career as a certified or licensed technician, service writer or manager, or automotive engineer—or even decide to open your own shop—hard work will give you the opportunity to become one of the 840,000 automotive professionals providing and maintaining safe and efficient automobiles on our roads. As a member of a technically proficient, cutting-edge workforce, you'll fill a need, and, even better, you'll have a career to feel proud of.

Best of luck in your studies,

The Editors of Thomson Delmar Learning

About the Author

Elisabeth H. Dorries has over twenty years of experience in the automotive industry. She "turned a wrench" full-time for the first ten of those years on both foreign and domestic vehicles in dealerships and independent repair facilities. Betsy has been a full-time professor in the automotive department at Vermont Technical College for the past thirteen years. She has led training courses internationally on the OBDII system and emissions-related repairs for technicians and emissions testing program managers. Betsy is a cofounder and director of the Vermont Center for Emissions Repair and Technician Training (VCERTT). She writes curricula, offers training courses, and trains trainers who deliver OBDII repair training to technicians across the state. In her spare time she enjoys being outside in Vermont with her family.

Betsy holds a number of ASE certifications. She is a:

Master Recertified Automobile Technician

Master Recertified Medium/Heavy Truck Technician

Master Recertified Engine Machinist

Recertified Advanced Engine Performance Specialist

Recertified Advanced Truck Electronic Diesel Engine Diagnosis Specialist

Recertified Alternative Fuel: CNG Specialist, and

Certified School Bus Technician

Betsy is also a member of the Society of Automotive Engineers (SAE), the Service Technicians Society (STS), and the International Automotive Technicians Network (IATN), and she is an active member in the North American Council of Automotive Teachers (NACAT).

Acknowledgments

My students have driven this textbook; it would not exist without them. Their persistent questions and curiosity over the past thirteen years have inspired endless conversations and forced me to come up with multiple sets of straightforward answers and explanations. This book is part of that ongoing conversation with students about how engines make power and how we can repair them. I offer my thanks to Thomson Delmar Learning for giving me a chance to write it down.

I would also like to thank Vermont Technical College for offering support for my professional endeavors, including the development of this textbook. The time and funding they provide for professional development has allowed me to be involved in many exciting and productive learning opportunities and projects.

Last, but certainly not least, I want to acknowledge my family. I give deep thanks to Dana, Kate, Alex, William, and Charlie for their patience and support. I owe them many sledding parties.

I'd like to thank the following reviewers, whose technical expertise was invaluable in creating this text:

Dave Crowley
Community College of Southern Nevada
N. Las Vegas, Nevada

Larry Leavitt
Pennsylvania College of Technology
Williamsport, Pennsylvania

Michael Longrich
Cuyahoga Community College
Cleveland, Ohio

Donny Seyfer
Seyfer Automotive
Wheat Ridge, Colorado

John Thorp
Illinois Central College
East Peoria, Illinois

Mitchell Walker
St. Louis Community College
St. Louis, Missouri

John Wood
Ranken Technical College
St. Louis, Missouri

Features of the Text

TechOne includes a variety of learning aids designed to encourage student comprehension of complex automotive concepts, diagnostics, and repair. Look for these helpful features:

Section Openers provide students with a **Section Table of Contents** and **Objectives** to focus the learner on the section's goals.

Interesting Facts spark student attention with industry trivia or history. Interesting facts appear on the section openers and are then scattered throughout the chapters to maintain reader interest.

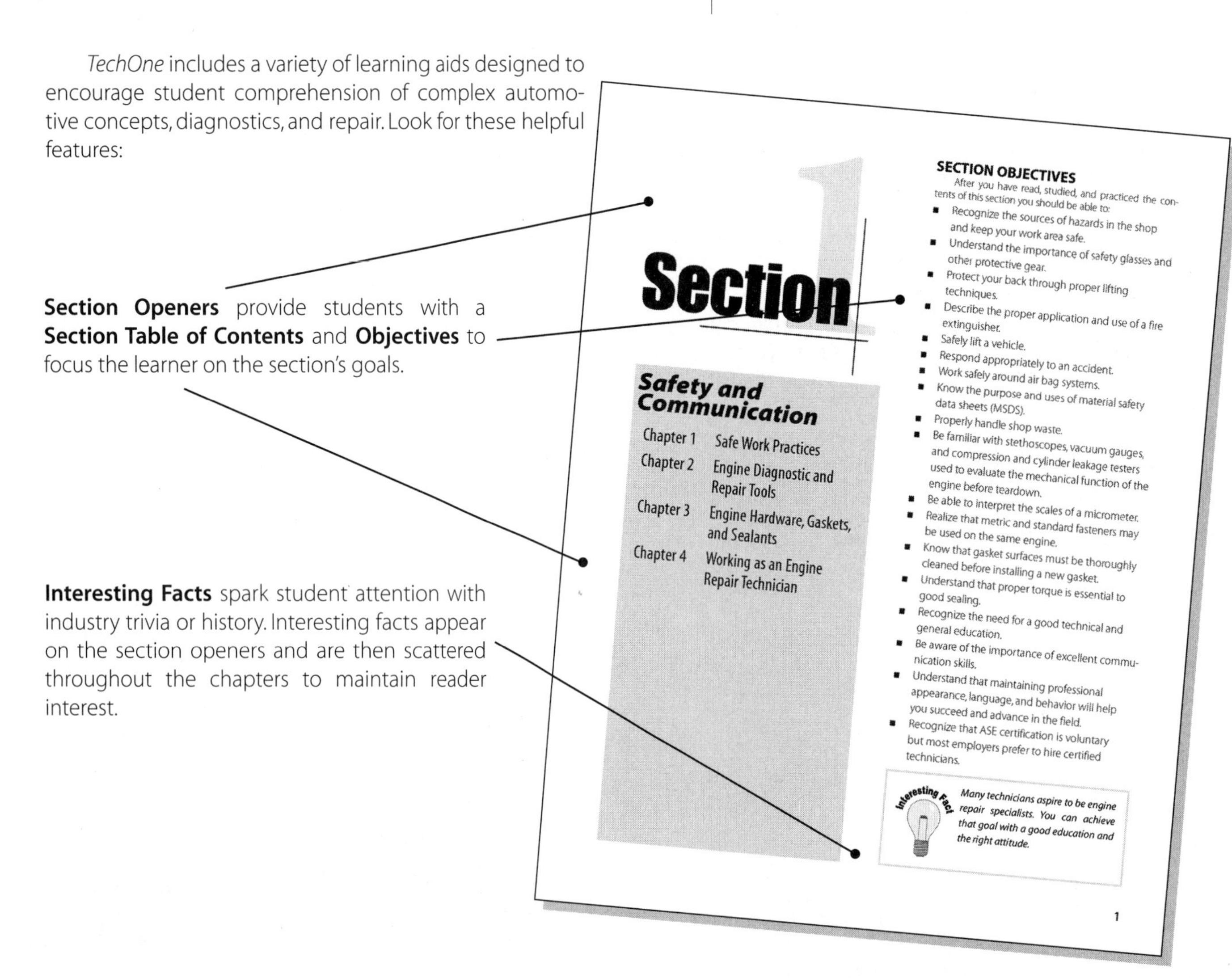
Section 1

Safety and Communication

Chapter 1 Safe Work Practices
Chapter 2 Engine Diagnostic and Repair Tools
Chapter 3 Engine Hardware, Gaskets, and Sealants
Chapter 4 Working as an Engine Repair Technician

SECTION OBJECTIVES

After you have read, studied, and practiced the contents of this section you should be able to:

- Recognize the sources of hazards in the shop and keep your work area safe.
- Understand the importance of safety glasses and other protective gear.
- Protect your back through proper lifting techniques.
- Describe the proper application and use of a fire extinguisher.
- Safely lift a vehicle.
- Respond appropriately to an accident.
- Work safely around air bag systems.
- Know the purpose and uses of material safety data sheets (MSDS).
- Properly handle shop waste.
- Be familiar with stethoscopes, vacuum gauges, and compression and cylinder leakage testers used to evaluate the mechanical function of the engine before teardown.
- Be able to interpret the scales of a micrometer.
- Realize that metric and standard fasteners may be used on the same engine.
- Know that gasket surfaces must be thoroughly cleaned before installing a new gasket.
- Understand that proper torque is essential to good sealing.
- Recognize the need for a good technical and general education.
- Be aware of the importance of excellent communication skills.
- Understand that maintaining professional appearance, language, and behavior will help you succeed and advance in the field.
- Recognize that ASE certification is voluntary but most employers prefer to hire certified technicians.

Interesting Fact

Many technicians aspire to be engine repair specialists. You can achieve that goal with a good education and the right attitude.

1

Timing Mechanism Inspection

Introduction

In order to prepare for valve timing mechanism service you must carefully inspect the components and decide which need replacement. Even when the timing mechanism is being replaced during a major engine overhaul you should carefully inspect for component damage to determine whether there was a particular cause of the wear. Belts, chains, and gears must be inspected for wear. Tensioners, sprockets, and guides must also be thoroughly evaluated to make a good decision about which components should be replaced. You want to provide repairs in a cost-effective manner but without the risk of premature failure of your work. In many cases you will replace multiple components in the timing mechanism. You should use the manufacturer's specific procedures when that information is available or if you have any doubt about your diagnosis. There are also industry-standard procedures for determining the condition of belts, chains, and gears that we will discuss here. This chapter will give you an overview of the inspection process.

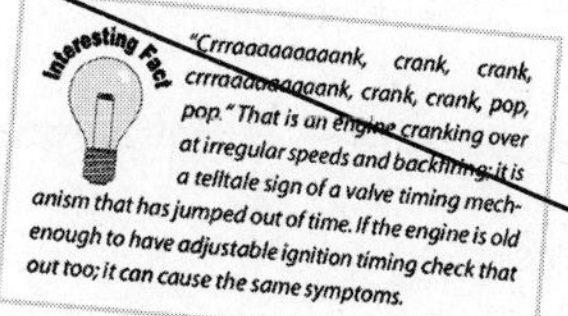

"Crrraaaaaaaaank, crank, crank, crrraaaaaaaaank, crank, crank, pop, pop." That is an engine cranking over at irregular speeds and backfiring; it is a telltale sign of a valve timing mechanism that has jumped out of time. If the engine is old enough to have adjustable ignition timing check that out too; it can cause the same symptoms.

SYMPTOMS OF A WORN TIMING MECHANISM

A worn timing chain may slap against the cover or against itself on a quick deceleration and make a rattling noise. Use a stethoscope in the area of the chain to confirm your preliminary diagnosis, and then proceed with the checks described in the next section to confirm the extent of the damage **(Figure 1)**.

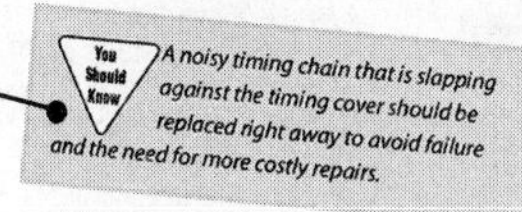

A noisy timing chain that is slapping against the timing cover should be replaced right away to avoid failure and the need for more costly repairs.

A customer may also report that the engine seems to have poor acceleration from startup but that it runs well, perhaps even better than usual, at higher rpms. When there is slack in the chain the camshaft timing is behind the crank. This is called retarded valve timing. This improves high-end performance while sacrificing low-end responsiveness.

Timing gears may clatter on acceleration and deceleration. The engine does not have to be under a load; just snap the throttle open and closed while listening under the cover for gear clatter. The customer is unlikely to notice the reduced low-end performance before the gears break from too much **backlash**.

Customers with worn or cracked timing belts will notice nothing until their engine stops running or begins running very poorly. What you and they should be paying close attention to is the mileage on the engine and the recommended belt service interval.

239

An **Introduction** orients readers at the beginning of each new chapter. **You Should Know** informs the reader whenever special safety cautions, warnings, or other important points deserve emphasis.

Technical Terms are bolded in the text upon first reference and are defined.

Chapter 3 Engine Hardware, Gaskets, and Sealants • 33

Summary

- Standard and/or metric specifications are given in engine repair procedures.
- Metric and standard fasteners may both be used on the same engine. They cannot be interchanged.
- Proper tightening of fasteners is critical to quality repairs.
- Torque to yield bolts cannot be reused.
- A tap and die set can be used to repair some fasteners.
- Gaskets may be made of paper, fiber, cork, synthetic rubber, steel, or any combination of the compounds.
- Gasket surfaces must be thoroughly cleaned before installing a new gasket.
- Seals are used when one of the parts rotates.
- Use a seal driver when installing seals to prevent premature failure.
- RTV sealant is an aerobic sealing compound used frequently on gasket seams or in the place of gaskets.
- Anaerobic sealants are used between two machined or cast surfaces.
- Thread sealant should be used on "wet" head or water pump bolts.

Review Questions

1. How many millimeters are in one inch?
2. To convert millimeters to inches, multiply millimeters by: ________.
3. What are the three common types of standard bolt threads?
4. What fastener can be used to hold a component on a rotating shaft?
   A. A stud
   B. A UNF bolt
   C. A key
   D. A nut
5. Technician A says that a torque to yield bolt is rotated after the initial torque specification. Technician B says that torque to yield bolts cannot be reused. Who is correct?
   A. Technician A only
   B. Technician B only
   C. Both Technician A and Technician B
   D. Neither Technician A nor Technician B
6. Technician A says that most head gaskets should be sealed with a continuous bead of RTV. Technician B says that RTV can form a gasket in some applications. Who is correct?
   A. Technician A only
   B. Technician B only
   C. Both Technician A and Technician B
   D. Neither Technician A nor Technician B
7. Technician A says that aerobic sealants can set before parts are assembled. Technician B says that anaerobic sealants set when all the air is squeezed out between the two components. Who is correct?
   A. Technician A only
   B. Technician B only
   C. Both Technician A and Technician B
   D. Neither Technician A nor Technician B
8. Technician A says to tap a seal in around its edge with a hammer. Technician B says that sometimes a socket can be used to safely install a seal. Who is correct?
   A. Technician A only
   B. Technician B only
   C. Both Technician A and Technician B
   D. Neither Technician A nor Technician B
9. Technician A says that modern head gaskets are often multilayered steel. Technician B says that some head gaskets have a top marking on them. Who is correct?
   A. Technician A only
   B. Technician B only
   C. Both Technician A and Technician B
   D. Neither Technician A nor Technician B
10. Technician A says that many oil pan gaskets are made of synthetic rubber. Technician B says to retighten a cork gasket to stop any leakage. Who is correct?
    A. Technician A only
    B. Technician B only
    C. Both Technician A and Technician B
    D. Neither Technician A nor Technician B

A **Summary** concludes each chapter in short, bulleted sentences. **Review Questions** are structured in a variety of formats, including ASE style, challenging students to prove they've mastered the material.

An **ASE Practice Exam** is found in the **Appendix** of every *TechOne* book, followed by a **Bilingual Glossary,** which offers Spanish translations of technical terms alongside their English counterparts.

Appendix A

ASE PRACTICE EXAM FOR ENGINE REPAIR

1. Technician A says that you should know the location of the eye wash station. Technician B says that you should wear work gloves when grinding or cleaning on a wire wheel. Who is correct?
   A. Technician A only
   B. Technician B only
   C. Both Technician A and Technician B
   D. Neither Technician A nor Technician B
2. You can use each of the following tools to evaluate an engine's mechanical condition *except* a:
   A. Brush hone
   B. Compression tester
   C. Vacuum gauge
   D. Stethoscope
3. Technician A says to use a micrometer to measure a component accurately to .0001 in. Technician B says that an error in measurement of .001 in. during an engine overhaul could cause serious problems. Who is correct?
   A. Technician A only
   B. Technician B only
   C. Both Technician A and Technician B
   D. Neither Technician A nor Technician B
4. Technician A says to torque bolts in a circular pattern around the pan or component. Technician B says that she has learned to gauge proper torque with an ordinary ratchet. Who is correct?
   A. Technician A only
   B. Technician B only
   C. Both Technician A and Technician B
   D. Neither Technician A nor Technician B
5. Technician A says that you may find both metric and standard fasteners on the same vehicle. Technician B says that you can use standard wrenches on either type of bolt. Who is correct?
   A. Technician A only
   B. Technician B only
   C. Both Technician A and Technician B
   D. Neither Technician A nor Technician B
6. Technician A says that good communication skills can make the difference between a good technician and an excellent technician. Technician B says that most of the diagnostic and repair skills needed to repair today's cars are learned on the job. Who is correct?
   A. Technician A only
   B. Technician B only
   C. Both Technician A and Technician B
   D. Neither Technician A nor Technician B
7. Technician A says that a bigger engine makes more power than a smaller one. Technician B says that the more air an engine can take in and exhaust, the more power it can make. Who is correct?
   A. Technician A only
   B. Technician B only
   C. Both Technician A and Technician B
   D. Neither Technician A nor Technician B
8. Technician A says that the intake and exhaust valves should be closed when the spark plug fires. Technician B says that both valves are open at the end of the exhaust stroke. Who is correct?
   A. Technician A only
   B. Technician B only
   C. Both Technician A and Technician B
   D. Neither Technician A nor Technician B

287

Bilingual Glossary

Air Bag System (SRS) A safety restraint system using inflatable bags to keep the passenger from hitting the dashboard or the driver from hitting the steering wheel.
Sistema de bolsa de aire (SBA) *Sistema de restricción que utiliza bolsas inflables para prevenir que el pasajero se golpee en el tablero o que el conductor se golpee en el volante.*

Air Conditioner Compressor A pump that cycles refrigerant throughout the air conditioning system.
Compresor del aire acondicionado *Bomba que hace circular el refrigerante por todo el sistema de aire acondicionado.*

American Petroleum Institute (API) The API rates the quality of motor oil and upgrades standards periodically to designate current oils with the highest rating.
Instituto Americano del Petróleo (IPA) *El IPA designa el grado de calidad del aceite para el motor y mejora los estándares periódicamente para designar los aceites actualizados con el máximo rendimiento.*

Anaerobic Sealant A sealant that will only cure in the absence of oxygen and takes up no space.
Sellador anaeróbico *Sellador que sólo cura si hay ausencia de oxígeno y que no ocupa espacio.*

Axle A shaft that connects the drive train to the wheel hub.
Eje *Árbol que conecta el tren de accionamiento al cubo de la rueda.*

Backlash The clearance between two gears.
Contrapresión *Espacio entre dos engranajes.*

Balance Shaft A shaft in the engine that rotates to reduce vibrations in the engine.
Eje de balance *Eje en un motor que da vueltas para reducir las vibraciones de la máquina.*

Ball Joint A flexible pivot to allow for turning a front wheel to the right or to the left and arcing up and down as the wheel encounters bumps.
Cojinete *Pivote flexible que permite dar vueltas a la llanta delantera hacia la derecha o la izquierda y arquearse hacia arriba o hacia abajo cuando la llanta se encuentra con topes.*

Bearing Crush The extension of the bearing half beyond the seat that is crushed into place when the bearing cap is tightened.
Agolpamiento del cojinete *Extensión del balero a la mitad del asiento que se agolpa en su lugar cuando se aprieta la tapa del balero.*

Bearing Inserts An interchangeable type of bearing. The engine bearings are a set of steel backed half rounds with a soft face to protect the shaft that rides in them.
Encartes de cojinete *Un tipo de cojinete intercambiable. Los cojinetes del motor son un juego de media rueda con soporte de acero con cara suave para proteger el árbol montado sobre los mismos.*

Bearing Spread The distance between the outside parting edges is larger than the diameter of the bore.
Extensión del cojinete *La distancia entre las dos orillas de separación es mayor que el diámetro de la barrena.*

Belt Tension Gauge Used to measure drive belt tension. The belt tension gauge is installed over the belt and indicates the amount of belt tension.
Manómetro de tensión de la banda *Se usa para medir la tensión de la banda de la trasmisión. El manómetro de tensión de la banda se instala sobre la banda e indica la cantidad de tensión que hay en ella.*

Bleeder Valve A valve that opens to allow removal of air from a hydraulic system.
Válvula purgadora *Válvula que se abre para permitir la salida del aire del sistema hidráulico.*

Blow-by The unburned fuel and combustion by-products that leak past the piston rings and enter the crankcase.
Soplado *Subproductos de combustión y de carburantes no quemados que escapan por los anillos del pistón y que entran en el cárter.*

295

A comprehensive **Index** helps instructors and students pinpoint information in the text.

Index

309

Safety and Communication

SECTION OBJECTIVES

After you have read, studied, and practiced the contents of this section you should be able to:

- Recognize the sources of hazards in the shop and keep your work area safe.
- Understand the importance of safety glasses and other protective gear.
- Protect your back through proper lifting techniques.
- Describe the proper application and use of a fire extinguisher.
- Safely lift a vehicle.
- Respond appropriately to an accident.
- Work safely around air bag systems.
- Know the purpose and uses of material safety data sheets (MSDS).
- Properly handle shop waste.
- Be familiar with stethoscopes, vacuum gauges, and compression and cylinder leakage testers used to evaluate the mechanical function of the engine before teardown.
- Be able to interpret the scales of a micrometer.
- Realize that metric and standard fasteners may be used on the same engine.
- Know that gasket surfaces must be thoroughly cleaned before installing a new gasket.
- Understand that proper torque is essential to good sealing.
- Recognize the need for a good technical and general education.
- Be aware of the importance of excellent communication skills.
- Understand that maintaining professional appearance, language, and behavior will help you succeed and advance in the field.
- Recognize that ASE certification is voluntary but most employers prefer to hire certified technicians.

Many technicians aspire to be engine repair specialists. You can achieve that goal with a good education and the right attitude.

Safe Work Practices

Introduction

Your safety is essential to your productive and profitable career as an automotive technician. A workplace injury can have devastating effects on your personal and professional life. You will be working in a shop with other technicians; you and they share a responsibility to keep the workplace safe. Most accidents and unsafe practices can be avoided. You need to be aware of the potential hazards in the shop and diligently protect yourself and your coworkers. This chapter will explain common workplace hazards and professional, safe practices.

Interesting Fact

Back injuries are common but largely preventable in the automotive industry. If you lift properly and use engine cranes and suitable levers you can protect yourself from injury. Engine work requires physical strength; protect your back.

SAFE WORK AREAS

Your work area should be kept clean and safe. The floor and bench tops should be clean, dry, and orderly. Any oil, coolant, or grease on the floor will make it slippery. Falling can result in serious injuries and time lost from work. To clean up oil use commercial oil absorbent. Keep all water off the floor. Water is slippery on smooth floors, and electricity flows well through water. Aisles and walkways should be kept clean and wide enough to move through easily.

Make sure all drain covers are snugly in place. Open drains or covers that are not flush with the floor can cause toe, ankle, and leg injuries.

Shop safety is the responsibility of everyone in the shop. Everyone must work together to protect the health and welfare of all who work in the shop. Be sure you hold up your part in the team. Report any unsafe equipment or practices to the shop manager.

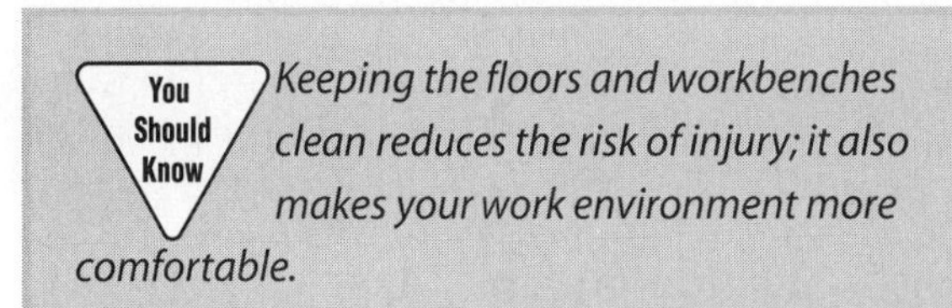

You Should Know

Keeping the floors and workbenches clean reduces the risk of injury; it also makes your work environment more comfortable.

PERSONAL SAFETY

Personal safety simply involves those precautions you take to protect yourself from injury. Your eyes can become infected or permanently damaged by many things in a shop. *Always* wear eye protection when you are working in the shop. There are many types of comfortable eye protection available. To provide adequate eye protection safety glasses have lenses made of safety glass. They also offer some sort of side protection. For nearly all services performed on the vehicle eye protection **(Figure 1)** should be worn.

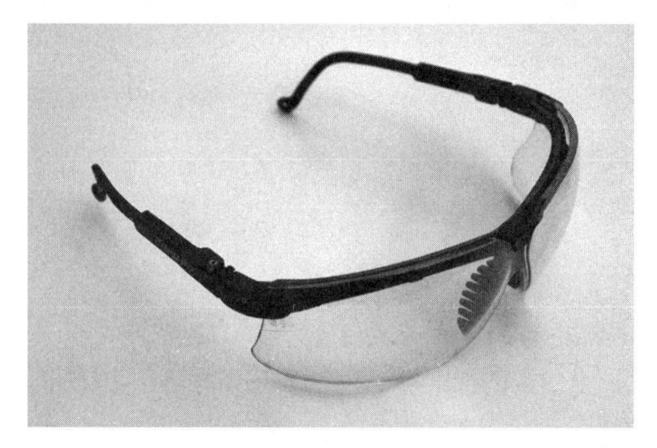

Figure 1. Find a comfortable pair of safety glasses that you will wear consistently.

> You Should Know *Each year hundreds of technicians suffer preventable eye injuries. A metal splinter projected while chiseling can easily implant itself in your eye. This causes serious pain, lost work, and potentially permanent damage.*

Some procedures may require that you wear other eye protection in addition to safety glasses. For example, when you are cleaning parts with a pressurized spray you should wear a face shield. A face shield not only provides added protection for your eyes but also protects the rest of your face. If you are welding or heating and cutting with torches be sure to use approved masks to prevent burning your eyes.

If chemicals such as battery acid, fuel, or solvents get into your eyes flush them continuously with clean water. Have someone call a doctor and get medical help immediately. Many shops have eye wash stations or safety showers that should be used whenever you or someone else has been sprayed or splashed with a chemical.

Your clothing should be well fitted and comfortable but made with strong material. If you have long hair tie it back or tuck it under a cap. Never wear rings, watches, bracelets, and neck chains. These can easily get caught in moving parts and cause serious injury.

Automotive work involves the handling of many heavy objects, which can be accidentally dropped on your feet or toes. Always wear leather or similar material shoes or boots with no-slip soles. Steel-tipped safety shoes can give added protection to your feet.

Good hand protection is often overlooked. A scrape, cut, or burn can limit your effectiveness at work for many days. A well-fitted pair of heavy work gloves should be worn during operations such as grinding and welding or when handling high-temperature components. Always wear approved rubber gloves when handling strong and dangerous caustic chemicals, which can easily burn your skin. Many technicians wear thin, surgical-type latex gloves whenever they are working on vehicles. These offer little protection against cuts but do offer protection against disease and grease buildup under and around your fingernails. These gloves are comfortable and are quite inexpensive.

> You Should Know *Not many years ago few technicians wore gloves; now many technicians wear them all the time. The major tool companies sell both heavy-duty work gloves for serious protection and latex gloves that make tasks as simple as oil changes simpler because you do not have to scrub your hands before getting in the customer's vehicle.*

LIFTING AND CARRYING

You will need to have a strong body for your whole career as a technician. When lifting a heavy object like an engine block use a hoist or have someone else help you. If you must work alone *always* lift heavy objects with your legs, not your back. Crouch down by bending your knees—do not lean over with your back **(Figure 2)**—and securely hold the object you are lifting; then stand up, keeping the object close to you. Trying to "muscle" something with your arms or back can result in severe damage to your back and may end your career and limit what you can do the rest of your life!

FIRE HAZARDS AND PREVENTION

Many items around a typical shop are a potential fire hazard. These include: gasoline, diesel fuel, cleaning solvents, and dirty rags. Each of these should be treated as a

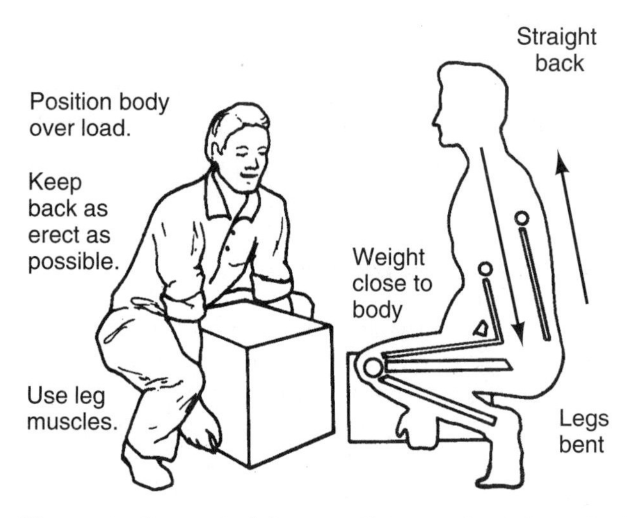

Figure 2. Learn to lift correctly; your back is critical to your prosperous career.

potential firebomb and handled and stored properly.

In case of a fire you should know the location of the fire extinguishers and fire alarms in the shop and should know how to use them. You should also be aware of the different types of fires and the fire extinguishers used to put out these types of fires.

Basically, there are four types of fires: Class A fires are those in which wood, paper, and other ordinary materials are burning; Class B fires are those involving flammable liquids, such as gasoline, diesel fuel, paint, grease, oil, and other similar liquids; and Class C fires are electrical fires. Class D fires are unique; they involve a burning metal. An example of this is a burning "mag" wheel; the magnesium used in the construction of the wheel is a flammable metal and will burn brightly when subjected to high heat. Many shop fire extinguishers work on Class A, B, and C fires. A separate extinguisher is provided for Class D fires **(Figure 3)**.

USING A FIRE EXTINGUISHER

Remember, during a fire never open doors or windows unless it is absolutely necessary; the extra draft will only make the fire worse. Make sure the fire department is contacted before or during your attempt to extinguish a fire. To extinguish a fire stand 6 to 10 feet from the fire. Hold the extinguisher firmly in an upright position. Aim the nozzle at the base, and use a side-to-side motion, sweeping the entire width of the fire. Stay low to avoid inhaling the smoke. If it gets too hot or too smoky get out. Remember: never go back into a burning building for anything. To help remember how to use an extinguisher remember the word "PASS."

Pull the pin from the handle of the extinguisher.
Aim the extinguisher's nozzle at the base of the fire.
Squeeze the handle.
Sweep the entire width of the fire with the contents of the extinguisher.

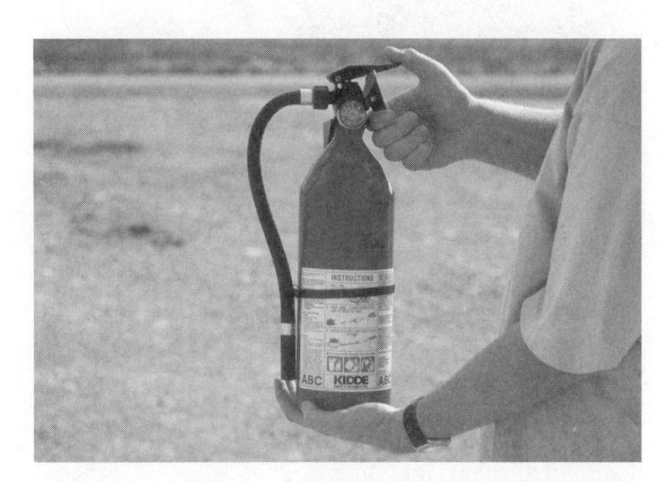

Figure 3. This fire extinguisher can be used on Class A, B, and C fires.

If there is no fire extinguisher handy a blanket or fender cover may be used to smother the flames. You must be careful when doing this because the heat of the fire may burn you and the blanket. If the fire is too great to smother move everyone away from the fire, and call the local fire department. A simple under-the-hood fire can cause the total destruction of the car and the building and can take some lives. You must be able to respond quickly and effectively to avoid a disaster.

SAFE TOOLS AND EQUIPMENT

Whenever you are using equipment make sure that you use it properly and that it is set up according to the manufacturer's instructions. All equipment should be properly maintained and periodically inspected for unsafe conditions. Frayed electrical cords or loose mountings can cause serious injuries. All electrical outlets should be equipped to allow for the use of three-pronged electrical cords. All equipment with rotating parts should be equipped with safety guards that reduce the possibility of the parts coming loose and injuring someone.

Do not depend on someone else to inspect and maintain equipment. Check it out before you use it! If you find the equipment unsafe put a sign on it to warn others, and notify the person in charge.

Never use tools and equipment for purposes other than those for which they are designed. Using the proper tool in the correct way will not only be safe but will also allow you to do a good job.

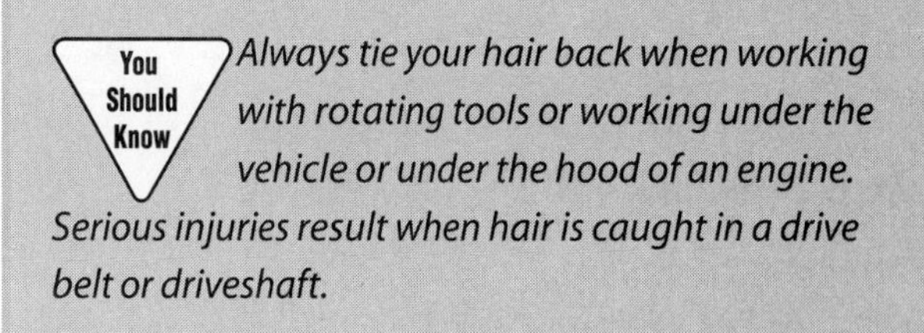

Always tie your hair back when working with rotating tools or working under the vehicle or under the hood of an engine. Serious injuries result when hair is caught in a drive belt or driveshaft.

LIFT SAFETY

Be careful when raising a vehicle on a lift or a hoist. Adapters and hoist plates must be positioned correctly on twin post and rail-type lifts to prevent damage to the underbody of the vehicle **(Figure 4)**. There are specific lift points. These points allow the weight of the vehicle to be evenly supported by the adapters or hoist plates. The correct lift points can be found in the vehicle's service manual or information CD or DVD. Always follow the manufacturer's instructions. Before operating any lift or hoist carefully read the operating manual and follow the operating instructions.

Once you know that the lift supports are properly positioned under the vehicle, raise the lift until the supports

Figure 4. Position the lift pads correctly so the vehicle will be secure on the lift.

contact the vehicle. Then check the supports to make sure they are in full contact with the vehicle. Shake the vehicle to make sure it is securely balanced on the lift, and next raise the lift to the desired working height. Then lower the lift until the mechanical locks are engaged—do not leave the vehicle supported by hydraulic pressure alone.

You Should Know *When you begin working at a shop or change shops be sure to get proper operating instructions for the lift(s) on which you will be working. Different lifts have varying safety mechanisms that may have to be set or released to operate the lift safely.*

JACK AND JACK STAND SAFETY

A vehicle can also be raised off the ground with a hydraulic jack **(Figure 5)**. The lifting pad of the jack must be positioned under an area of the vehicle's frame or at one of the manufacturer's recommended lift points. Never place the pad under the floor pan or under steering and suspension components: these are easily damaged by the weight of the vehicle. Always position the jack so the wheels of the vehicle can roll as the vehicle is being raised.

You Should Know *Never use a lift or jack to move something heavier than it is designed for. Always check the rating before using a lift or jack. If a jack is rated for 2 tons do not attempt to use it for a job requiring 5 tons. It is dangerous for you and for the vehicle.*

Figure 5. This hydraulic floor jack can lift passenger cars and light duty trucks.

Figure 6. Always use jack stands to support the vehicle; don't rely on the jack!

Safety or **jack stands** are supports of different heights that sit on the floor. They are placed under a sturdy chassis member, such as the frame or axle housing, to support the vehicle **(Figure 6)**. Once the jack stands are in position the hydraulic pressure in the jack should be slowly released until the weight of the vehicle is on the stands. Like jacks, jack stands also have a capacity rating. Always use a jack stand with the correct rating.

Never move under a vehicle when it is supported only by a hydraulic jack. Rest the vehicle on the safety stands before moving under the vehicle. The jack should be removed after the jack stands are set in place. This eliminates a hazard, such as a jack handle sticking out into a walkway. A jack handle that is bumped or kicked can cause a tripping accident or cause the vehicle to fall.

BATTERIES

When possible you should disconnect the battery of a car before you disconnect any electrical wire or component. This prevents the possibility of a fire or electrical shock. It also eliminates the possibility of an accidental short, which can damage the car's electrical system and electrical components. This is especially true of newer cars that are equipped with many electronic and computerized controls. To properly disconnect the battery disconnect the negative or ground cable first **(Figure 7)**, then disconnect the positive cable. Since electrical circuits require a ground to be complete removing the ground cable eliminates the possibility of a circuit accidentally becoming completed. When reconnecting the battery connect the positive cable first, then the negative.

You Should Know

The active chemical in a battery, the **electrolyte,** *is basically sulfuric acid. Sulfuric acid can cause severe skin burns and permanent eye damage, including blindness, if it gets in your eye. If some battery acid gets on your skin wash it off immediately and flush your skin with water for at least five minutes. If the electrolyte gets into your eyes immediately flush them out with water, then see a doctor right away.* Never *rub your eyes, just flush them well and go to a doctor. Wearing safety glasses when working with and around batteries is an absolute must.*

Figure 7. Disconnect the negative battery cable first so you won't accidentally create a spark near the battery.

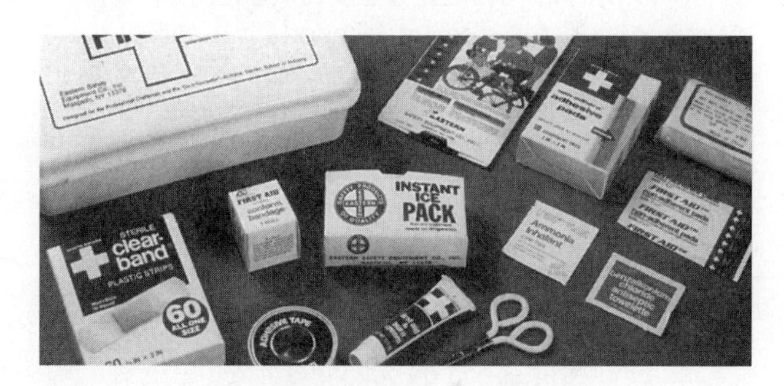

Figure 8. A basic first aid kit.

ACCIDENTS

Make sure you are aware of the location and contents of the shop's first-aid kit **(Figure 8)**. There should be an eye wash station in the shop so that you can rinse your eyes thoroughly should you get acid or some other irritant in them. If there are specific first-aid rules in your school or shop make sure you are aware of them and follow them. Some first-aid rules apply to all circumstances and are normally included in each shop's rules.

If someone is overcome by carbon monoxide get that person fresh air immediately. Burns should be cooled right away by rinsing them with water. Whenever there is severe bleeding from a wound try to stop the bleeding by applying pressure with clean gauze on or around the wound, and get medical help. Never move someone who may have broken bones unless the person's life is otherwise endangered. Moving that person may cause additional injury. Call for medical assistance.

Your supervisor should be informed immediately of all accidents that occur in the shop. It is a good idea to keep an up-to-date list of emergency telephone numbers posted next to the telephone. The numbers should include a doctor, a hospital, and the fire and police departments.

ELECTRICAL SYSTEM REPAIRS

Some electronic replacement parts are very sensitive to **static electricity**. These parts will be labeled as such. Whenever you are handling a part that is sensitive to static electricity you should follow these guidelines to reduce any possible electrostatic charge buildup on your body and the electronic part:

1. Do not open the package until it is time to install the component.
2. Before removing the part from the package ground the package to a known good ground on the car.
3. Always touch a known good ground before handling the part. Repeat this periodically while handling the part and more frequently after sliding across the seat, sitting down, or walking a distance.
4. Never touch the electrical terminals of the component.
5. Use static straps if they are available **(Figure 9)**.

AIR BAG SAFETY AND SERVICE WARNINGS

The dash and steering wheel contain the circuits that control the **air bag system**. Whenever you are working on or around air bag systems it is important to follow some safety warnings. There are safety concerns with both **deployed** and live (undeployed) air bag modules.

1. Wear safety glasses when servicing the air bag system.
2. Wear safety glasses when handling an air bag module.
3. Wait at least 10 minutes after disconnecting the battery before beginning any service on or around the air bag system. The reserve energy module is capable of storing enough power to deploy the air bag for up to 10 minutes after battery voltage is lost.
4. Handle all air bag sensors with care. Do not strike or jar a sensor in such a manner that deployment may occur.
5. When carrying a live air bag module face the trim and bag away from your body.
6. Do not carry the module by its wires or connector.

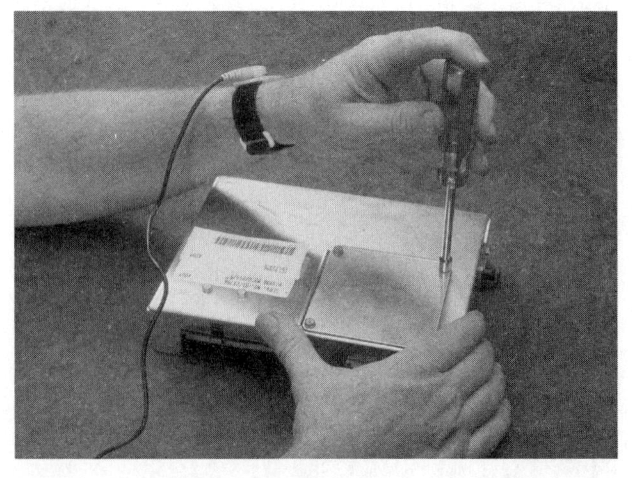

Figure 9. Use a grounding strap to prevent a static electricity discharge from damaging electronic components.

7. When placing a live module on a bench face the trim and air bag up.
8. Deployed air bags may have a powdery residue on them. Sodium hydroxide is produced by the deployment reaction and is converted to sodium carbonate when it comes into contact with atmospheric moisture. It is unlikely that sodium hydroxide will still be present after deployment. However, wear safety glasses and gloves when handling a deployed air bag. Wash your hands immediately after handling the bag.
9. A live air bag module must be deployed before disposal. Because the deployment of an air bag is an explosive process improper disposal may result in injury and in fines. A deployed air bag should be disposed of in a manner consistent with **Environmental Protection Agency (EPA)** guidelines and manufacturer procedures.
10. Do not use a battery- or AC-powered voltmeter, ohmmeter, or any other type of test equipment not specified in the service manual. Never use a test light to probe for voltage.

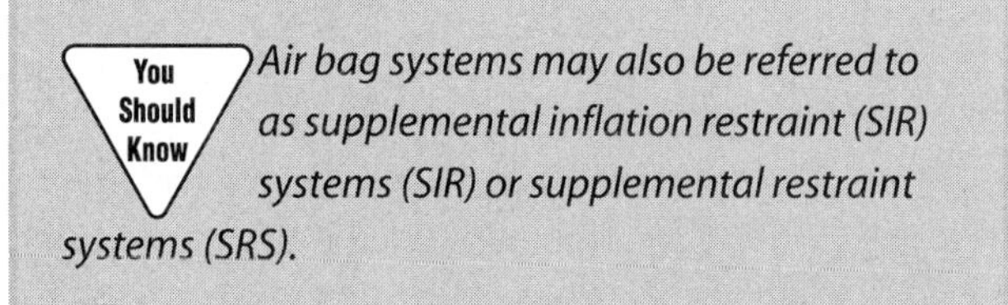

Air bag systems may also be referred to as supplemental inflation restraint (SIR) systems (SIR) or supplemental restraint systems (SRS).

HAZARDOUS MATERIALS

Many solvents and other chemicals used in an auto shop have warning and caution labels. Read the warnings so you know the proper precautions and understand the dangers **(Figure 10)**. Most automotive solvents and chemicals are considered hazardous materials. Also, many service procedures generate what are known as **hazardous wastes**. Dirty solvents and liquid cleaners are good examples of these.

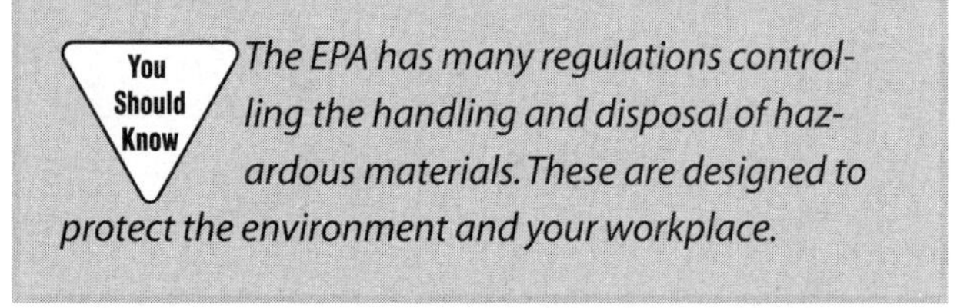

The EPA has many regulations controlling the handling and disposal of hazardous materials. These are designed to protect the environment and your workplace.

Every employee in a shop is protected by **"Right-to-Know" Laws** concerning hazardous materials and wastes. The general intent of these laws is for employers to provide a safe workplace as it relates to hazardous materials. All employees must be trained about their rights under the

Figure 10. Many common automotive solvents can cause serious health risks. Know what you are using.

legislation, including the nature of the hazardous chemicals in their workplace, the labeling of chemicals, and the information about each chemical listed and described on **Material Safety Data Sheets (MSDS)**. These sheets are available from the manufacturers and suppliers of the chemicals. They detail the chemical composition and precautionary information for all products that can present health or safety hazards.

Employees must be familiar with the intended purposes of the material, the recommended protective equipment, accident and spill procedures, and any other information regarding the safe handling of hazardous materials. This training must be given annually to employees and provided to new employees as part of their job orientation. The Canadian equivalents of the MSDS are called Workplace Hazardous Materials Information Systems (WHMIS).

You Should Know *When handling any hazardous material always wear the appropriate safety protection. Always follow the correct procedures while using the material, and be familiar with the information given on the MSDS for that material.*

All hazardous materials should be properly labeled, indicating what health, fire, or reactivity hazard they pose and what protective equipment is necessary when handling each chemical. The manufacturer of the hazardous material must provide all warnings and precautionary information, which must be read and understood by all users before they use it. Pay attention to the label information; by doing so you will use the substance in the safest possible manner.

A list of all hazardous materials used in the shop should be posted for the employees to see. Shops must maintain documentation on the hazardous chemicals in the workplace, proof of training programs, records of accidents or spill incidents, and records of satisfaction of employee requests for specific chemical information via the MSDS. A general right-to-know compliance procedure manual must also be utilized within the shop.

There are many government agencies charged with ensuring that all workers have safe work environments. These include the **Occupational Safety and Health Administration (OSHA)**, Mine Safety and Health Administration (MSHA), and National Institute for Occupational Safety and Health (NIOSH). These, in addition to state and local governments, have instituted regulations that must be understood and followed. Everyone in a shop has the responsibility to adhere to these regulations.

OSHA

In 1970 OSHA was formed by the federal government to "assure safe and healthful working conditions for working men and women; by authorizing enforcement of the standards developed under the Act; by assisting and encouraging the States in their efforts to assure safe and healthful working conditions by providing research, information, education, and training in the field of occupational safety and health."

Safety standards have been established that will be consistent across the country. It is the employers' responsibility to provide a place of employment that is free of all recognized hazards and that will be inspected by government agents knowledgeable in the laws of working conditions. OSHA controls all safety and health issues of the automotive industry.

OSHA and the EPA have other strict rules and regulations that help to promote safety and environmentally friendly practices in the automotive shop. These are described throughout this text whenever they are applicable.

HANDLING SHOP WASTES

Maintaining a vehicle involves handling and managing a wide variety of materials and wastes. Some of these wastes can be toxic to fish, wildlife, and humans when improperly managed. No matter the amount of waste produced it is to the shop's legal and financial advantage to manage the wastes properly and, even more important, to prevent pollution.

The following is a list of the recommended procedures for disposal and handling of various waste materials. Always refer to the local, state, and federal regulations that affect you.

- Recycle engine oil. Set up equipment, such as a drip table or screen table with a used oil collection bucket, to collect oil dripping off parts. Place drip pans underneath vehicles that are leaking fluids. Never allow oil to enter into the shop's sewage or drain system **(Figure 11)**.
- Used oil filters should be drained for at least 24 hours, crushed, and recycled.
- Old batteries should be recycled by sending them to a reclaimer or back to the distributor.
- Collect metal filings when machining metal parts. Keep them separate from other dirt, and recycle the metal, if possible. Prevent metal filings from falling into a storm sewer drain.
- Recover and/or recycle refrigerants during the service and disposal of motor vehicle air conditioners and refrigeration equipment. It is illegal to knowingly vent refrigerants to the atmosphere. Recovery and/or recycling during service must be performed by an EPA-certified technician using certified equipment and following specified procedures.
- Replace hazardous chemicals with less toxic alternatives that have equal performance. For example, substitute water-based cleaning solvents for petroleum-based solvent degreasers. Hire a hazardous waste management service to clean and recycle solvents. Store solvents in closed containers to prevent evaporation. Properly label spent solvents.
- Store materials such as scrap metal, old machine parts, and worn tires under a roof or tarpaulin to protect them from the elements and to prevent the potential to create contaminated runoff.
- Collect and recycle coolants from radiators. Store transmission fluids, brake fluids, and solvents containing chlorinated hydrocarbons separately, and recycle or dispose of them properly.
- Keep waste towels in a closed container marked for dirty rags. To reduce costs and liabilities associated with disposal of used towels, which can be classified as hazardous waste, investigate using a laundry service that is able to treat the wastewater generated from cleaning the towels **(Figure 12)**.

You Should Know *It is illegal to throw a used oil filter from the vehicle into the ordinary trash. Follow your shop's procedures for draining, crushing, and disposing of used oil filters.*

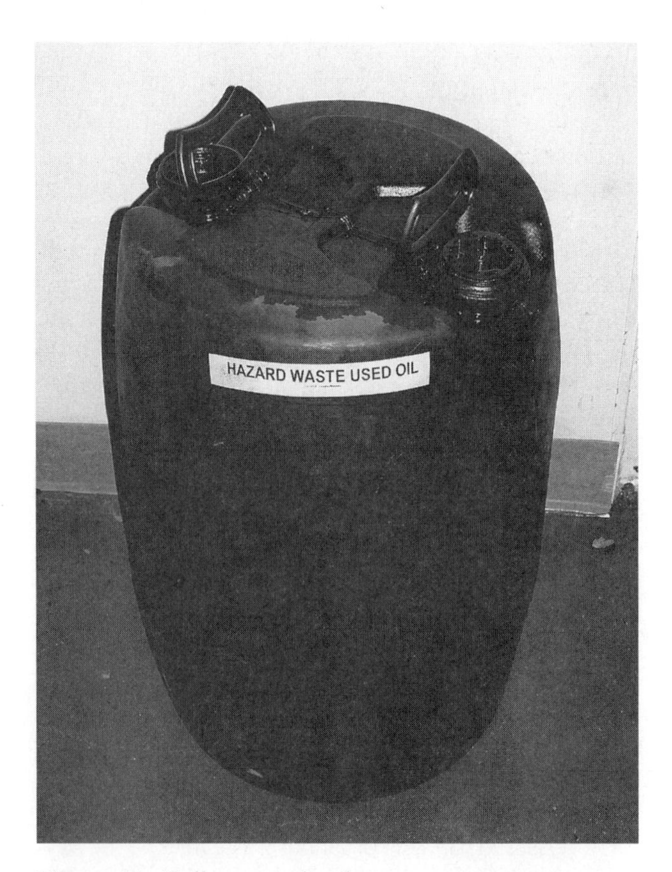

Figure 11. Collect used oil in an appropriate container until pick up for recycling.

Figure 12. Follow your shop's practices for proper handling of used rags.

Summary

- True safe work practices are based on common sense and knowledge of safety equipment.
- For a work area to be safe it needs to be kept clean and orderly.
- Shop safety is the responsibility of everyone in the shop, and everyone must work together to protect the health and welfare of everyone in the shop.
- Eye protection should be worn whenever you are working in the shop.
- If chemicals or solvents get into your eyes flush them continuously with clean water. Have someone call a doctor, and get medical help immediately.
- Dress to protect yourself and to avoid accidents. Wear well-fitted and comfortable clothing and foot and hand protection. Never wear anything that might get caught in moving parts or can conduct electricity.
- When lifting heavy objects use a hoist or have someone help you, and always lift with your legs not your back.
- Gasoline, diesel fuel, cleaning solvents, and dirty rags should be treated as potential firebombs and handled and stored properly.
- Know where the fire extinguishers and fire alarms are located in the shop, and know how to use them (think of the word PASS!).
- Inspect all tools and equipment before using them. Check them for unsafe conditions.
- Never use tools and equipment for purposes other than those they are designed for.
- Make sure the contacts of a lift are properly positioned on the vehicle before lifting it.
- Use jack stands to secure a vehicle after you raise it with a jack.
- When possible disconnect the battery of a car before you begin to work around it or before you disconnect any electrical wire or component.
- Follow the proper procedures for eliminating static electricity when handling parts that are sensitive to static electricity.
- Know the location and contents of the shop's first-aid kits, eye wash stations, and list of emergency telephone numbers.
- Whenever working on or around air bag systems it is important to follow certain procedures.
- Many solvents and other chemicals used in an auto shop have warning and caution labels that should be read and understood by everyone who uses them.
- Handle and dispose of all hazardous wastes according to federal and local laws and your own common sense.
- Know where the MSDS can be found in the shop, and know how to quickly find the appropriate information on them.

Review Questions

1. List three simple ways to avoid shop accidents.
2. Describe the correct procedure for disconnecting a battery.
3. When should you wear eye protection?
4. What is the correct procedure for putting out a fire with an extinguisher?
5. When a chemical gets into a technician's eye get a clean rag and try to lightly rub the chemical out. True or False?
6. Technician A says the manager should take full responsibility for ensuring a safe work area for technicians. Technician B says the apparel and work habits of technicians can help prevent accidents. Who is correct?
   A. Technician A only
   B. Technician B only
   C. Both Technician A and Technician B
   D. Neither Technician A nor Technician B
7. Technician A says you should avoid knocking any air bag system sensors. Technician B says you should disconnect the battery to disarm the air bag system. Who is correct?
   A. Technician A only
   B. Technician B only
   C. Both Technician A and Technician B
   D. Neither Technician A nor Technician B
8. Technician A says that most shop fires are Class D fires. Technician B says that you should open the doors when a fire occurs to provide adequate ventilation. Who is correct?
   A. Technician A only
   B. Technician B only
   C. Both Technician A and Technician B
   D. Neither Technician A nor Technician B
9. Technician A says to throw oil filters into the trash after they are drained. Technician B says that used oil must

be drained into a hazardous waste container. Who is correct?

A. Technician A only
B. Technician B only
C. Both Technician A and Technician B
D. Neither Technician A nor Technician B

10. Technician A says that the purpose of OSHA is to set standards to protect the environment. Technician B says that the EPA sets regulations about collection and disposal of hazardous wastes. Who is correct?

A. Technician A only
B. Technician B only
C. Both Technician A and Technician B
D. Neither Technician A nor Technician B

Engine Diagnostic and Repair Tools

Introduction

When servicing or rebuilding an automotive engine you will need to be familiar with various types of tools. Many tools are common to different aspects of automotive repair; other tools are specialized to the repair of an engine. Special engine evaluation tools are used to locate the source of engine problems before engine repairs are made. Many engine repair procedures require the use of specialized tools. Tools are needed to properly install components and to recondition parts of the engine. Close measurement of engine components is critical to a high-quality service. Instruments must be used properly and treated appropriately in order to guarantee accurate measurements. In the modern automotive engine many measurements must be accurate to within a ten thousandth of an inch, .0001 in., which is roughly one twentieth of the thickness of a hair. This chapter will look at many of the tools required for engine diagnosis, repair, and major service.

Some engine overhaulers use the same tools they bought 40 years ago. Many measuring tools have not changed at all, though the engines certainly have. Taking good care of high-quality tools really can make them last a lifetime.

MEASURING TOOLS

Repair and overhaul of an engine requires many specialty measuring tools. Precision is critical in the professional rebuild of an engine. A mistake in measurement as small as .001 in. can cause a serious problem in a modern engine where tolerances are very small. Measuring tools must be used, treated, and stored carefully to ensure accuracy.

Machinist's Rule

A **machinist's rule** is a steel ruler with increments as small as 1/128th of an inch. They typically provide decimal equivalents on the opposite side. Some machinist's rules are graduated in decimal intervals of .1 in. (one tenth of an inch), .02 in. (two hundredths or twenty thousandths of an inch) and .001 in. (one thousandth of an inch). A machinist's rule may also be metric; these are divided into .5 mm (millimeter) increments. One use of a machinist's rule is to measure valve seat width during a valve job.

Feeler Gauge

A **feeler gauge set** holds many different steel strips, each made to a precise thickness. Feeler gauges may be metric or standard; many sets offer both measurements on each blade **(Figure 1)**. The gauge set typically provides gauges in sizes from .0015 in. up to .024 in. or .040 in. The smaller sizes come in increments of .001 in.; the larger sizes are provided in .002 in. increments. Some feeler gauge sets are go/no go gauges. These feeler gauges have a smaller size on the tip and a size .002 in. bigger further down on the strip. For example, if you were trying to size a .007 in. gap you would use a .006/.008 go/no go gauge. Adjust the gap so the .006 in. part of the strip fits but the .008 in. part of the strip does not. Feeler gauges are used to measure clearances. Valve adjustment may require the precision measurement of the clearance between the valve and the rocker or the valve and the follower. Cylinder heads and

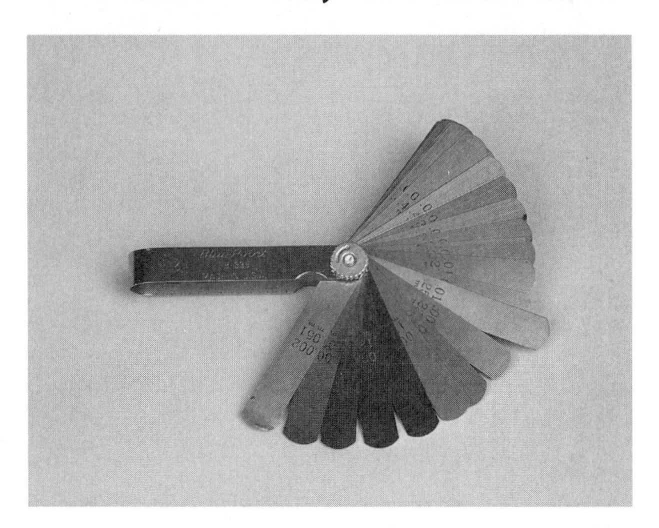

Figure 1. A feeler gauge set with standard and metric designations.

blocks may be measured for straightness using a feeler gauge and a straightedge.

Dial Caliper

A **dial caliper** can be used to measure outside diameter, inside diameter, or depth **(Figure 2)**. It may be metric or standard. The caliper has jaws that will fit within a bore or on the outside diameter of a component. A depth gauge provides an arm that scrolls down using a thumbwheel. Many dial calipers are accurate to .0005 in. or .02 mm.

Micrometer

The **micrometer** is used to measure component diameter or thickness. Micrometers may be accurate to .001 in. or to .0001 in. Metric micrometers may be accurate to .01 mm. Newer engines typically require more precise measurement. Some micrometers offer a digital readout, but many require interpretation of the scales. The micrometer includes the frame, the spindle, the anvil, the barrel, and the thimble. You will measure pistons, cylinder bores, main and rod journals, and lifters using a micrometer. Accurate measurement requires practice and a good feel to achieve repeatability. Some micrometers use a slip gauge at the end of the thimble so you cannot overtighten the micrometer on the piece being measured.

To interpret the measurement, first identify the number of lines showing on the barrel **(Figure 3)**. Each line represents .025 in.; every fourth line is longer and shows an increment of .1 in. Seven lines showing would indicate a reading of .175 in.

Next read the number on the thimble as it lines up to the zero mark. The number on the thimble is a reading of thousandths of an inch **(Figure 4)**. If the 17 on the thimble lines up with the zero line, this indicates .017 in. A reading of 8 would indicate .008 in. The reading on the thimble is added to the reading on the barrel. For example, .175 in. plus .017 in. equals .192 in. Finally, add the number that lines up on the vernier scale to the end of the measurement; this is in ten thousandths of an inch. A reading of .0006 in. on the vernier scale would give a total measurement of .1926 in.

Different sized micrometers are available to measure varying diameters. A 0–1 in. micrometer, for example, would be used to measure lifters. A 3–4 in. micrometer is often used to measure piston diameter on light-duty gas engines.

An inside micrometer may be used to measure the size of a bore or hole. On these tools the measuring arms extend outward to the edge of the bore.

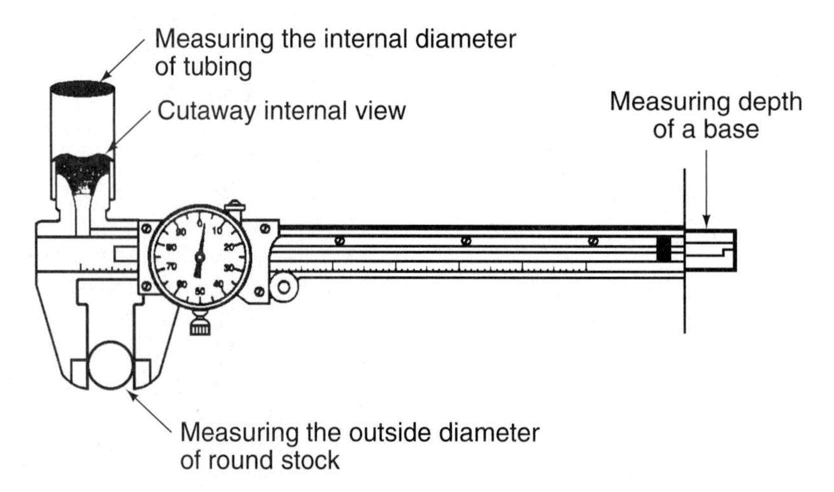

Figure 2. Use a dial caliper to make precision measurements of inside diameter, outside diameter, or depth.

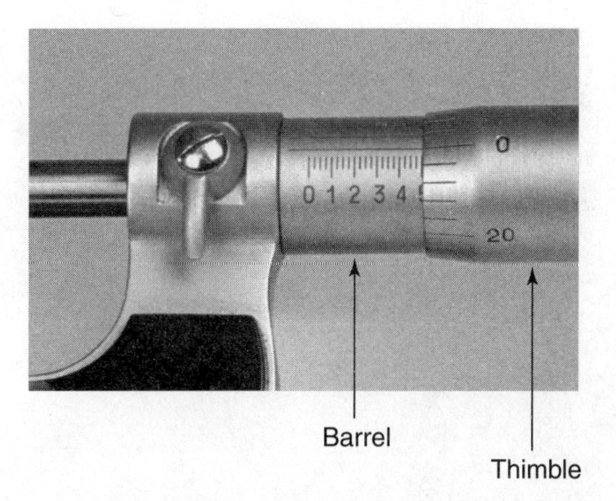

Figure 3. The numbered lines on the barrel mark .1" increments. The three smaller lines in between each represent .025".

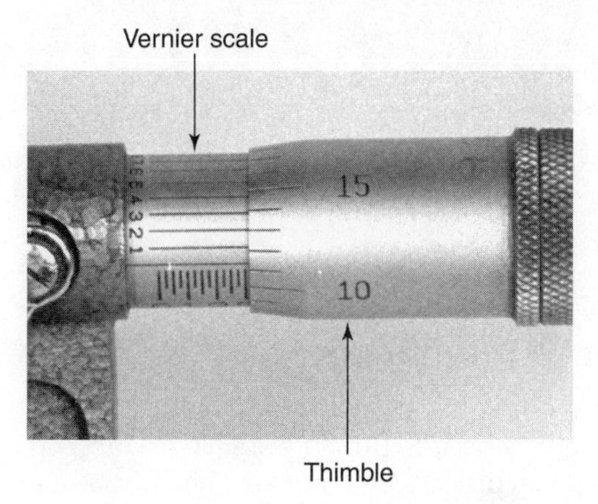

Figure 4. The marks on the thimble indicate .001" increments. The vernier scale displays .0001" divisions.

Telescoping or Snap Gauges

When an inside micrometer is not available, **telescoping gauges** and a micrometer can be used to measure the inside diameter of a bore **(Figure 5)**. This system is frequently used to measure the cylinder bore for size, out of round, and taper. To use it, place the gauge inside the bore and loosen the locking screw. The telescoping gauge will extend to the diameter of the bore. Rock the gauge lightly to straighten it out in the bore. Sometimes it is easier to set the gauge on the top of the piston to level it. Then lock the setscrew, and measure the diameter of the gauge with a micrometer; this gives you the bore diameter.

Small Hole Gauges

Small hole gauges are telescoping gauges for measuring small bores. They function identically to telescoping gauges; the only difference is their smaller size **(Figure 6)**.

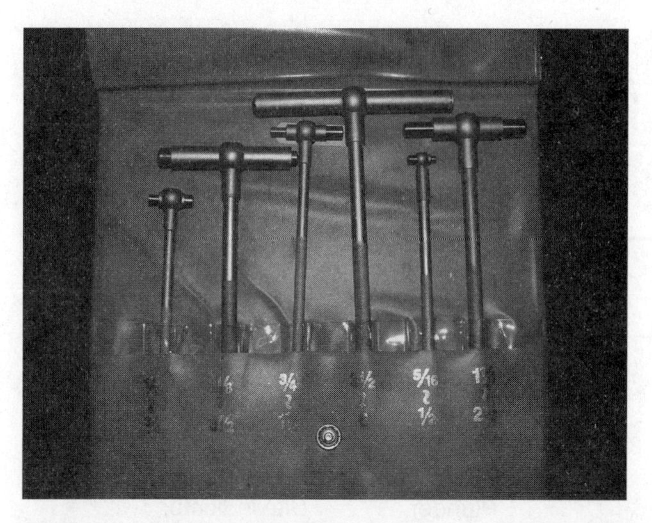

Figure 5. These telescoping gauges can be used with a micrometer to measure the inside diameter of bores.

Figure 6. A small hole gauge set.

Dial Indicator

A **dial indicator** is used to measure rotational variation, movement, or endplay of a component **(Figure 7)**. Crankshaft endplay, for example, must be greater than 0 in. to prevent binding and less than the specified maximum in order to provide smooth, trouble-free operation of the

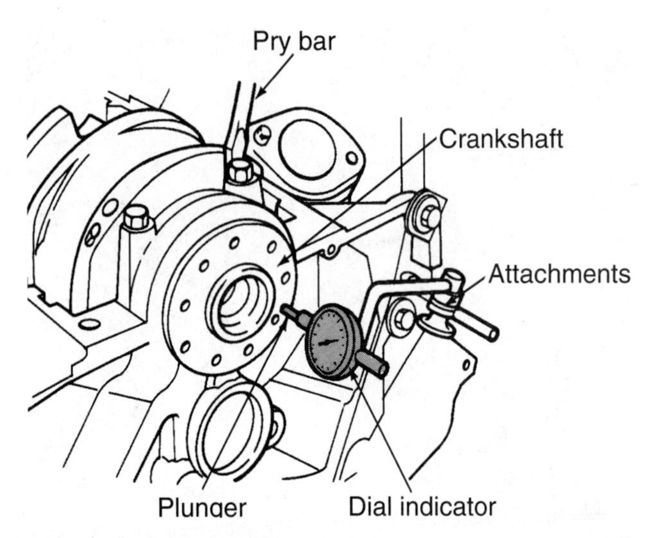

Figure 7. Use a dial indicator to measure crankshaft or camshaft end play, gear backlash, or runout.

crankshaft. The dial indicator mounts on an immovable piece, and the needle rests on the piece that moves. The distance the piece moves is reflected in the indicator gauge. Most dial indicators are accurate either to.001 in. or to .0005 in.

Dial Bore Gauge

A **dial bore gauge** is a dial indicator designed to measure the diameter of a bore. This is an excellent tool for measuring cylinder taper or out of round. The tool has a measuring tip and centering arms for ease of measurement. The bore gauge is moved up and down gently in the cylinder, and the needle deflection shows any changes in diameter. With a gentle touch and some practice an engine overhauler can make short work of cylinder bore measurements.

Oil Pressure Gauge

An **oil pressure gauge** is used to accurately check for adequate oil pressure in an engine. If the vehicle is equipped with a gauge on the instrument panel it should not be relied upon for engine diagnostic information; the sending unit, wiring, or gauge could be faulty **(Figure 8)**. A threaded adapter fits into a service port, usually where the oil pressure sending unit fits. A hose allows oil to run up to the gauge and deflect the needle relative to pressure. Inadequate oil pressure can rapidly damage an engine. Worn engine bearings or a weak oil pump are common culprits.

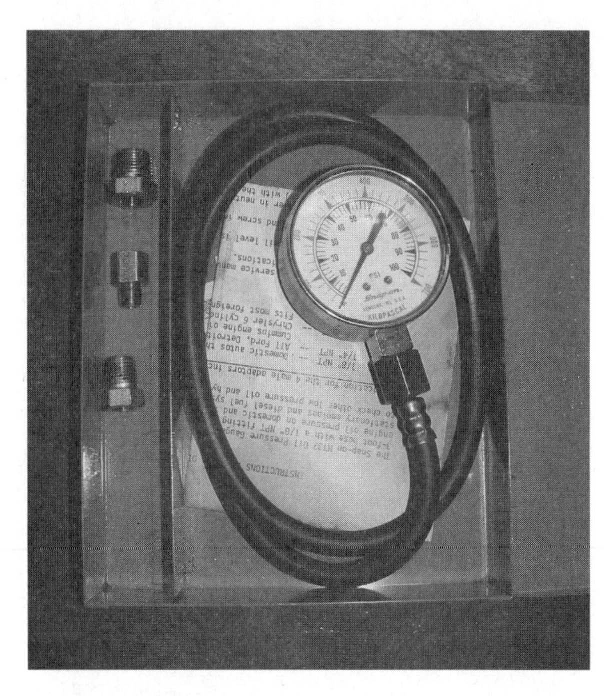

Figure 8. An oil pressure gauge set with threaded adaptors.

Belt Tension Gauge

Proper tightness of a timing belt or accessory drive belt can be achieved using a **belt tension gauge**. The mechanism deflects to fit over the tightened belt, and the gauge indicates its tension. A timing belt that is installed too tightly can snap and cause serious engine damage.

COOLING SYSTEM TOOLS

Maintaining and troubleshooting the engine cooling system require a few important special tools. These specialty tools are usually provided by the shop; you will typically need to purchase only a coolant hydrometer.

Coolant Hydrometer

A **coolant hydrometer** is a simple device that measures the antifreeze protection level of coolant. The proper ratio of water and coolant must fill the cooling system to prevent freezing during the winter. A frozen coolant system can crack the engine block or the cylinder head. The hydrometer draws some coolant into a tube. The tube contains either specially weighted balls or a needle. The specific gravity of the coolant is proportional to the antifreeze protection level. The balls or needle move depending on the specific gravity of the coolant and provide a reading of the antifreeze protection level. A coolant refractometer may be necessary to provide accurate readings with some newer coolants.

Coolant Refractometer

A **refractometer** can be used to measure the freezing point of coolant or the specific gravity of battery electrolyte. It is accurate on any type of coolant. Place a few drops of the coolant on the prism, and look through the eyepiece to read the results. Read the tool user's manual for complete instructions.

Cooling System Pressure Tester

A **cooling system pressure tester** can be used to check for leaks in the system and to evaluate the function of the radiator cap **(Figure 9)**. The correct adapter fits onto the top of the radiator or the reservoir. Then pressure is pumped into the system. You carefully inspect the system for leaks while the coolant is pressurized. With a different adapter you can pump pressure onto the radiator cap to be sure it holds its rated pressure. You can also apply vacuum to the cap to be sure it releases all vacuum.

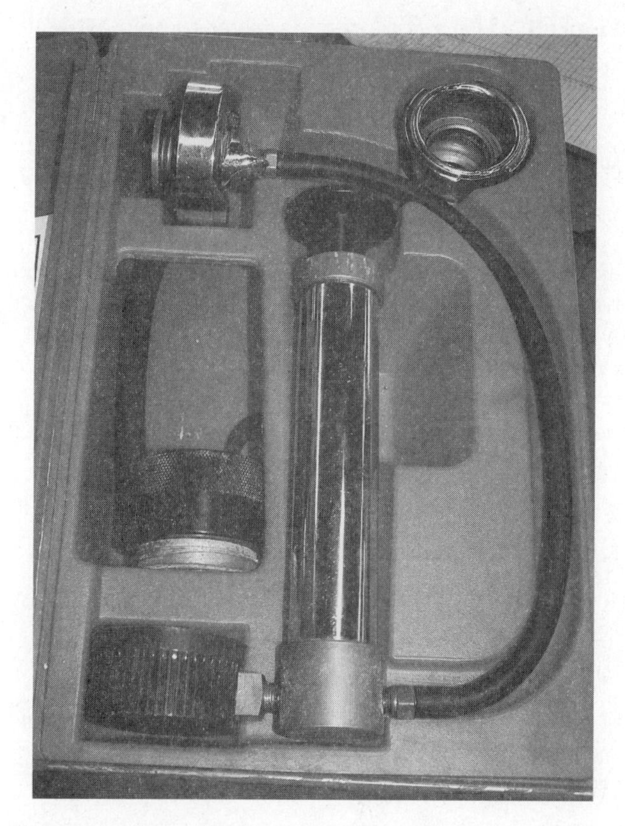

Figure 9. A cooling system pressure test set can test the cooling system for leaks and the cap for proper function.

Coolant Recovery and/or Recycling Station

Many shops now use a coolant recovery and/or recycling station to handle draining and refilling of cooling systems. Often these stations also have a means of power flushing the cooling system to remove residue from the cooling passages. The coolant is pulled into the recovery portion of the station. On recycling systems an internal recycler filters the used coolant to enable it to be used again rather than discarded as hazardous waste. Some manufacturers do not recommend the use of recycled coolant; check with the manufacturer before using this equipment. Coolant is refilled into the system under pressure to reduce the risk of air bubbles getting trapped in the engine.

ENGINE EVALUATION TOOLS

It is important to fully evaluate an engine before repairing, rebuilding, or replacing it. First, you want to be sure that an expensive repair is actually required. Second, the more information you have about the damage, the better able you are to estimate the job for the customer. An engine that has burned valves, for example, may require only a valve job rather than a full engine overhaul or replacement. The following tools can help determine the extent and location of engine damage.

Scan Tool

A **scan tool** is an electronic tool used to interface with a vehicle's on-board computer. It provides diagnostic information about various electrical sensors and actuators. It cannot directly diagnose mechanical engine problems, but it is very useful in helping to locate the source of engine drivability concerns.

Stethoscope

A common **stethoscope** is a useful tool in diagnosing engine noises **(Figure 10)**. Generally, a knocking noise from the top end of the engine is much less costly to repair than a knock from the bottom end. Place the tip of the stethoscope at different areas on the engine block and head. The stethoscope isolates the sound; when you near the source of noise it will be clear and loud. A stethoscope can also be used to locate worn generator, water pump, and other bearings, or to locate vibrations from loose components.

Vacuum Gauge

An inexpensive and easy tool to use to help diagnose mechanical and performance concerns in an engine is a

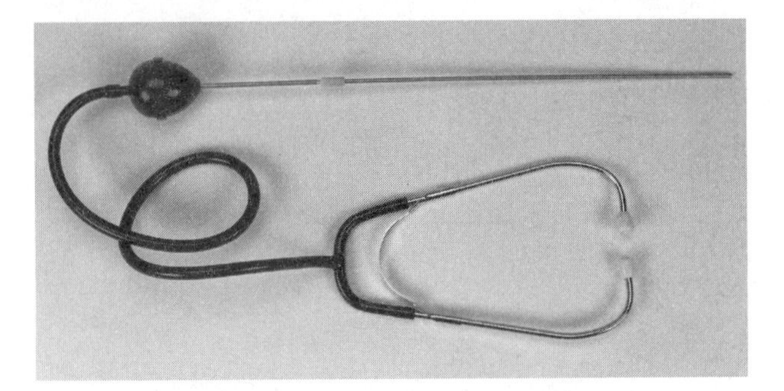

Figure 10. A stethoscope will help isolate an engine noise.

Figure 11. A vacuum gauge set.

vacuum gauge. It is used to measure engine vacuum **(Figure 11)**. A line off the gauge is hooked up to a vacuum source, most often off the intake manifold. Engine vacuum is read with the engine running. Variations in the expected steady reading of 18 in. hg vacuum (or higher) at idle and at 2500 rpm can help diagnose engine performance problems. Cranking vacuum can also be checked as a means of isolating engine mechanical failures on a vehicle that will not start.

Compression Tester

A **compression tester** measures the pressure developed inside a cylinder on the compression stroke **(Figure 12)**. This is a useful tool in assessing the mechanical condition of the engine. A threaded hose with a Schrader valve screws into the spark plug hole. The gauge attaches to the other end of the hose. Compression can be measured with the engine cranking or with the engine running. When compression pressures are lower than specified, it indicates mechanical engine problems such as worn rings, burned or leaking valves, or head gasket leakage.

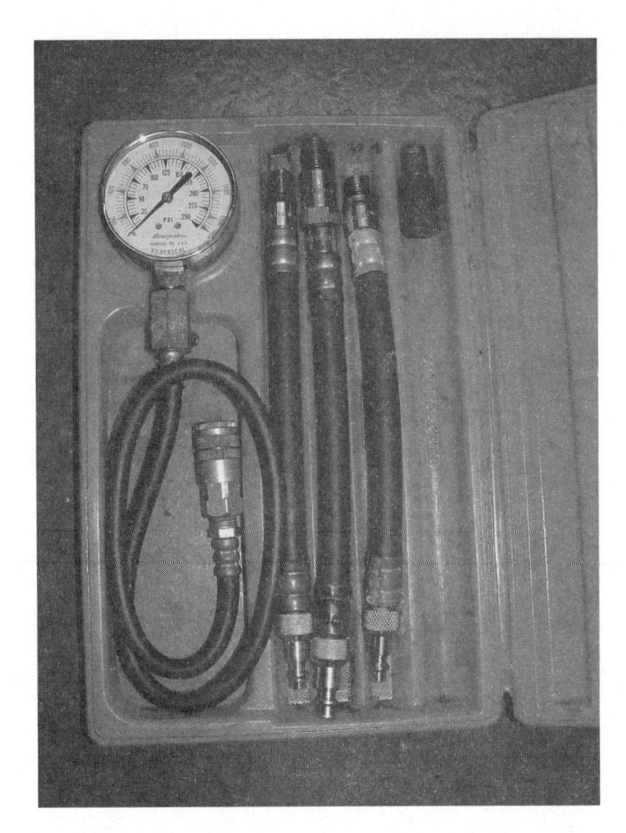

Figure 12. Use a compression tester to help evaluate an engine's mechanical condition.

Cylinder Leakage Tester

A **cylinder leakage tester** can be used to help pinpoint the cause of low compression readings **(Figure 13)**. It, too, uses a threaded adapter that fits in the spark plug hole. Then regulated air pressure is applied to the cylinder when the piston is at top dead center with the valves closed. If the cylinder sealing is compromised, air can be heard leaking from the different parts of the engine depending on the fault. Air heard escaping the tailpipe, for example, indicates a leaking exhaust valve.

ENGINE RECONDITIONING TOOLS

There are numerous specialty tools you will use when performing major repairs or overhauling an engine. Many of them are essential to professional work. Often many of these are shop-owned tools, but engine overhaul specialists are likely to purchase some of their own tools as well. We will fully discuss the proper use of these tools later in the book as we review the steps of engine repair, overhaul, and replacement.

Portable Engine Crane

An **engine crane** is used to pull the engine out of the vehicle and roll it through the shop to where you will ser-

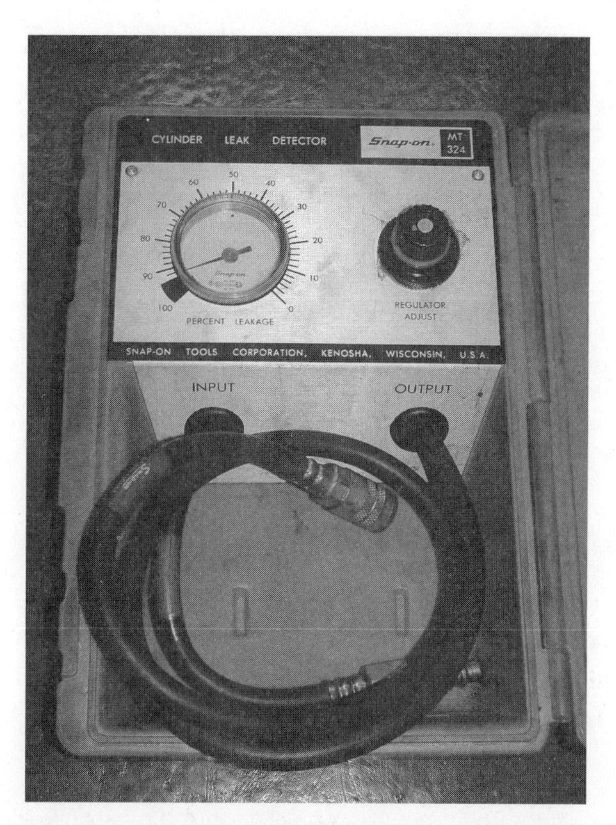

Figure 13. This cylinder leakage tester can pinpoint a problem with combustion chamber sealing.

Figure 14. This Chevy 350 is properly mounted on an engine stand and ready for an overhaul.

vice it. Most engines have lifting hooks on them. Secure a chain to the lifting hooks. The crane has a hook at the end of its beam to pick up the center of the chain. The crane uses a hydraulic unit to lift the beam easily even with the weight of the engine on it. You should not attempt to perform repairs on the engine while it is hanging from the crane.

Engine Stand

The best way to disassemble, repair, and reassemble an engine is on an **engine stand (Figure 14)**. The stand, usually on wheels, allows you to hold the engine from the block at the flywheel end. A universal mounting bracket bolts onto the end of the engine in the **bell housing** bolt holes. Then the post off the end of the mounting plate slides into the engine stand holding tube. Once the engine is securely mounted, you can rotate the engine on the post to conveniently work on any part of the engine.

Pullers

An assortment of gear, pulley, bushing, and bearing **pullers** is often needed to disassemble parts of the engine. A puller set contains a variety of pullers with yokes, flanges, or jaws to attach to the component. These universal puller sets are not made for manufacturer-specific tasks. You will have to find the proper setup for the different jobs you may run into. The crankshaft pulley, for example, is often press fit onto the crankshaft snout. When removing the crankshaft pulley you would likely choose a flange-type puller **(Figure 15)**. The flange mounts to the pulley itself. Then the center screw (with a round button

Figure 15. This flange-type puller can be used to remove the crankshaft pulley.

on the end to prevent thread damage) is driven in against the crankshaft snout to pull the pulley off the crank. The bearing splitter is another form of puller used to remove bearings races from shafts. Tighten the splitter's tapered jaws behind the race to pull it forward on the shaft. Once the jaws are secure behind the race, a puller can be screwed into the separating plate to pull the race from the shaft. A slide hammer is another tool used to pull components. Jaws or an expanding collet are placed at the end of a shaft. On the other end of the tool is a large movable weight. Once the slide hammer is attached to the part to be removed, slide the weight firmly against its stop to drag the component out of its bore or off its shaft. A slide hammer is commonly used to remove the pilot bearing from the crankshaft or flywheel.

You Should Know *Always wear safety glasses when using pullers. The tension applied by the puller can send a part flying or even splinter a component if the puller is not fully seated on the part.*

Figure 16. Use a ridge reamer before removing the pistons to avoid damaging them.

Bushing and Seal Pullers and Drivers

A seal puller is generally a lever with a handle at one end and two different shaped hooks at the other end. They are used as a lever to remove a seal from a bore. They will catch the back of the seal and pry it out; this destroys the seal. Never reuse seals; they will undoubtedly leak. Bushing and **seal drivers** are discs of assorted diameters used to remove or install a bushing or a seal. Select the disc that is just slightly smaller than the diameter of the bushing or seal. This allows you to drive the component in or out while applying force equally around the circumference. The driver will prevent the edge of the bushing or seal from being damaged when it is installed. Using a punch on just one spot would likely deform a bushing and surely destroy a seal.

Ridge Reamer

Before removing pistons from a worn cylinder, the ridge formed at the top of ring travel must be removed to prevent damaging the pistons **(Figure 16)**. A ring ridge removing tool, a **ridge reamer**, cuts the ridge of carbon and metal away so that the pistons can be removed without being scratched or broken. The tool sits on top of the piston and is centered in the bore by movable arms that expand out equally to reach the cylinder walls. A carbide cutter is adjusted to just touch the cylinder wall below the ring ridge. Then you use a wrench to thread the shaft holding the cutter upward. The cutter rotates around the cylinder as it slowly moves up through the top of the cylinder. The tool does not provide a fine finish; you will repair that later through honing or brushing.

You Should Know *We have had more than one experience of breaking a piston in our shop while trying to remove a piston past a "small" ridge. Play it safe and remove* any *ring ridge before taking the pistons out.*

Ring Expander

Once the pistons are removed from the cylinders the rings are removed from the pistons. A **ring expander** should be used to prevent scratching the sides of the piston when removing the rings. Any vertical scratches on the piston can allow oil to leak by and be burned in the combustion chamber. The ring expander is like a set of pliers. The ends of the jaws have a lip on them to catch the ring ends. Squeeze the ring expander grips to spread the piston ring. The expander is used again when the time comes to install new rings.

Ring Groove Cleaner

A **ring groove cleaner** has various sized bits on it to clean different sizes of ring grooves. You install the proper size bit on the end of the tool and rotate the tool around the piston in the groove to thoroughly clean the grooves of carbon and oil residue. Some technicians snap one of the old rings and use it to scrape the groove clean. This is an appropriate, though less efficient, technique when the correct size bit is not available for the ring groove cleaner.

Ring Compressor

When the pistons with new rings are ready to be installed in the engine, you use a **ring compressor (Figure 17)**. A sleeve is clamped around the piston and rings so that the assembly will fit into the bore. Then, rest the sleeve on the deck and carefully drive the well-oiled piston assembly into the bore. This just takes gentle pressure from a light plastic or wooden mallet.

Cylinder Hone

A **cylinder hone** can be used to properly refinish a freshly bored cylinder or to restore a slightly worn cylinder. On newer engines only fixed hones should be used and only after the cylinder has been bored. The tolerances in modern engines are too small to allow the variations in size left by flexible honing. If, for example, the cylinder is out of round, a flexible hone will not make it round again; it will follow the wear pattern already established. Most boring and honing of cylinders is done at a specially equipped machine shop.

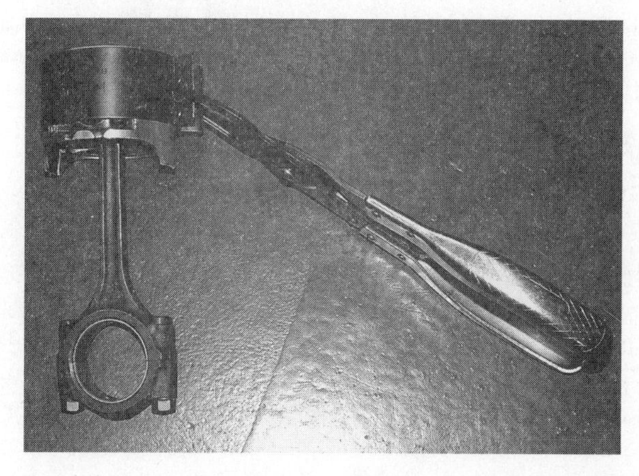

Figure 17. The ring compressor clamps the rings fully into the grooves to allow piston installation.

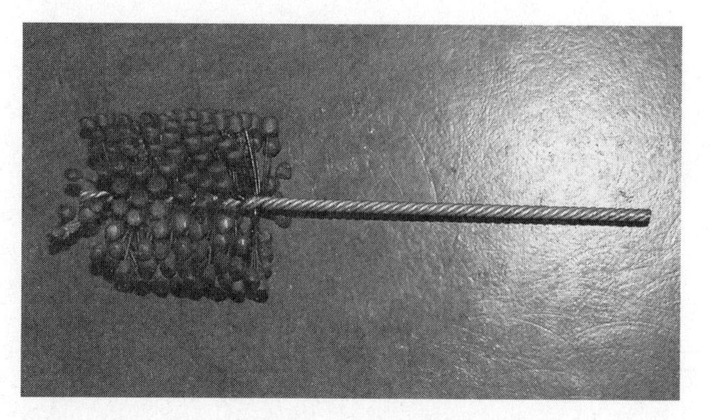

Figure 18. A brush hone can provide a good cylinder surface for the new rings to break into.

Cylinder Brush or Deglazer

A **cylinder brush** or deglazer is a long stem that fits into a drill with multiple flexible arms with abrasive balls on the end **(Figure 18)**. The balls form the circumference of the tool and are fitted into the cylinder. The correct diameter tool must be used for the cylinder. The tool takes very little material off the cylinder walls. It can be used when the cylinder measurements are within specification and only deglazing of the walls is required. The cylinder walls become glazed over time as the rings wear down and glide over the surface of the walls making them smooth. It is important to remove this glaze when installing new rings to help the new rings break in properly. You run the brush up and down in the cylinder on a drill at a speed that produces a proper crosshatch pattern. The small diagonal scratches formed by the abrasive balls hold a little oil to lubricate the rings.

Straightedge

The engine block's deck and the cylinder head surface must be very close to flat to allow proper sealing by the head gasket. You check this by laying a machined **straightedge** on the surfaces. Then, try to slide feeler gauges under any part of the straightedge. In this way you can measure the amount of warpage and determine whether the surface needs to be refinished.

You Should Know *A straightedge is only as good as its treatment. If a straightedge is dropped it should be remachined true. Always store a straightedge clean; acids can eat away at the machined surface.*

Vee Blocks

Machined **vee blocks** are used to measure the straightness of the camshaft and crankshaft. Lay the crank or cam on its main journals in the vee blocks. Carefully rotate the shaft and measure the variations in straightness using a dial indicator.

Cam Bearing Driver

A **cam bearing driver** is used to install cam bearings for an in-the-block camshaft **(Figure 19)**. Because access to the bearings is limited, a long tapered shaft with driving discs on it is used to tap the bearings into place. This process is often performed at the machine shop after engine cleaning.

Oil Primer

An oil priming tool may be used before the reassembled engine is started. This tool pumps oil throughout the engine so that oil pressure develops immediately on engine startup. An adapter threads into the oil pressure sending unit port, and oil is pumped throughout the lubrication system.

CYLINDER HEAD REPAIR TOOLS

The top end of the engine can be repaired independently or in conjunction with an engine overhaul. A "valve job" involves refurbishing the cylinder head, valves, and valve train components. Several specialized tools are required for in-shop cylinder head repairs. Some shops send all their valve jobs out to a machine shop or simply replace the head with a refurbished unit.

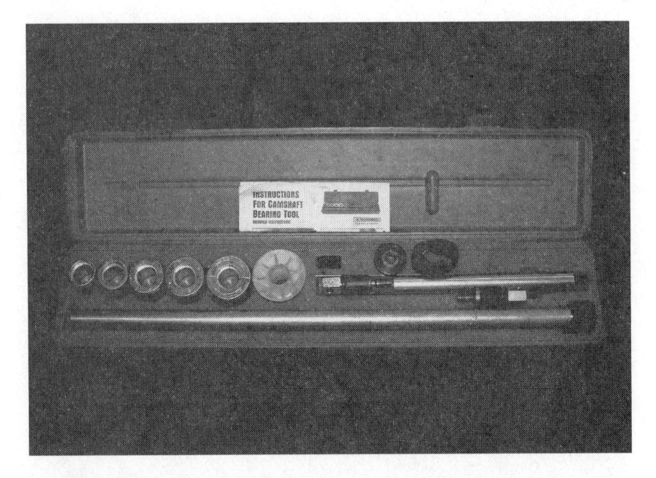

Figure 19. This camshaft bearing tool can be used to install newcam bearings in the block.

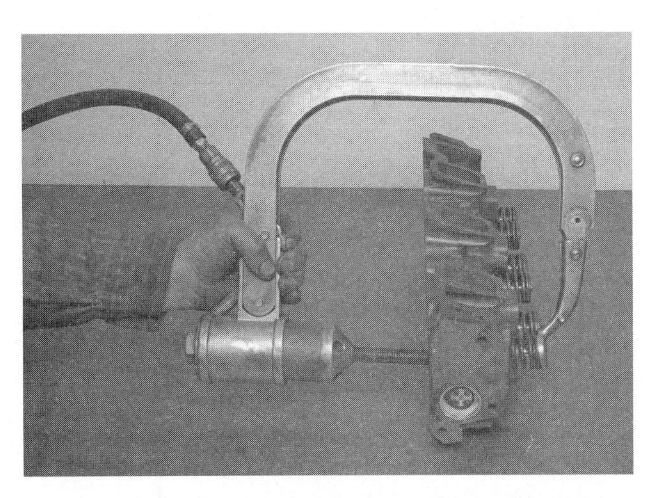

Figure 20. This valve spring compressor uses shop air to compress the spring for valve removal.

Valve Spring Compressor

Two types of **valve spring compressors** are used in engine repair. One is a small spring-loaded screw-type unit used to compress valve springs with the cylinder head on the engine and fully assembled. This can be used to remove the springs to replace valve seals without removing the cylinder head. Two or three jaws grab the spring at its base. Then the top of the tool screws down on top of the spring retainer to compress the spring.

The other type of spring compressor is used to disassemble the cylinder head when the head is removed from the engine. This spring compressor uses a large clamp with a lever or air pressure to compress the spring against the spring seat **(Figure 20)**. One end of the clamp fits on the valve head, and the other end pushes down on the valve spring retainer. Once it is compressed, you remove the keepers and loosen the tension on the compressor. Then the valve train components can be removed.

Valve and Valve Seat Resurfacing Equipment

This tool set includes valve face and tip grinding equipment and valve seat refinishing tools **(Figure 21)**. To refinish the valve face the valve is rotated in the chucks of an electric motor. You adjust the valve very close to a rotating cutting stone. Then you slowly move the valve into the stone as you pass the valve back and forth across the stone to clean up the valve face. Often you can restore a valve adequately to reuse it.

The valve seats must also be refinished in order to ensure good valve sealing. First the valve guides must be repaired or replaced if needed. Once the guides are accept-

Figure 21. This valve grinding bench can restore used valves effectively.

able, a pilot is placed in the guide. A cutting stone with the appropriate angle is attached to a special motor and placed over the guide. The stone is allowed to spin around on the seat to cut it; no pressure is applied. Again, the goal is to just clean the seat while removing as little material as possible.

Valve Spring Tester

A **valve spring tester** is used to measure the spring's tension. Place the spring on the table of the tool, and adjust the height gauge to the spring height. Place a dial-indicating torque wrench in the end of the tensioning bar. Compress the spring until the tone sounds on the tool. Read the torque value on the gauge, and multiply it by two to obtain the valve spring tension reading.

Summary

- Stethoscopes, vacuum gauges, and compression and cylinder leakage testers can be used to evaluate the mechanical function of the engine before teardown.
- Many measuring tools are used to assess engine wear. These tools may be as accurate as .0001 in.
- Micrometers may measure thickness, diameter, and depth. Metric and standard micrometers are available.
- Specialized cooling system tools are used to maintain and troubleshoot the system.
- Block reconditioning tools allow the technician to perform many repairs in the automotive shop.
- Valve and seat refinishing tools are used to perform a valve job.

Review Questions

1. Name three tools that may be useful in diagnosing engine mechanical failures before disassembly.
2. A machinist's rule is usually accurate to this fraction of an inch: ___________.
3. List three components you might measure with a micrometer.
4. Technician A says that a dial indicator can be used to measure cylinder diameter. Technician B says that telescoping gauges and a micrometer can be used to measure cylinder bores for wear. Who is correct?
   A. Technician A only
   B. Technician B only
   C. Both Technician A and Technician B
   D. Neither Technician A nor Technician B
5. What tool should be used to remove the buildup at the top of the cylinder before removing the pistons?
   A. A ring groove cleaner
   B. A ring compressor
   C. A flange puller
   D. A ridge reamer
6. Technician A says to use a cooling system pressure tester to check the radiator cap. Technician B says to use a hydrometer to check for system leaks. Who is correct?
   A. Technician A only
   B. Technician B only
   C. Both Technician A and Technician B
   D. Neither Technician A nor Technician B
7. Technician A says that you must remove the cylinder head to replace valve seals. Technician B says that you need a spring compressor to replace valve seals. Who is correct?
   A. Technician A only
   B. Technician B only
   C. Both Technician A and Technician B
   D. Neither Technician A nor Technician B
8. Technician A says that a valve face may need to be reconditioned. Technician B says that worn valve guides must be replaced before reconditioning the valve seats. Who is correct?
   A. Technician A only
   B. Technician B only
   C. Both Technician A and Technician B
   D. Neither Technician A nor Technician B

9. Each of the following is a precision measuring tool, *except* a:
   A. Micrometer
   B. Telescoping gauge
   C. Dial caliper
   D. Machinist's rule
10. Technician A says to use a cylinder brush to deglaze cylinders before installing new rings. Technician B says that a crosshatch pattern helps rings break in and holds oil for ring lubrication. Who is correct?
   A. Technician A only
   B. Technician B only
   C. Both Technician A and Technician B
   D. Neither Technician A nor Technician B

Chapter 3 Engine Hardware, Gaskets, and Sealants

Introduction

In an engine the proper fasteners must be used in the correct location. Before reusing nuts and bolts, check their threads; they must be in good condition to hold adequately. Nuts and bolts must be tightened to the correct specifications to ensure a quality engine repair. One loose connecting rod bolt could destroy an expensive overhaul. Engine gaskets seal everything from oil to combustion pressure. They must be installed properly to function as designed. The most common cause of gasket failure is improper installation. Sealants should be used only when specified by the manufacturer. Appropriate application ensures that the quality of your repair work will not be marred by a leak.

A freshly rebuilt engine was brought into the shop because it had been damaged by a loss of oil pressure. During the inspection process the technician noted an excess of room temperature vulcanizing (RTV) sealant around the oil pan. Once the oil pan was removed the reason for the failure was clear. The technician had used so much RTV when sealing the oil pan that much of it squeezed out during tightening and wound up plugging the oil pick-up tube screen.

MEASURING SYSTEMS

The United States still uses two measuring systems, International Systems, known as **metric**, and United States Customary System, known as the **standard** or English system. Most other countries use the metric measuring system. You should be prepared to use both systems. When you are working on engines you are likely to encounter both measuring systems on the same vehicle; some fasteners may be metric and others standard. Some engine measurements are given in either metric, standard, or both. Some manufacturers still use the standard system when providing specifications for engine repairs, although the trend is toward the international metric measurements. Some European and Asian manufacturers provide only metric specifications.

The standard or English system measures fasteners, dimensions, and clearances in portions of inches. A component dimension may be measured to the inch (1.0 in.), the tenth of an inch (.1 in.), the hundredth of an inch (.01 in.), the thousandth of an inch (.001 in.), or the ten thousandth of an inch (.0001 in.). The inch may also be broken down by fractions as read on a machinist's rule: 1/2, 1/4, 1/8, 1/16, 1/32, 1/64. When converting inches to millimeters, multiply the standard measurement of inches by 25.4 to find the metric equivalent in millimeters.

In the metric system units are divided into meters. One meter is equal to 3.28084 feet. When measuring automotive components and fasteners we typically use millimeters, (mm), or thousandths of a meter. One millimeter is a very small fraction of an inch, under 3/64 in. To convert millimeters to inches multiply the millimeters by .03937. For example, 3/4 of an inch or .750 in. times .03937 = 19.05 mm.

Since vehicles use both metric and standard fasteners, you will need to have a full set of metric and standard wrenches and sockets.

FASTENERS

Many types of fasteners are used to hold components together. The most common is a nut and a bolt; this is called a threaded fastener. Threaded fasteners come in many different forms, sizes, and designs. Standard bolts and screws are available in Unified National Coarse (UNC), Unified National Fine (UNF), Unified National Extra Fine (UNEF), and Unified National Pipe Thread (UNPT or NPT).

Some components are held together using studs and nuts. This fastening system is used on rocker arms, for example. A stud is threaded or pressed into a bore in the cylinder head. Then the rocker arm is placed over the stud, and a nut holds the rocker arm in place.

Set screws may be used to prevent a part from rotating on a shaft. Typically these are countersunk screws (having no protruding head) or Allen head screws.

Keys are machined pieces that fit into a groove (keyway) on a rotating member to prevent spinning. The crankshaft pulley is typically held on the crankshaft with a key. The key should be square on the edges if no movement has been occurring between the two pieces. Replace a rounded key or have the keyway remachined to fit an oversize key if damage has occurred.

Bolt Identification

Bolts are properly identified by the diameter of the threaded area—the **shank**, not the **head**. A 6 mm bolt, for example, typically has a 10 mm head. Therefore, a 6 mm bolt usually requires a 10 mm wrench to loosen it. The length of the bolt is measured by its threaded shank, from the bottom of the head to the end of the bolt. On standard bolts, the **pitch** of the bolt is measured in threads per inch. A 3/8-in. bolt may be UNF with 24 threads per inch or a UNC with 16 threads per inch. In the metric system the distance between two threads determines the pitch of a bolt. Most bolts run between 1.0 and 2.0 mm between threads. A common 8-mm bolt with a 13-mm head is an 8- × 1.25-mm bolt. The 1.25-mm measurement is the distance between threads.

Bolt strength or **grade** is marked on top of the bolt head. On standard bolts there are typically three lines for regular grade 5 fasteners. Six lines indicate the higher tensile strength of grade 8 bolts. Never exchange a higher-grade bolt with a lower-grade replacement. On metric bolts the strength is identified by the numbering; a common bolt is marked 8.8, while a higher-grade bolt reads 10.9 on its head. Nuts are not always marked, but they are graded as well. Always use a high-grade nut with a high-grade bolt; failure to do so compromises the strength of the fastener **(Figure 1)**.

Always pay careful attention to the hardware. The correct strength and length bolt must always be used. Many technicians use small bags to hold groups of nuts and bolts during disassembly procedures and label their locations. This is much more efficient than digging through a big can of nuts and bolts looking for the timing cover bolts, for example. Inspect the threads for damage, and repair or replace any imperfect fasteners. Also check a bolt for stretch as shown in **Figure 2**. Do not reuse a stretched bolt; it is likely to break and will not hold adequately. Many manufacturers specify replacement of fasteners whenever they are removed during repair work. Read the service information carefully to avoid a costly mistake.

Fastener Tightening

Fasteners must be tightened correctly to meet the holding demands of the application. During an engine repair, replacement, or overhaul, almost every bolt and nut must be

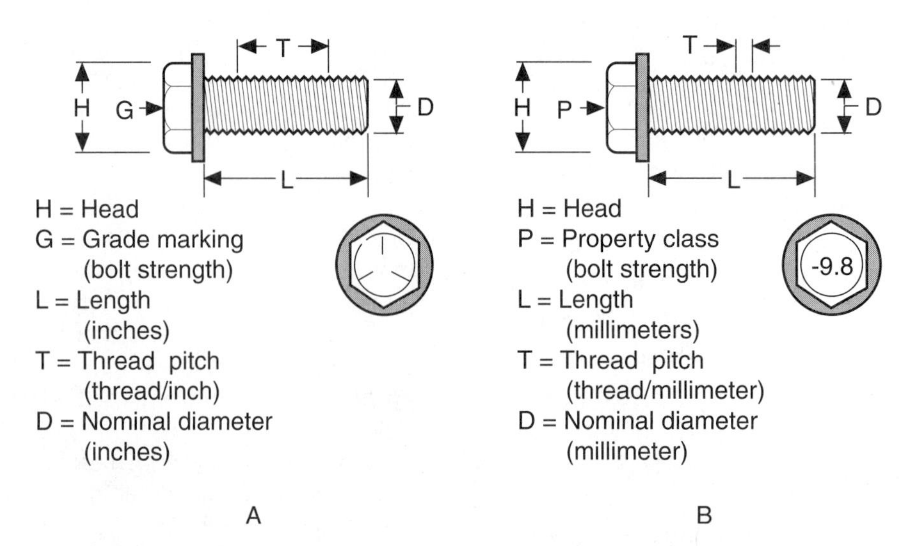

Figure 1. English standard (A) and metric (B) terminology.

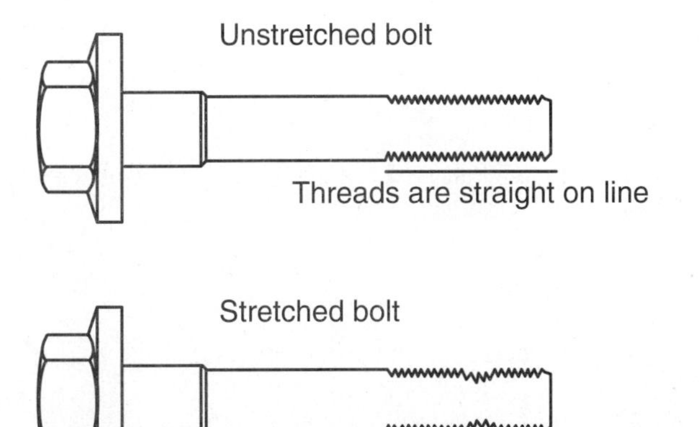

Figure 2. Check all fasteners for stretch or thread damage before reusing them.

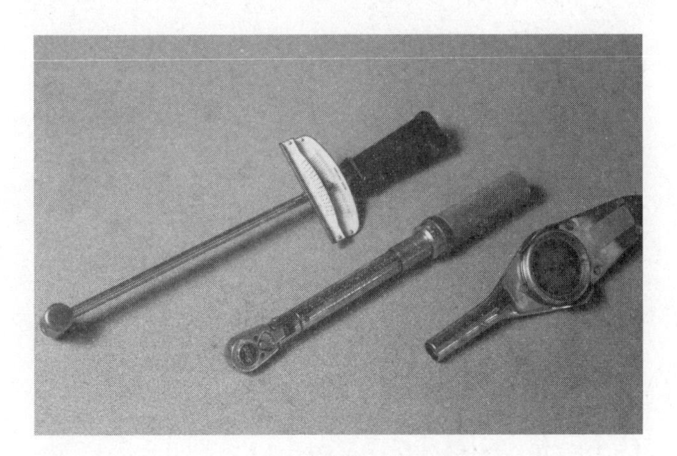

Figure 3. Various types of torque wrenches.

torqued to its specification to ensure longevity of the engine. **Torque** is twisting force as measured in foot-pounds or newton-meters. A **torque wrench** is used to tighten bolts and nuts to their torque specification **(Figure 3)**. The appropriate socket fits on the end of the torque wrench. Some torque wrenches use a dial gauge where the needle deflects as the bolt is tightened. Tighten the bolt until the gauge displays proper torque. Other torque wrenches are click type or breakaway type. To use this type of tool, set the desired torque on the handle of the wrench. Then tighten the bolt until you feel a click or a slip.

Torque to Yield Bolts

Manufacturers often use **torque to yield bolts** to improve consistency of clamping forces on cylinder heads and other critical components. Traditional tightening of a bolt is performed using a torque wrench. You twist each bolt head to a specified torque in a particular pattern. This procedure, however, actually leaves significant variations in the clamping forces at the threads of different bolts torqued to the same specification. To improve cylinder head sealing, for example, engineers now often specify torque to yield bolts. In these cases the bolts are torqued to a specified tightness and then rotated an additional number of degrees. You can use a **torque angle gauge** to measure the specified rotation. The gauge fits over an extension or socket and displays the degrees of rotation. This brings the bolt to its yield point where clamping forces equalize. A crucial point to remember is that torque to yield bolts cannot be reused. During the tightening process they stretch beyond their elastic limit. Follow the manufacturer's procedures carefully for bolt replacement and tightening.

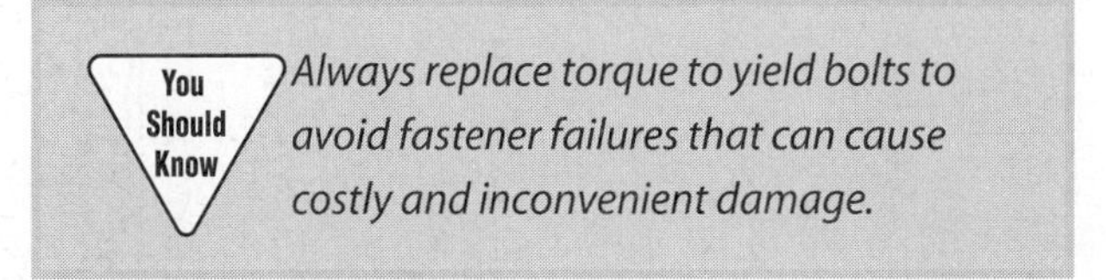

Always replace torque to yield bolts to avoid fastener failures that can cause costly and inconvenient damage.

Thread Repairs

Thread repairs are accomplished using a thread file, a tap and die set, or a **helicoil** threaded insert. If very minor imperfections are seen on a bolt, you can use a thread file to remove the displaced metal. Sometimes the very end of the bolt is lightly damaged and a thread file can clean up the threads easier than a die.

When more serious defects are seen in the threads, a tap or die should be used. A die is used to repair the threads of a bolt or screw. Use a thread pitch gauge to find the proper thread pitch **(Figure 4)**. Then pick a die with the same diameter and pitch as the bolt. Use the die holder to gently screw the die over the bolt. If the die becomes hard to tighten, back the die off and clean the threads. Continue moving forward and backward slowly until you reach the

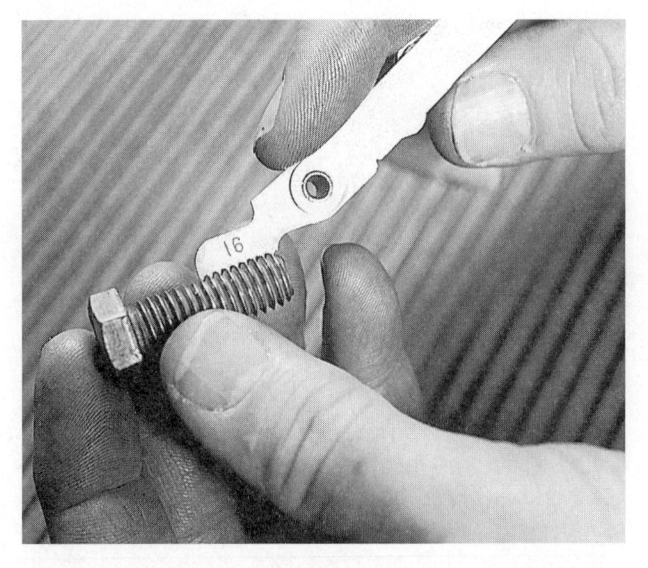

Figure 4. Use a thread pitch gauge to select the correct die.

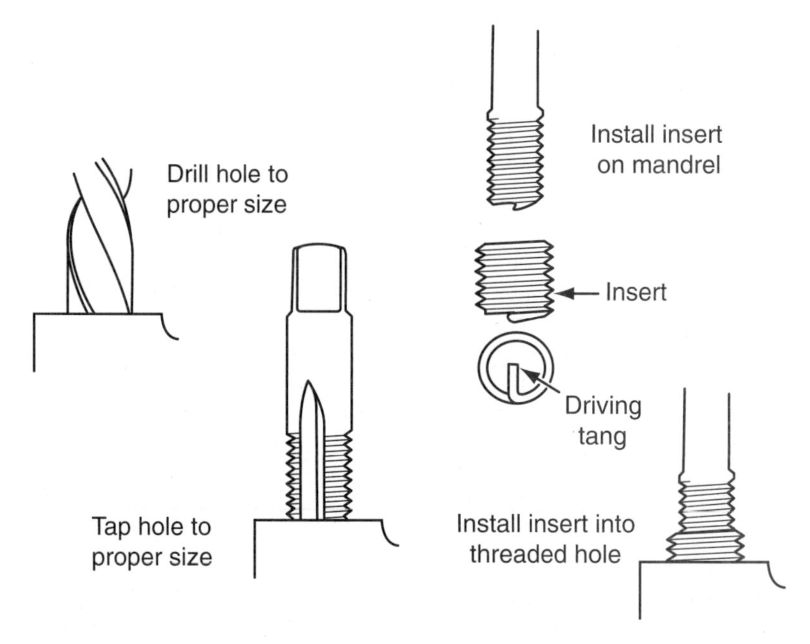

Figure 5. Follow the steps to install a helicoil.

end of the threads. A die is not intended to rethread a bolt; it should only be used to clean up existing threads. If the bolt is damaged too significantly, the die will remove too much material from the bolt to make it salvageable. Whenever possible, replace a damaged bolt. A tap is used to restore internal threads. The same careful procedure should be used to run the tap down the bore. Be sure to clean all the metal filings out of the hole before refitting the bolt.

If the threads in a hole are stripped beyond repair, a threaded insert or helicoil can be installed **(Figure 5)**. First, drill the hole to the size indicated on the threaded insert kit. Use the indicated tap to rethread the new hole. Then use the insert tool to screw the helicoil into the threaded hole. This repair is commonly needed on aluminum cylinder heads or engine blocks.

GASKETS

Gaskets are used to seal the joint between two stationary components. Gaskets can be made of paper or fiber, steel, cork, synthetic rubber, and many combinations. Timing cover gaskets and water pump gaskets are frequently made of paper but can also be molded or synthetic rubber **(Figure 6)**. Steel or metal combination gaskets are used to seal the cylinder head, the exhaust manifold, and the exhaust gas recirculation valve. Some valve cover gaskets are still made of cork, though many parts manufacturers offer a synthetic rubber option. Synthetic rubber gaskets are often used to seal valve covers, intake manifolds, some thermostat or water pump housings, and oil pans. All gaskets need to be installed properly to provide a good, durable seal. The old gasket material and any sealant must be completely removed from the sealing surfaces. Read the manufacturer's service information for proper installation, and use no sealants unless specified. Always tighten a gasket cover to its specified torque in the proper sequence. Most gaskets should be replaced whenever they are removed. This is inexpensive insurance compared to the labor of having to perform the procedure again to correct a leak.

Cylinder Head Gasket

The head gasket provides critical sealing for the engine. It must contain combustion pressures over 2500 psi and withstand temperatures over 2000°F. It must also keep

Figure 6. This timing cover gasket is made of steel with a rubber sealing bead.

engine oil and coolant from leaking into the cylinders or out of the engine. Newer multilayer steel or rubber-coated embossed head gaskets are made of several materials and multiple layers. The base is made of steel for gasket strength and resistance to deformation. The steel is covered with composite facing material to allow the gasket to compress to the surface irregularities of the cylinder head and block deck. An antifriction coating of Teflon or silicone is used to prevent abrasion as the cylinder head and block expand and contract at different rates. The slippery coating allows the head gasket to slide between the two surfaces without tearing or abrasion. Many gaskets have elastomer sealing beads around fluid passages to prevent leakage. The fire ring is the part of the head gasket circling the cylinder bores. This stainless steel ring must remain intact while being exposed to the heat and pressure of combustion. Some older or cheaper head gaskets may be made of composite fiber with steel fire rings. Other head gaskets may be perforated steel core. These gaskets typically have fiber and clay facing on a thicker perforated core **(Figure 7)**. The gasket is designed to compress significantly to seal surface imperfections. The perforated core makes these gaskets somewhat weaker than mulitlayered steel gaskets.

Head Gasket Installation

Use a good quality gasket that meets or exceeds original equipment specifications. Check service information for possible updates to the head gasket; many manufacturers have revised their recommendations since an engine was produced. Clean the cylinder head and engine block surfaces completely. Be careful when you scrape or grind the old head gasket away not to remove material from the engine parts. Many manufacturers state that only plastic scrapers may be used. Parts stores carry abrasive discs that can be used on cast iron and some aluminum cylinder heads; make sure you are using the correct type. Be careful to read the appropriate service information; many manufacturers do not allow the use of these abrasive discs on aluminum surfaces. They can damage the surface finish of the head and leave abrasive particles in the engine. Often, chemicals are recommended to assist in removing the old gasket material **(Figure 8)**. Most modern head gaskets are installed dry without any additional sealants. Only use sealers or coating if the head gasket manufacturer specifies them. Look carefully at the gasket before installation. Many gaskets have a front and/or top marking; if not, look carefully to be sure all the cooling and oil passages line up with a proper-size passage in the head gasket. Similarly, some head gaskets are different for the left and right heads of a vee engine. Improper placement may block cooling passages and cause overheating.

The cylinder head must be properly torqued onto the block to ensure that the head gasket seals properly. Check to be sure whether the old head bolts can be reused or if new torque to yield head bolts are specified. The head bolts are torqued in stages in a particular sequence. **Figure 9** shows a typical torque pattern. The torque procedure may specify tightening the head bolts to 30 ft.-lb in sequence, then to 55 ft.-lb, and finally an additional 60°. Many head gaskets come with the torquing instructions in the package. Always locate the specified torque procedure; guessing will result in head gasket failure. Torque

Figure 7. This perforated steel core gasket has an elastomer sealing ring around the exterior.

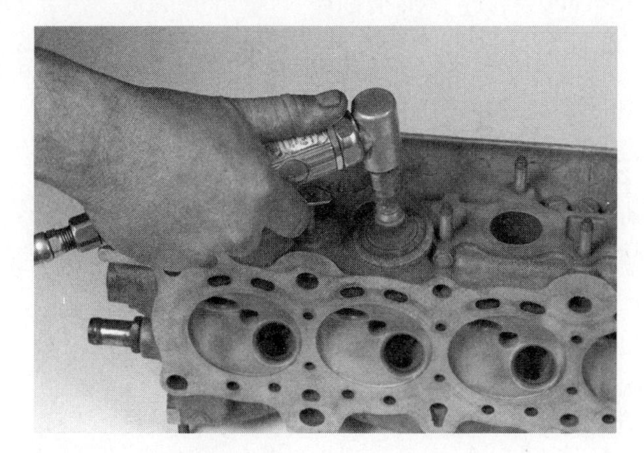

Figure 8. Use the proper disc to avoid damaging the head.

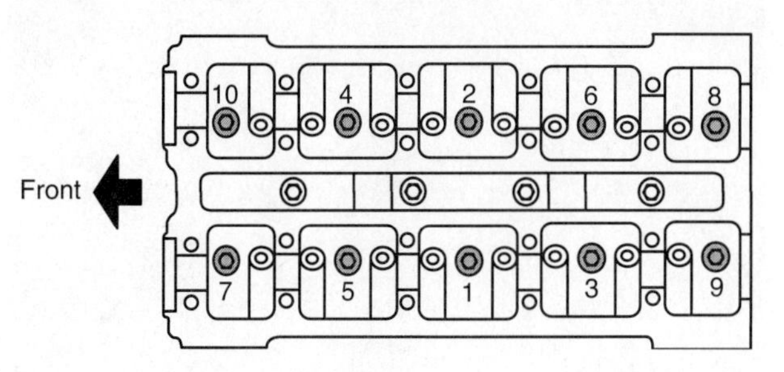

Figure 9. Follow the torque stages and sequence exactly to ensure good sealing.

specifications are given for bolts with clean and lightly lubricated threads.

Paper and Fiber Gaskets

Paper and fiber gaskets are often used between two machined surfaces to conform to and seal any surface imperfections. The timing cover, for example, often uses a thin paper gasket. These gaskets rarely require additional sealant, but sometimes a thin layer of spray adhesive or aviation sealant can be used to hold the gasket in place for installation. Be sure to torque covers down in a diagonal pattern even when a specific sequence is not given. Start at the center of the cover, and work your way out to the ends diagonally to prevent distortion of the cover or the gasket. Use the proper torque specification to ensure equal clamping force around the gasket.

Cork Gaskets

Occasionally you will still run into a cork gasket. New designs have replaced cork applications with synthetic rubber for better sealing. Cork gaskets are very soft and absorbent. They are easily distorted and regularly weep some fluid. Be careful to use a diagonal pattern when installing a cork gasket. Do not overtighten the cover, and use the proper torque specification to prevent irregular thicknesses from occurring. Never retighten a cork gasket; it was compressed as much as it should have been when it was originally installed. Retightening will usually result in greater, not less, leakage.

Synthetic Rubber Gaskets

Many gaskets are now made of synthetic rubber or molded synthetic rubber. These gaskets provide superior sealing qualities. They distort to the surfaces but are not actually compressed; they attempt to return to their original shape. This means that after they are torqued into place they try to revert to their natural form, and this improves their sealing effect **(Figure 10)**.

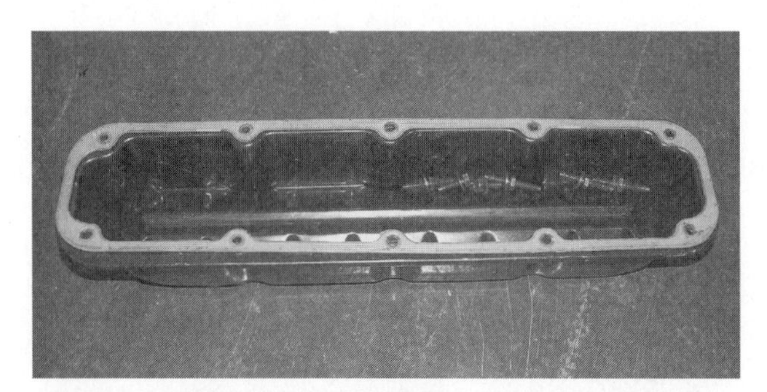

Figure 10. This molded synthetic rubber valve cover gasket will seal much better than a cork gasket.

Many oil pan and valve cover gaskets are made of synthetic rubber. Proper torque in a diagonal pattern is still essential to good sealing. Do not use any sealers or adhesives on synthetic rubber gaskets; they can prevent the gasket from sealing well. If the valve cover or oil pan is made of thin stamped steel (rather than aluminum), be sure that the bosses around the bolt holes have not been distorted during a previous installation. If the cover is misshapen, place the gasket surface on a flat, hard surface and tap the bosses back into place **(Figure 11)**.

SEALS

Seals are used when one of the components rotates. When overhauling an engine you will always replace the front and rear main engine seals and the camshaft seal if one is used. These seals are typically radial lip seals **(Figure 12)**. The seal has a flexible lip that rides against the rotating part. Behind the lip there is a space for hydraulic pressure to push against the lip and often a garter spring to add some tension to the sealing lip **(Figure 13)**.

Figure 11. You can straighten the sealing lip of a steel cover; lay it on a flat surface and tap it into shape.

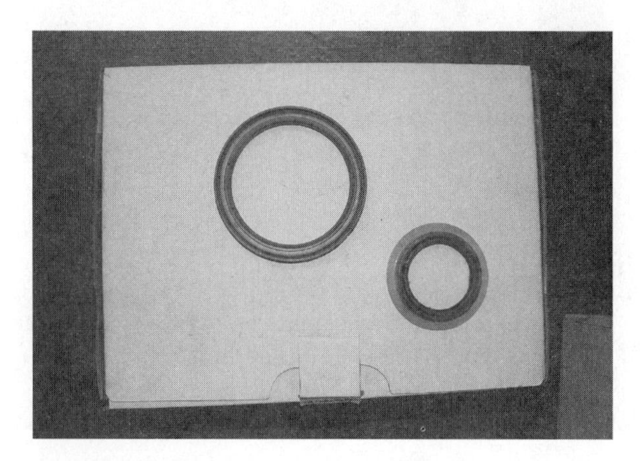

Figure 12. A soft radial lip seal (left) and one with a steel casing (right).

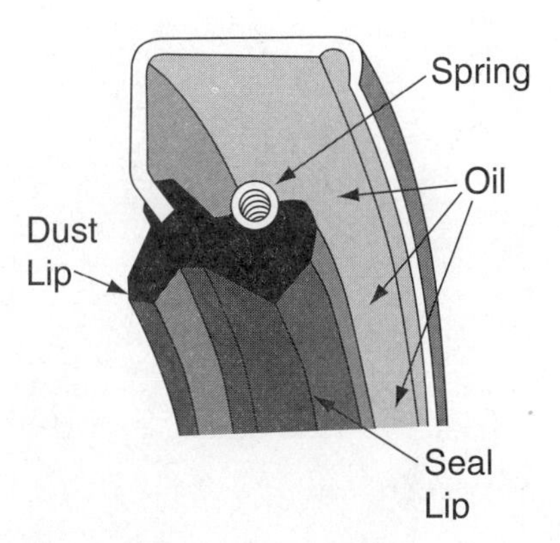

Figure 13. The parts of a radial lip seal work to keep dust out and fluid in.

Figure 14. Use a seal driver to install the front main seal into the timing cover.

Proper installation will ensure a good seal. Some seals use silicone or an adhesive on the outer ring of the seal. Use only what is specified in the service information. The lip should be lubricated with whatever fluid it is sealing **(Figure 14)**. In the case of the engine seals, run a layer of engine oil around the surface of the lip. This will prevent tearing during initial start-up. Use a seal driver to install the seal; this applies equal pressure around the circumference of the seal. Never use a hammer directly on a seal; you will damage it.

To replace a leaking rear main seal, you often need to remove the transmission or transaxle. Be sure you perform the job properly the first time. Before installing a new front main seal, inspect the crankshaft carefully. On an engine with high mileage the old seal may have worn a groove into the crankshaft, preventing a new seal from working properly. If the crankshaft is worn, thin sleeves are available to provide a smooth new sealing surface. As with gaskets, the most common cause of failure is improper installation.

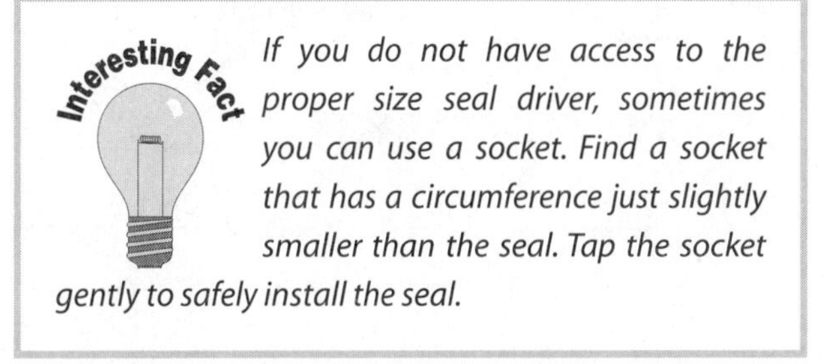

If you do not have access to the proper size seal driver, sometimes you can use a socket. Find a socket that has a circumference just slightly smaller than the seal. Tap the socket gently to safely install the seal.

SEALANTS AND ADHESIVES

Manufacturers sometimes specify the use of sealants or adhesives to assist a gasket or seal, to seal or lock a bolt, or to form a gasket. Use these compounds only when specified by the manufacturer; improper use can destroy a seal or damage a component. More is not better when using sealants and adhesives. Too much sealant can get into the engine and clog up oil passages. This can rapidly destroy an engine. Proper application is essential to good performance.

Room Temperature Vulcanizing Compound

Room temperature vulcanizing (RTV) sealer is a silicone compound **(Figure 15)**. It is frequently used at gasket seams and as a gasket forming material. It should not be used to coat gaskets. Some oil pans are sealed not with a gasket but with a continuous bead of RTV. Run a bead around each of the bolt holes as you make your way

Figure 15. Ultra black high temperature RTV.

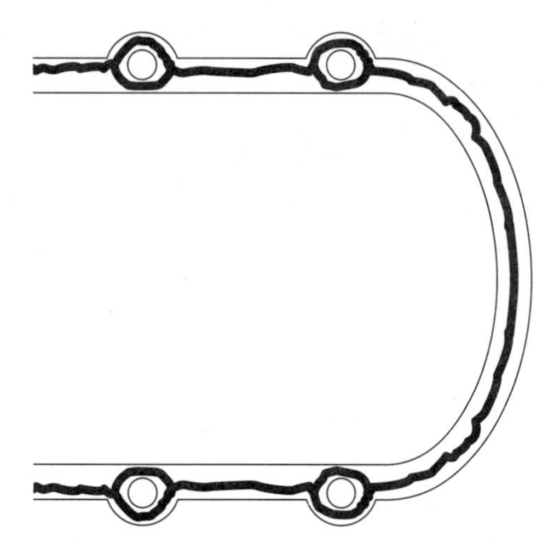

Figure 16. Run a continuous bead around the cover and circle the bolt holes.

around the oil pan **(Figure 16)**. RTV is an aerobic sealant, meaning it seals in the presence of air. The moisture in the air helps the compound set. RTV begins to set in about fifteen minutes; torque the components into place before the sealant hardens. RTV comes in a variety of colors that denote the temperature range they will work in. Be careful to use the correct RTV for the application. Check to be sure the RTV you use advertises either that it is safe for oxygen sensors or that it is free of acetic acid. Most automotive RTV compounds are now made with a type of silicone that does not damage oxygen sensors.

Anaerobic Sealants

Anaerobic sealants are typically used between two machined or cast surfaces **(Figure 17)**. They cure in the absence of air; they will not set until the two pieces have been torqued together. The surfaces must be thoroughly cleaned for the sealant to work properly. Loctite formulas are commonly used anaerobic sealants. Loctite numbers

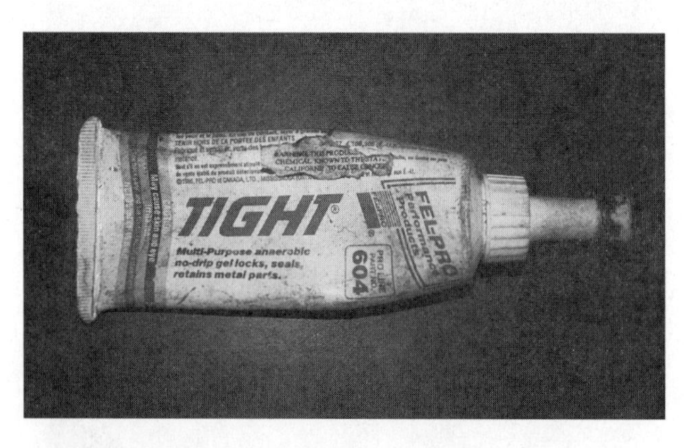

Figure 17. A multi-purpose anaerobic sealant.

Figure 18. Use a threadlocker when specified to lock bolts in place.

its products, and each formula has a distinct application. Loctite formulas can be used to form a seal on some timing covers. A different formula is used on threads to lock some bolts into place once they are torqued. This makes the bolts very difficult to remove later **(Figure 18)**. Use anaerobic sealants only when specified.

Thread Sealants

Some bolt threads must be sealed with Teflon thread tape or brush-on **thread sealer (Figure 19)**. Many head bolts or water pump bolts are turned into threads that end in a coolant passage. These bolts must be properly sealed or they will leak. Either thread tape or brush-on thread sealer can be used on "wet" head or pump bolts. Read the service information carefully to determine which bolts, if any, need sealant.

Figure 19. Wrap thread tape or use Teflon paste on the threads of a bolt that runs into a cooling passage.

Summary

- Standard and/or metric specifications are given in engine repair procedures.
- Metric and standard fasteners may both be used on the same engine. They cannot be interchanged.
- Proper tightening of fasteners is critical to quality repairs.
- Torque to yield bolts cannot be reused.
- A tap and die set can be used to repair some fasteners.
- Gaskets may be made of paper, fiber, cork, synthetic rubber, steel, or any combination of the compounds.
- Gasket surfaces must be thoroughly cleaned before installing a new gasket.
- Seals are used when one of the parts rotates.
- Use a seal driver when installing seals to prevent premature failure.
- RTV sealant is an aerobic sealing compound used frequently on gasket seams or in the place of gaskets.
- Anaerobic sealants are used between two machined or cast surfaces.
- Thread sealant should be used on "wet" head or water pump bolts.

Review Questions

1. How many millimeters are in one inch?
2. To convert millimeters to inches, multiply millimeters by:__________.
3. What are the three common types of standard bolt threads?
4. What fastener can be used to hold a component on a rotating shaft?
   A. A stud
   B. A UNF bolt
   C. A key
   D. A nut
5. Technician A says that a torque to yield bolt is rotated after the initial torque specification. Technician B says that torque to yield bolts cannot be reused. Who is correct?
   A. Technician A only
   B. Technician B only
   C. Both Technician A and Technician B
   D. Neither Technician A nor Technician B
6. Technician A says that most head gaskets should be sealed with a continuous bead of RTV. Technician B says that RTV can form a gasket in some applications. Who is correct?
   A. Technician A only
   B. Technician B only
   C. Both Technician A and Technician B
   D. Neither Technician A nor Technician B
7. Technician A says that aerobic sealants can set before parts are assembled. Technician B says that anaerobic sealants set when all the air is squeezed out between the two components. Who is correct?
   A. Technician A only
   B. Technician B only
   C. Both Technician A and Technician B
   D. Neither Technician A nor Technician B
8. Technician A says to tap a seal in around its edge with a hammer. Technician B says that sometimes a socket can be used to safely install a seal. Who is correct?
   A. Technician A only
   B. Technician B only
   C. Both Technician A and Technician B
   D. Neither Technician A nor Technician B
9. Technician A says that modern head gaskets are often multilayered steel. Technician B says that some head gaskets have a top marking on them. Who is correct?
   A. Technician A only
   B. Technician B only
   C. Both Technician A and Technician B
   D. Neither Technician A nor Technician B
10. Technician A says that many oil pan gaskets are made of synthetic rubber. Technician B says to retighten a cork gasket to stop any leakage. Who is correct?
    A. Technician A only
    B. Technician B only
    C. Both Technician A and Technician B
    D. Neither Technician A nor Technician B

Chapter 4 Working as an Engine Repair Technician

Introduction

To succeed as an engine repair technician you will need to prove yourself first as an excellent general service technician. Most employers want to determine that you consistently produce high-quality work before they rely on you to perform major engine work or an overhaul. It would be unusual to be chosen as an engine repair specialist at the very beginning of your career, though it is expected that you are capable of replacing, rebuilding, or making repairs to an engine. Being an engine repair specialist is a coveted position and one that holds a tremendous amount of responsibility. You will regularly generate repair bills in the thousands of dollars. Your work is expected to hold up for 100,000 miles. A seemingly small error can cause a catastrophic engine failure. To become a successful engine repair technician you must be well educated, professional, and conscientious and pay attention to detail.

You may be the technician most technically capable of performing an engine overhaul but if you do not communicate and present yourself professionally you may never get the chance.

EDUCATION

Your training to become a professional automotive technician is critical to your success. You must have a solid technical education and a well-rounded general education. You are required to understand a huge amount of technical information and to regularly interpret new information. As soon as you leave school you will need to find ways to update your training **(Figure 1)**. Specialized skills as an engine repair technician must be accompanied by a thorough understanding of automotive systems.

An ability to perform mathematical functions is essential; an error of one thousandth of an inch (.001 in.) could destroy an engine repair. Accuracy in measurements, interpretation of specifications, and conversion of numbers are all required skills. Additionally, the problem solving and analytical thinking used in math will be useful in almost every repair job.

Vehicle designs change constantly; new systems appear every year. The muscle cars of the 1960s had no electronics, antilock brake systems were nonexistent in the 1970s, air bags were brand new in the 1980s, a federally mandated **on-board diagnostic system generation II (OBDII)**

Figure 1. Reading professional periodicals is essential to keep your knowledge up to date.

radically changed vehicle diagnosis in the mid-1990s, and the new millennium brought production hybrid and electric vehicles into our service bays. These systems and hundreds of others are based on principles learned in physics and electronics. Your education will allow you to understand new systems as they are implemented. This will help keep your skills relevant even when 2040's vehicles make 2000's look unsophisticated.

You also need excellent communication skills to be successful. When performing a **warranty repair** it is your responsibility to effectively describe the customer concern, the cause of the problem, and all the necessary corrections. The manufacturer relies on the information you provide on the repair order to determine whether it will reimburse the dealer for the parts and labor required for the job. Failure to properly document your engine work could cost your employer thousands of dollars. In addition, repair information is kept on file so that vehicle history is available when other repairs are needed. It is also extremely important that you can communicate effectively with your service advisors, service manager, and customers. You will need to be able to explain technical information in a way that lay people can understand. Your service advisors rely on your description of your work to provide information to the customer. Often, particularly with large jobs such as an engine overhaul or replacement, the customer will want to speak directly with the technician **(Figure 2)**. You must be able to professionally and effectively describe your work to an anxious customer.

PROFESSIONALISM

From the moment you begin repairing vehicles, either as a student or as a paid professional the image you portray will help define your role in the shop, your rate of pay, your opportunities for advancement, and your overall success in the automotive industry. Clean work uniforms, proper language and behavior, and excellent communication skills will help develop your image as a professional service technician. Your presentation helps you as an individual and the service industry as a whole. When you dress, behave, and speak like a professional you will be treated and paid accordingly. There will be times when you are frustrated with a job. Take a short break and walk away before losing your cool. Profanities will not loosen a bolt or help your career. Your employer, a potential employer such as a manufacturer's representative, or a customer could be in earshot at any time. It will damage your image and your potential if they hear a string of foul language coming from you.

Keep yourself and your work area clean; you should always be prepared to present yourself to a customer. Neither your employer nor your customers want a filthy technician in a clean vehicle. Customers' vehicles are typically their second largest investment; they want service provided by a professional. You can perform the best engine repair but if you leave a spot of grease on the driver's seat, the customer will be unhappy with the whole job **(Figure 3)**.

> **You Should Know** *Some employers will let a perfectly competent technician go if she acts unprofessional in the shop or with management or customers.*

Your ability to communicate well with your service advisors, your employer, and your customers is essential to your career. If you cannot describe your work professionally you simply will not get that work. The more complex and in depth the job, the more critical it is that you can explain

Figure 2. Explain your repair fully and professionally to gain respect and trust from your customers and coworkers.

Figure 3. This customer would not appreciate a smudge of grease on her spotless interior.

your diagnosis and repair to management and your customers. When performing engine repairs you will need to explain why the problem occurred, how you found it, and what you did to correct it. If you are discussing an engine rebuild you are trying to "sell" thousands of dollars of work. Your employer relies on your explanations to get authorization for the job. Imagine the different effects the following two statements might have on a customer deciding whether to have work performed at your shop. "The engine blew; I'll have to totally overhaul it. It will probably cost around two thousand dollars." "Two cylinders have low compression caused by worn piston rings. This is the cause of your lack of power. Usually when the rings are worn there are other components in the engine that will need service or replacement. Once we disassemble and evaluate the engine we can offer you specific repair options. In the worst case a thorough engine overhaul or replacement may be needed; this could cost upwards of two thousand dollars." Who is the customer more likely to trust? If you were the boss which technician would you choose to perform the work? Offer your customers and your management respectful and professional communication; they will respond in kind.

COMPENSATION

You have the opportunity to enter a rewarding profession. Your skills are in high demand; rarely will a competent technician have any difficulty securing work. Skilled technicians can make a very comfortable wage. Your pay will increase as your skill level improves.

Most technicians start out working **"straight time."** This means that they are paid a dollar amount for every hour that they are at work. This gives them a chance to learn the shop practices and develop some speed in repairing vehicles. Some technicians will work for an hourly wage throughout their careers. A technician who performs primarily diagnostic work or electrical work will often be paid straight time. Sometimes engine overhaulers are paid straight time too. The employer understands that the speed of the overhaul is not nearly as critical as thoroughness and attention to detail.

Many shops will change your method of payment to **"flat rate"** pay when they determine that you will be productive on this system. When you are paid flat rate it means you get paid by the job hour. If, for example, a vehicle needs a head gasket the service advisor will look up the flat rate time and quote that amount of labor to the customer. When you do the job in six hours rather than the quoted eight hours you will be paid for eight hours. Your billable hours for the week are added up to calculate your pay. If you produced fifty-six hours while working for forty-five hours you would be paid for fifty-six hours. In a strictly flat rate shop this can also work to your disadvantage. If a few jobs took you much longer than the posted flat rate time you might actually be paid for fewer hours than you worked that week. This could happen if you replace a timing belt on a vehicle one week and it comes in the next week with a snapped timing belt. Many shops will not pay you again to repair a "come back"; if your service fails you can be held accountable and repeat the job without getting paid hours for it. This is just one of the many reasons why it is critical to perform quality work.

Some shops use a combination of the two forms of payment. They may guarantee you a certain number of base hours per week and pay you extra for increased production. This ensures that the technician is adequately compensated every week regardless of the type or quantity of work available. It also serves as an incentive to be as productive as possible.

ASE CERTIFICATION

The **National Institute of Automotive Service Excellence (ASE)** offers a well-recognized national certification program for automotive technicians. The standardized ASE tests are offered in eight basic areas of automotive repair, as well as in several specialized and advanced areas. When you pass all eight basic tests you become an ASE certified master automotive technician. Engine repair is one of the eight core tests needed to become a master technician.

The ASE tests are written tests offered twice a year at test sites all across the country. The tests check your practical knowledge of automotive systems and of diagnostic and repair techniques. To become certified you must pass the test and prove that you have two years of work experience. If you complete a two-year college program in automotive technology it will count for one year of experience. Once you are certified you will receive a certificate, a wallet card, and a shoulder patch documenting your achievement **(Figure 4)**. Technicians must retake the tests every five years to keep their certifications current.

Figure 4. This ASE certified technician wears his ASE patch proudly; it is a sign of accomplishment.

The engine repair ASE test asks questions about general engine diagnosis, cylinder head service, engine block service, and cooling and lubrication system repair, as well as a few questions on fuel, electrical, ignition, and exhaust systems. A free preparation booklet is available from ASE. It lists the tasks covered in the test and offers some sample questions. Many of the questions are in the format we use in this text:

Technician A says that a compression test can locate a weak cylinder. Technician B says that a cylinder leakage test can pinpoint the cause of a low compression reading. Who is correct?

A. Technician A only
B. Technician B only
C. Both Technician A and Technician B
D. Neither Technician A nor Technician B

In this question both technicians are correct, as you will learn in upcoming chapters. Successfully answering these questions requires a thorough knowledge of the subject and careful reading. While many technicians find the questions tricky, if you really know the information the correct answer will usually be clear. The engine repair test is made up of sixty multiple-choice questions.

The ASE tests are written by industry experts; technicians; technical service experts; vehicle, parts, and tool manufacturers; and automotive instructors. Each question must be accepted by each member of the board and then tested on a sample of technicians. The tests do a good job of checking your technical knowledge. Certification is voluntary, but most employers prefer to hire ASE certified technicians. ASE certification shows your employer that you are knowledgeable and motivated to excel **(Figure 5)**.

SKILLS REQUIRED OF AN ENGINE REPAIR TECHNICIAN

Special skills are required to succeed as an engine repair technician. You must thoroughly understand engine operation and design to effectively diagnose problems. For example, you will need to remember how critical engine breathing is to proper performance when diagnosing a low-power condition as a restricted exhaust system. Your study and practice of the information and procedures described in this text will provide you with a solid base of engine operation, diagnosis, and repair knowledge.

Figure 5. This tech and her shop boast the blue seal of excellence.

You must learn to thoroughly diagnose an engine problem before attempting a repair. Do not attempt to guess; test. We describe many different diagnostic procedures to pinpoint the cause of engine problems. Study and practice these tests now so you will be competent at them in the shop. The customer should receive a realistic quote for engine work; you should not get halfway through the job and *then* realize that more significant work is needed. Performing the proper tests before disassembly will allow you and your service advisor to estimate the required work. They will also help you locate the root cause of the problem so that your repair is 100 percent effective.

Your ability to access and interpret service information is essential to performing quality work. Manufacturers and aftermarket information companies provide service information on the internet, DVD, CD Roms, and in books. Service information from electronic sources is the norm; you will need basic computer skills to be able to maneuver your way through the information programs. When you diagnose an engine drivability concern you should check your service information source for applicable technical service bulletins (TSBs). These notices are developed by the manufacturer when a condition exists on many vehicles that is not identified in the regular service information. A TSB may describe a noise that occurs in the engine just after start up only when the engine is cold. The TSB will outline the correct repair procedure, such as installing a revised bracket to solve the noise concern. If you failed to search for applicable technical service bulletins it is unlikely that you would find the root cause of the problem or that you would successfully repair it. TSBs can save time and help you resolve concerns effectively. You will also need to locate the specifications for a particular engine when you are assessing its condition or making repairs. You will learn how to measure an engine's main bearing clearances in this text. When performing an engine overhaul you will need specific information to compare these measurements to the manufacturer's specifications.

Service information programs require that you correctly identify the vehicle you are working on. You must know the vehicle year, make, model, and engine size to look up information pertinent to the vehicle you are working on. Sometimes you will need to refer to the vehicle identification number (VIN) to correctly identify the model year, engine, or other pertinent information. The VIN is located at the left front corner of the dashboard and is visible from the outside of the driver's windshield. It is also stamped in several other places on the vehicle in case the VIN plate on the dash has been tampered with. The eighth character of the VIN identifies the engine and the tenth character denotes the model year. Use service information to decode the characters. This method of identification is the definitive one. You may need to provide the VIN when ordering parts or when searching for vehicle service history. When ordering engine parts the parts specialist may also ask you for an engine code which will be stamped on the engine block. Again, service information will help you locate and interpret the code.

You must master many repair procedures using special tools in order to become a skilled engine repair technician. An engine overhaul requires thorough preliminary diagnosis, accurate measurement with special tools, precision machining and refinishing, and careful torque and assembly techniques. You must be competent at all the techniques to complete the overhaul. In many cases a significant portion of these repairs will be performed at an engine machine shop. We will focus more on the repairs normally performed in general automotive repair shops, either dealerships or independent garages **(Figure 6)**.

In many cases your diagnosis will uncover engine problems that may be more easily and cost-effectively resolved with a rebuilt engine. These "crate" engines are delivered to your shop fully assembled except for all the accessories such as water pumps, generators, and electronic components. This type of repair is clearly the trend in the industry due to the high cost of labor and individual parts. The crate engines come with a warranty so the shop would not be responsible for the cost of repairs if the rebuild were to fail. This makes your job of diagnosis even more critical; you certainly would not want to install a "new"

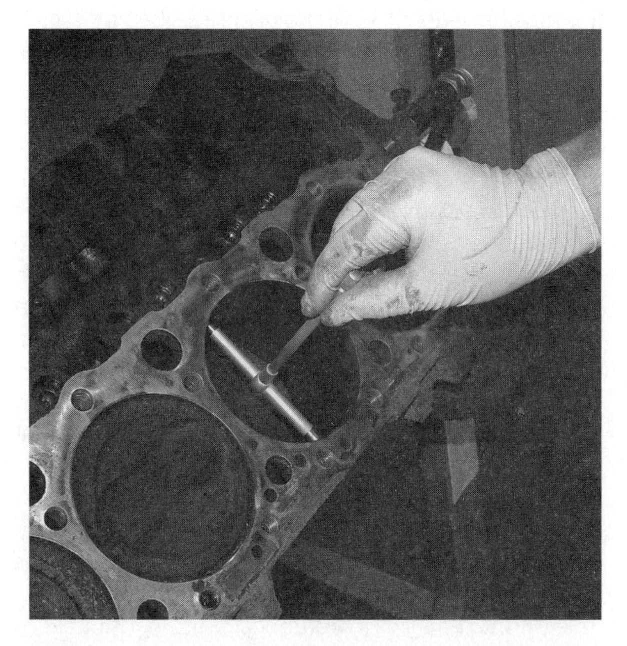

Figure 6. Accurate measuring, patience, and good organization are needed to succeed as an engine repair technician.

engine if the problem were not actually an internal engine failure. When you work as an engine repair technician you will have some of the responsibility of determining whether the engine should be repaired or replaced. It is essential that you perform quality work in finishing assembly of the engine and installing it.

Equally as important as your technical skills is your attitude. An engine repair technician must be patient enough to be extremely thorough. You cannot rush through critical engine measurements and expect accurate results. You need to evaluate the problem thoroughly and complete each step required for repair. Your attention to detail is essential to your success. Locating a cracked cylinder head requires close inspection. Forgetting to analyze the engine bearing wear patterns could cause you to miss a problem that would damage your freshly rebuilt engine. Aside from taste all your senses are useful during engine diagnosis. Listen for abnormal noises before disassembly, smell the engine oil for signs of contamination or overheating, look carefully at each part for clues about the causes of failure, and touch the cylinder walls, crank journals, and valve stems to feel for wear. Use logical steps to analyze and repair faults. A haphazard approach to engine repair will likely cause you to miss something critical. You will learn all the steps required to fully analyze and overhaul an engine; keep notes on the steps to refer to when you perform your first several overhauls. If you forget one step, such as measuring crankshaft endplay, the engine could come apart within the first few hundred miles of operation.

Keep yourself well organized. Label brackets and wires and hoses when you remove the engine or cylinder head. Many technicians put the bolts removed during disassembly loosely back in position or label them in bags. Locating the proper bolts for the timing cover could become a time-consuming and frustrating task if you just dumped all the hardware into one bin. Organize your engine measurements into a "blueprint" so that you can look at all the data to determine what repairs need to be made. If you do not write your results down as you go along you will forget some information and have to repeat the procedure. You will need documentation to justify your repairs for warranty purposes or for the customer. As an engine repair technician you will need patience to work your way through the big job of diagnosing and repairing an engine mechanical failure.

Honesty is essential when building your reputation as an engine repair technician. No one wants to pay for unneeded work. Your employer does not want to be caught in a lie to the customer. Be realistic about your estimation of what is required for the repair. If you minimize the work you will have to come back to the customer later to get authorization for the additional work. If you overestimate the necessary repairs the customer could get a more realistic quote elsewhere. That customer will never return. If you make an honest mistake offer an honest explanation.

Do not try to cover up mistakes with unprofessional work. All technicians make mistakes; you will too. What separates the excellent technician from the shoddy one is how she or he deals with her or his mistakes.

Mastery as an engine repair technician requires technical knowledge, hands-on experience, and good work habits. If you enjoy engine work learn and practice the diagnostic and repair techniques presented here. Then, always remember how essential your professional attitude is to your success. Engine repair is a challenging and rewarding part of automotive service; aim for excellence.

Summary

- You will need a good technical and general education to excel in the automotive industry.
- Math, physics, and electronics will help you repair vehicles today and throughout your career.
- Excellent communication skills are an essential part of your job as an engine repair technician.
- Maintaining professional appearance, language, and behavior will help you succeed and advance in the field.
- Competent technicians will always find good employment opportunities.
- Technicians may be paid "straight time" or hourly for the time they are at work, or they may be paid "flat rate" or by the billable job hours produced that week. Some shops use a combination of the two for a guaranteed salary with the incentive to be more productive.
- ASE certification is voluntary, but most employers prefer to hire certified technicians.
- The engine repair ASE test checks your knowledge of general engine diagnosis; cylinder head service; engine block service; cooling and lubrication system repair; and fuel, electrical, ignition, and exhaust systems.
- Careful study of this text and practice on the procedures presented will prepare you to repair engines professionally.
- Thorough diagnosis is required before engine repairs are attempted.
- Engine repair technicians must be patient, thorough, logical, organized, and honest.

Review Questions

1. Explain three attributes required of a professional technician.
2. What special skills are required of an engine repair technician?
3. Why is effective communication with your employer and your customers essential to your success?
4. What areas are covered on the engine repair certification test?
5. What should you do when you become very frustrated trying to loosen a hard-to-reach nut?
6. Your explanation on a repair order is essential for:
   A. Communication with the customer
   B. Warranty claims
   C. Vehicle history
   D. All of the above
7. Physics and electronics courses are particularly helpful in:
   A. Replacing brakes
   B. Installing a water pump
   C. Understanding a new system
   D. Talking to the customer
8. Technician A says that "straight time" pay means you get paid for each hour you are at work. Technician B says that "flat rate" pay can pay you for more hours a week than you were actually at work. Who is correct?
   A. Technician A only
   B. Technician B only
   C. Both Technician A and Technician B
   D. Neither Technician A nor Technician B
9. Technician A says that achieving ASE certification requires knowledge of diagnostic and repair techniques. Technician B says you cannot be ASE certified without hands-on experience. Who is correct?
   A. Technician A only
   B. Technician B only
   C. Both Technician A and Technician B
   D. Neither Technician A nor Technician B
10. Technician A says that a compression test can locate a weak cylinder. Technician B says that a cylinder leakage test can pinpoint the cause of a low compression reading. Who is correct?
    A. Technician A only
    B. Technician B only
    C. Both Technician A and Technician B
    D. Neither Technician A nor Technician B

Section 2

Engine Operation and Support Systems

SECTION OBJECTIVES

After you have read, studied, and practiced the contents of this section you should be able to:

- Explain the operation of a four stroke/cycle internal combustion engine.
- Calculate engine displacement in cubic inches and in liters.
- Describe causes of power loss in the internal combustion engine.
- Explain methods that are used to improve engine torque and horsepower.
- Explain the process of normal combustion.
- Explain the effects and causes of abnormal combustion.
- Identify the three major pollutants emitted from gas engines.
- Describe the operation and components of the lubrication system.
- Perform an engine oil pressure test.
- Identify reasons for and effects of low oil pressure.
- Perform proper maintenance on the lubrication system.
- Describe the operation and components of the cooling system.
- Perform proper maintenance procedures on the cooling system.
- Diagnose and repair faults causing engine overheating and overcooling.
- Describe the important functions of the intake and exhaust systems.
- Perform tests to identify problems in the intake and exhaust systems.
- Describe the purpose and operation of forced induction systems.
- Diagnose and repair problems in a turbocharging or supercharging system.

A bigger engine does not necessarily generate more power than a smaller one. For example, a turbocharged 2.3 liter dual overhead cam engine can develop more horsepower than some 3.3 liter engines!

Engine Operation

Introduction

Understanding the fundamentals of engine operation is critical to your ability to diagnose and repair engine concerns. Basic engine operation is relatively simple; it has not changed much in fifty or even one hundred years. But the materials used and the manufacturing processes employed to build modern engines have improved dramatically. Modern engines are lighter, more powerful, and more durable than ever. Many current systems used to improve engine performance, such as variable intake manifolds, variable valve timing, and turbocharging, build on a basic theory of engine operation; the better an engine can breathe the more power it can produce. When new systems are added to engines you will be able to understand their purpose once you have a good grasp of the principles of the four-stroke internal combustion gas engine. Study this chapter carefully; it describes essential information on engine operation **(Figure 1)**.

Figure 1. This modern engine uses many improvements to squeak more horsepower out of a small engine.

Some automotive enthusiasts like to call our automotive engines big air pumps. It is true that the more air they can take in, the more power they can put out.

INTERNAL COMBUSTION FOUR-STROKE ENGINES

Automotive gas engines are referred to as **internal combustion (IC) engines** because the combustion occurs in a combustion chamber within the engine. Some early automotive and later locomotive engines had a separate chamber where combustion took place. The automotive IC engine is also most commonly a four-stroke engine. This means that the piston must complete four **strokes** (movements up and down in the cylinder) in order to create one useful stroke, the power stroke. When the engine completes its four strokes—intake, compression, power, and exhaust—it has completed one full cycle. This is described in the phrase "four strokes per cycle" internal combustion engine. We will discuss the engine cycle further in the upcoming paragraphs.

ENGINE CONSTRUCTION

The engine has two main structures, the **engine block** and the **cylinder head** that sits on top of the block. The engine block has **cylinder bores** in which the **pistons** are driven up and down. The cylinder bores are precisely machined to nearly perfect round bores of the intended diameter, often between 2.5 and 4.25 inches on common production passenger cars and light-duty trucks. The number of cylinder bores describes the engine: a three-, four-, five-, six-, eight-, ten-, or twelve-cylinder engine. The cylinder head seals the top of the cylinders and forms the top of the **combustion chamber**. The piston is usually about one-thousandth of an inch (.001 in.), or half the thickness of a strand of hair, smaller than the cylinder bore. The **piston rings** seal that gap. The piston and rings form the bottom of the combustion chamber. The combustion chamber is sealed by the piston and rings, the engine valves, the spark plug, and the seal between the block and head, the head gasket **(Figure 2)**. The **valves** seal the top of the combustion chamber. They open to allow fresh air and fuel in (intake valves) and to allow spent exhaust gases out (exhaust valves). They close when the air and fuel mixture is being compressed to make it easier to burn and during the **combustion** process when the air-fuel mixture is ignited and the heat and expansion of gases is pushing down on the piston. The power of combustion is contained within the combustion chamber so that all the force can be exerted on the top of the piston.

The piston is connected to the **crankshaft** by a **connecting rod**. The piston connects to the connecting rod through a **wrist pin** (sometimes called a piston pin) at the small end of the connecting rod. The piston moves up and down, spinning the crankshaft around. The connecting rod attaches to the crankshaft on **journals** that are offset from the centerline of the crank, forcing the crank to twist as the piston pushes the connecting rod down on the crank. The big end bore of the connecting rod holds a soft-coated rod bearing that rotates around the hard, finely machined crankshaft rod journal **(Figure 3)**.

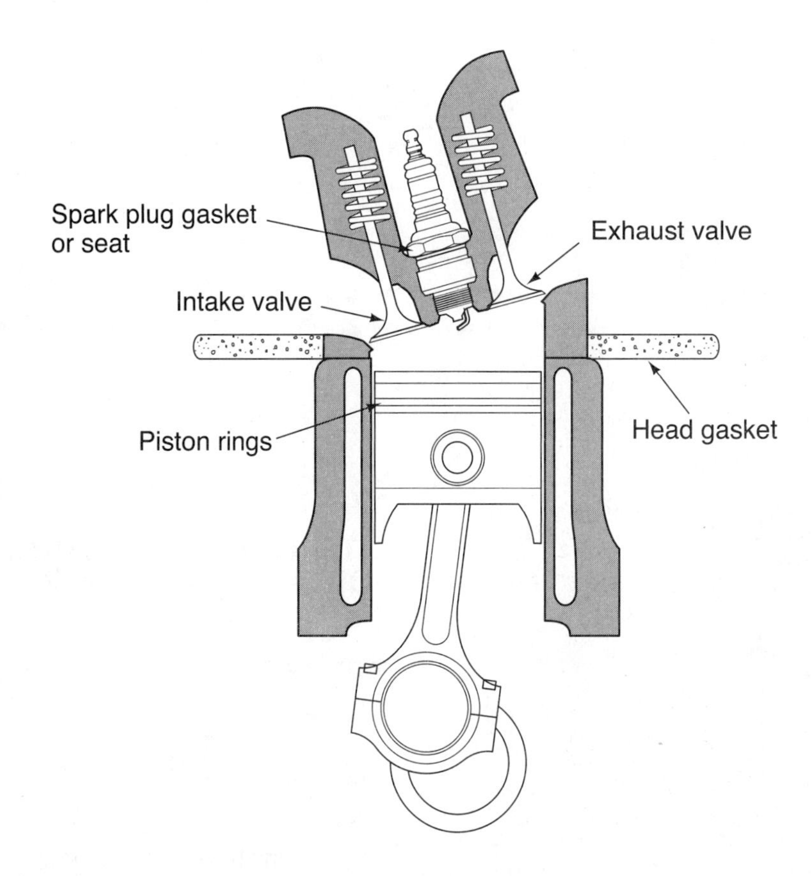

Figure 2. The combustion chamber must be sealed well for the engine to put out its maximum power.

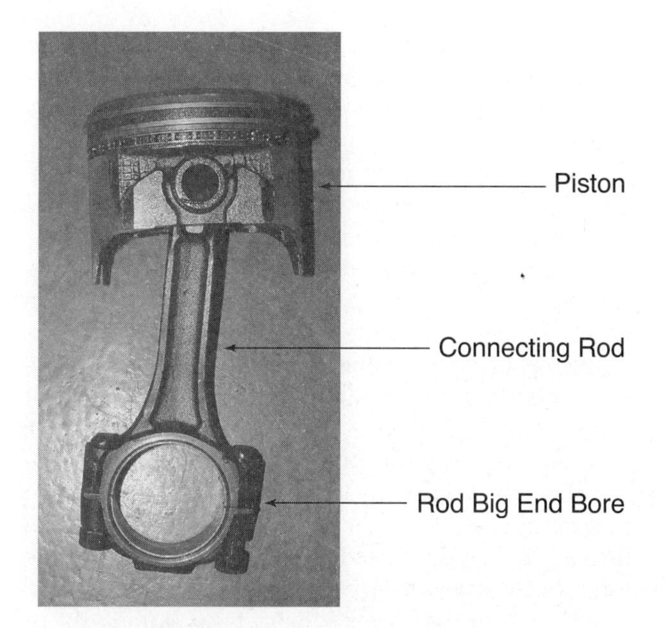

Figure 3. The forces of combustion push the piston down in the cylinder. The connecting rod bolts to the crankshaft and transmits this force.

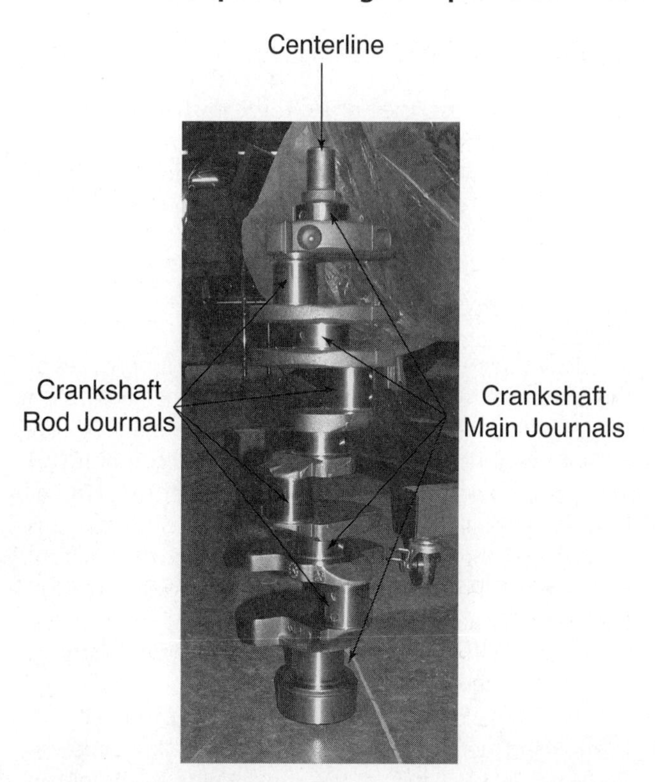

Figure 4. The main journals are on the centerline of the crankshaft. The rod journals are offset so the rod turns the crank as it pushes down.

The crankshaft converts the reciprocating (up and down) motion of the pistons into a rotary motion to spin the **transmission**. The transmission then turns the wheels, indirectly from the force on the piston. The crankshaft has main journals and rod journals **(Figure 4)**. The main journals form the centerline of the crankshaft. The crankshaft is mounted to the engine block with bearings laid in the main bore of the block. Half of the main bore is integral to the block, and the other half is formed by a main cap. You can remove the main caps to take the crankshaft out and replace the bearings. Sometimes the main caps are formed into a modular housing and all the bore halves are removed as a unit. The connecting rod journals are spread off the centerline of the crankshaft. The distance they are offset times two determines the distance the piston moves up and down in the cylinder, the **stroke**. The larger the offset, the greater the leverage the piston has to twist the crankshaft. This force turning the crank is the torque of the engine. The rod journals hold the connecting rods for each piston. The connecting rod holds bearings between the rod and the crank journals. There is just enough clearance between the main and rod bearings and the journals that a thin layer of pressurized oil keeps the journals from actually riding directly on the bearings. The correct clearance and adequate oil pressure prevent damage to the bearings or crank.

The cylinder head holds the engine valves and forms the top half of the combustion chamber **(Figure 5)**. Many engines now use four or five valves per cylinder, unlike the older engine designs that almost always used two valves per cylinder. Another common trend is overhead **camshaft**(s), though many engines still use a camshaft mounted in the block. The camshaft opens the valves at the correct time through lobes ground onto the shaft.

Figure 5. This cylinder head has dual overhead camshafts.

The camshaft operates the valves through valve lifters or followers and sometimes pushrods and rocker arms. The valve springs close the valves. The camshaft rotation is timed to correspond perfectly with the crankshaft rotation. This is called **valve timing**. The camshaft is connected to the crankshaft through timing gears, a belt, or a chain.

ENGINE OPERATION

To complete one full cycle of the engine four strokes of the piston are required. A stroke is the distance the piston travels from the bottom of its throw to the top of its throw and vice versa. The piston is at **bottom dead center (BDC)** when at its lowest point and at **top dead center (TDC)** at its highest point. The distance between those two points is a stroke. Another stroke occurs as the piston travels from TDC to BDC. The strokes are typically expressed in the following order: intake, compression, power, and exhaust. You will need to remember the strokes in that order including what occurs during each stroke.

The intake stroke fills the combustion chamber with fresh air and fuel. The intake valve(s) open to allow airflow in from the intake manifold. The stroke starts as the piston begins moving down from TDC. As the piston moves down in the cylinder the volume of the cylinder expands. This creates a low pressure area or a **vacuum**. When the intake valve opens external air at atmospheric pressure of 14.7 pounds per square inch (psi) rushes in to fill the vacuum created by the piston moving down. The fuel injector squirts a small amount of fuel into the rushing air stream in the intake manifold, and it mixes well with the air. To fully utilize the momentum of airflow the intake valve actually stays open a little beyond BDC. Now the combustion chamber is full (or as full as time allowed) of fresh air and fuel.

Next is the compression stroke; this adds heat and pressure to the mixture to help one short spark begin an efficient burning process: combustion. The piston moves from BDC to TDC with the intake valve soon closed and the exhaust valve fully closed. The air and fuel molecules become tightly packed under compression pressure.

Just near the end of the compression stroke the spark plug ignites a spark to begin the combustion process. A small flame front begins around the spark plug and then quickly propagates (spreads) across the chamber. The burning of the mixture increases the heat and pressure inside the chamber tremendously. This controlled burning is called combustion. Contrary to the common misconception it is not an explosion; it is much slower. This power stroke begins just after the piston crosses TDC and begins moving down. The spark is started at a time that will allow the peak pressure on top of the piston to occur at about 10° of crankshaft rotation after TDC (ATDC). The faster the engine is spinning, the earlier the spark has to occur to achieve that goal. That is called **ignition timing** advance. When proper combustion occurs at the correct time a force of over 2000 lbs. is exerted on the top of the piston.

That is the power stroke; the piston is forced from top dead center to bottom dead center. The impressive force on the piston spins the crankshaft. This is the only stroke that makes power. The other three strokes prepare the engine to create this movement; the power stroke harnesses the force of the incredible heat and pressure created by combustion. This is the reason that four-stroke IC engines are still only slightly above twenty-five percent efficient. One in four strokes creates power.

Near the end of the power stroke the volume of the chamber has increased significantly. This reduces the amount of pressure left. At roughly 10° to 40° of rotation before BDC (BBDC), depending on the engine design, the exhaust valve opens to let out the spent gases. The residual pressure left in the chamber helps force all the exhaust gases out. The piston moves from BDC to TDC during the exhaust stroke to clean out (scavenge) the chamber.

The exhaust valve stays open for 10° to 40° ATDC during the beginning of the intake stroke. The intake valve opens BTDC during the end of the exhaust stroke. This is called **valve overlap**, when both valves are open at the same time. The idea is to scavenge all the old gases out and pull in as much fresh charge as possible. The airflow near the end of the exhaust stroke creates a strong stream. It helps fully clean the chamber to let the flow continue a little ATDC. When the intake valve opens just BTDC the slow-moving intake charge gathers speed along with the rushing exhaust flow. This helps start the flow of intake air faster and more efficiently. The amount of overlap is carefully planned so that the intake charge does not rush out the exhaust. The engineers also design carefully to control reversion, which occurs when exhaust gases flow back into the intake.

The purpose of valve overlap is to help the engine fill the chamber with a full, fresh charge of air and clean all the spent gases out. The better the engine can breathe the more power it can make. This is why **turbochargers** and **superchargers** are sometimes used. They pressurize the intake air, forcing more air into the chambers. They can improve engine power dramatically.

The cycle begins again: intake, compression, power, exhaust. The crankshaft has to rotate two complete revolutions, or 720°, to complete the four-stroke cycle. The intake stroke takes the first 180° from TDC to BDC, and the compression stroke finishes that revolution, moving the piston from BDC to TDC. Then the power stroke occurs, forcing the piston down for 180° of the second revolution, and finally the crank comes full circle as the cycle finishes with the exhaust stroke. The camshaft rotates only one time during this cycle. We want the cylinder #1 intake valve to open only once during the whole cycle, and the same must be true for the exhaust valve. So the crankshaft spins at twice the speed of the camshaft. To complete one cycle the crank rotates twice (720°) and the cam rotates once (360°). Look at the timing circle of the crankshaft, showing typical opening and closing times for the engine valves. Notice that valve opening is defined in degrees of crankshaft rotation **(Figure 6)**.

ENGINE CLASSIFICATIONS

There are many different types of engines; classifications help describe the engine size, configuration, firing order, camshaft location and number, valves, timing type and mechanism, intake type, and fuel and ignition systems. As a professional you should be able to describe an engine

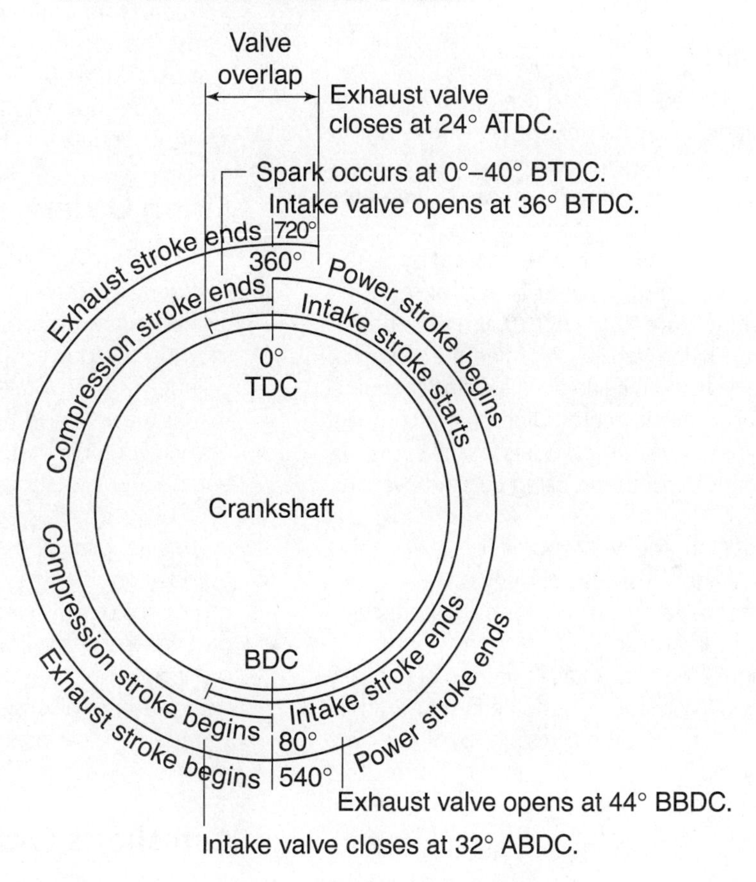

Figure 6. Start in the inner circle with the intake stroke. Follow the lines to see typical valve positions throughout one complete engine cycle.

accurately to a customer and to the parts person when ordering engine repair parts. Can you explain to a customer what a 4.6L, V8, DOHC, 32-valve, VVT, SFI, COP power plant is?

Cylinder Number and Arrangement

The number of cylinders is described when discussing an engine. If the engine has eight cylinder bores with eight pistons it is an eight-cylinder engine. Generally, though certainly not always, an engine with more cylinders produces more power.

An engine may have the cylinders arranged in one line, an inline three-, four-, five-, six-, or eight-cylinder engine, for example. The cylinders are in a row, one behind the other. Most, though not all, three-, four-, and five-cylinder engines are inline. The cylinders may be upright or slanted slightly sideways to allow a lower-profile hood line.

Many manufacturers started making **vee-type engines** so that the front of the vehicle did not have to be so long. An inline eight-cylinder must be covered by quite a long hood. A vee-style engine has cylinders offset at an angle to each other **(Figure 7)**. V6, V8, V10, and V12 engines are all used in production vehicles. Imagine the length of the nose of a vehicle with an inline twelve-cylinder engine! The vee configuration allows more cylinders in a compact space. Many engines have the cylinders offset by 90° to each other, but manufacturers are using 60° and even 15° vee engines.

A newer twist on the vee engine is a W engine designed by Volkswagen. It is essentially two vee engines cast together in steeper angles. This allows for more cylinders to fit in an even smaller space.

Another design is the horizontally opposed or pancake engine. Porsche and VW made this engine famous in their early cars, but it is still in common use today by Subaru. The cylinders, typically four, lie horizontally to the ground, two opposite the other two. This can help keep the weight of the engine very low and allow for a low-profile hood.

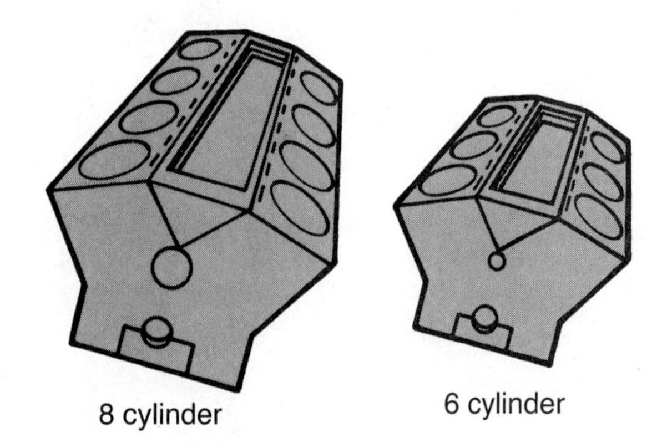

Figure 7. These 90° vee engines are a common design.

Cylinder Numbering

Cylinders are numbered starting with number one at the very front of the engine. The front of the engine is the end with the crankshaft pulley and accessory drive belt(s). The back of the engine is the end to which the transmission is bolted. Even on a vee engine the #1 cylinder is slightly ahead of the #2 cylinder. Most manufacturers number their vee engines with the #1 bank or side of the block counting #1, #3, #5, and #7, and the side of the engine with the #2 cylinder holding cylinders #2, #4, #6, and #8, for example. The cylinders are numbered in ascending order from the front of the engine, so you count back and forth across the engine toward the rear. Ford, on the other hand, numbers the bank with cylinder #1 with cylinders #1, #2, #3, and #4. The opposing side of the engine houses cylinders #5, #6, #7, and #8.

Firing Order

The **firing order** of an engine is the order in which the cylinders go through the power stroke. A typical four-cylinder firing order is 1-3-4-2. This means that as cylinder #1 is on the power stroke, cylinder #3 is on compression, cylinder #4 is on intake, and cylinder #2 is on exhaust. That means that every 180° of crankshaft rotation a power stroke occurs: 720° divided by 4 = 180°. In an eight-cylinder engine a power stroke occurs every 90° of crank rotation: 720° divided by 8 = 90°. That is what makes an eight-cylinder engine operate so much smoother than a four-cylinder engine. The crankshaft is being pushed around more consistently. It is impossible to remember all the firing orders because there are a great number in use. A V8 firing order may be 1-5-4-8-6-3-7-2, 1-8-7-2-6-5-4-3, 1-5-4-2-6-3-7-8, or a host of other orders. The firing order is typically stamped on the intake manifold and is always available in service information.

Camshafts and Valves

A description of an engine often includes the camshafts and valves. Our example of the 4.6 liter V8 had dual overhead camshafts (DOHC) opening 32 valves, four per cylinder. The camshaft(s) may be described as in the block, overhead (OHC), or dual overhead (DOHC). Usually the valves are expressed as the total number the engine uses, such as 16 valves on a four-valve-per-cylinder four-cylinder engine.

Timing Type and Mechanism

An engine uses timing gears, a timing belt, or a timing chain to link the camshaft(s) to the crankshaft. This is called the **timing mechanism**. Some engines also use one or more balance shafts to help smooth the engine; these are also driven by the timing mechanism **(Figure 8)**.

Figure 8. This intake manifold has five runners coming off the plenum, one to feed each cylinder.

Figure 9. These three coils sit right on top of the spark plugs in this coil on plug ignition system.

Many modern engines are using **variable valve timing (VVT)** systems. These are designed to improve the engine's performance at different engine rpms. A VVT system changes the timing of the camshaft in relation to the crankshaft at different engine speeds and operating conditions. This allows the camshaft time to more effectively fill the combustion chambers with air, improve power, and minimize fuel consumption and emissions.

Intake System

The intake of an engine may use **forced induction** or be **naturally aspirated**. Forced induction intakes use either a turbocharger or supercharger to pressurize the intake air. This helps fill the cylinders with air and greatly improves engine performance. Engines without a forced induction system are termed naturally aspirated, meaning they breathe without assistance. Some intake systems also use variable intakes. The intake manifold may have two sets of runners (tubes) for air to flow into the engine. A longer set of runners may be used at lower rpms to take advantage of the momentum of airflow to help push air into the cylinders. Then, at a higher rpm, the mechanism opens the shorter set of runners to help increase flow when there is a shorter time for cylinder filling.

Fuel System

The fuel system describes the type of fuel and fuel injection used. Some of the fuels used are gasoline, diesel fuel, compressed natural gas, propane, and M85. Some engines are bi-fuel and can run on either gasoline or M85. A gas engine may use MPFI, multipoint **fuel injection**, where each cylinder has an injector that delivers fuel into the intake manifold very close to the intake valve. Most common now is SFI, sequential fuel injection, where each injector squirts in the firing sequence during the cylinder's intake stroke. Some manufacturers are introducing gasoline direct injection engines, where the fuel is injected under higher pressures directly into the combustion chamber of the engine. A few vehicles use CPI, central port injection. Older vehicles used TBI, throttle body injection, where one or two injectors squirted fuel into the throttle body. Some manufacturers may use variations on these examples to describe their fuel systems.

Ignition System

Ignition systems have changed dramatically in the past several years. Now COP, coil on plug, ignition is quite common **(Figure 9)**. This system uses one coil per cylinder, and it sits right on the spark plug to eliminate the spark plug wire. There is also coil near plug ignition, where a high-tension wire delivers each coil's spark to the cylinder. Other manufacturers use waste spark ignition systems, where one coil provides spark for two cylinders. All of these systems are distributorless ignition systems and are the most common types of ignition. The proper term for these is EI, or **electronic ignition** systems. Ignition systems that still use a distributor are officially called DI, distributor ignition systems. This can be somewhat confusing, however, because several manufacturers refer to their distributorless ignition systems as DIS systems.

ENGINE IDENTIFICATION

The engine is casually identified by its size and configuration, a 5.7 liter V8, for example. When ordering parts for engine repair more specific information may be required. The **vehicle identification number (VIN)** is located in several locations on the vehicle but is easily read on the driver's side of the dashboard through the front window. The eighth character of the VIN defines the engine type. The tenth character defines the model year of the vehicle. Often there are several versions of the same size engine; the eighth character of the VIN can distinguish the model. Service information will identify the different engines described by the VIN.

Most engines also have identifying tags or stamps. Their location is described in the service information. They may contain the date or location of manufacture, the engine group or version, and other important identifying information. These codes are interpreted in the service information. When ordering parts for an engine overhaul your parts person may request this information.

Summary

- The gas IC four-stroke/cycle engine is the most commonly used automotive engine.
- The crankshaft converts the reciprocating motion of the piston to usable rotary motion.
- The cylinder head, valves, head gasket, and piston and rings seal the combustion chamber.
- The camshaft opens the valves; the valve springs close the valves.
- The piston is connected to the crankshaft by the connecting rod.
- The piston stroke is the distance the piston travels from BDC to TDC.
- The engine goes through four strokes—intake, compression, power, and exhaust—to complete one full cycle.
- Combustion is a controlled burning that causes expansion of gases inside the combustion chamber.
- Combustion creates over 2000 pounds of force on top of the piston.
- Valve overlap occurs at the end of the exhaust stroke and at the beginning of the intake stroke. Both valves are open to help fill the chamber with a fresh charge.
- The crankshaft rotates twice per cycle; the camshaft rotates once.
- Engines may be inline, vee, W, or horizontally opposed.
- Engines are classified by their cylinder number and arrangement, firing order, camshaft and valves, timing mechanism, and fuel and ignition systems.

Review Questions

1. Describe the four strokes in the order in which they are commonly expressed, and describe the valve position and purpose of each stroke.
2. The __________ connects the piston to the crankshaft.
3. Define overlap, and describe its purpose.
4. A coil on plug ignition system is a form of __________, while a system using a distributor is a __________ ignition system.
5. Technician A says that during the power stroke the intake valve is just closing. Technician B says that the piston moves from TDC to BDC during the power stroke. Who is correct?
   A. Technician A only
   B. Technician B only
   C. Both Technician A and Technician B
   D. Neither Technician A nor Technician B
6. Technician A says that combustion is an explosion of gases. Technician B says that combustion pressure should peak at about 40° before top dead center at the end of the compression stroke. Who is correct?
   A. Technician A only
   B. Technician B only
   C. Both Technician A and Technician B
   D. Neither Technician A nor Technician B
7. Technician B says that the spark usually occurs at the end of the compression stroke. Technician B says that the exhaust valve opens at the end of the power stroke. Who is correct?
   A. Technician A only
   B. Technician B only
   C. Both Technician A and Technician B
   D. Neither Technician A nor Technician B
8. Each of the following is a common engine configuration, *except*:
   A. Inline
   B. Vee
   C. Layer
   D. Pancake
9. A 4.6L, V8, DOHC, 32-valve, SFI, COP is:
   A. An eight-cylinder engine with a cam in the block and a carburetor on the plenum

B. A 4.6 liter engine with two timing chains and single point fuel injection
C. A 4.6 liter engine with four cams, four valves per cylinder, and coil on plug ignition
D. An eight-cylinder engine with two valves per cylinder and sequential fuel injection

10. Technician A says that the firing order of an engine is the order in which the cylinders fire. Technician B says that the firing order used for V8 engines is 1-8-4-5-3-6-2-7. Who is correct?
A. Technician A only
B. Technician B only
C. Both Technician A and Technician B
D. Neither Technician A nor Technician B

Engine Measurements and Ratings

Introduction

As an engine repair technician you should understand some of the design characteristics of various engines. Knowing the size and ratings of an engine may be important when talking with customers. You will need an engine's size and measurements when ordering parts or looking up service information. It is also interesting information to discuss with your friends. If you ever hope to work in a performance shop or upgrade your engine you will need to know what these descriptions mean. This information will even help you understand trade magazines.

> **Interesting Fact**
>
> *You can increase the size of a 350-cubic-inch displacement (cid) engine to a 360 cid during an overhaul by having the cylinders bored .060 in. oversize and replacing the pistons. Another popular modification is to increase the stroke by changing the crankshaft, connecting rods, and pistons to create a 383-cubic-inch displacement engine.*

BORE AND STROKE

The engine size is determined by the diameter of the cylinder bore and the length of the stroke. The engine is described by stating the bore followed by the stroke. A 4.00-in. (101.6-mm) by 3.5-in. (88.9-mm) engine, for example, means the bore is 4.0 in. and the stroke is 3.5 in. The bore is simply a measurement of the diameter of the cylinder. You will measure the bore diameter during an engine overhaul in part to determine whether the engine has ever been bored. **Boring** an engine means machining the bore to a greater size to make the cylinder round and fresh again **(Figure 1)**.

The stroke is the distance the piston travels from BDC to TDC. The distance the rod journals are offset from the centerline of the crankshaft determines the stroke. You can also measure stroke by multiplying the rod journal offset by two.

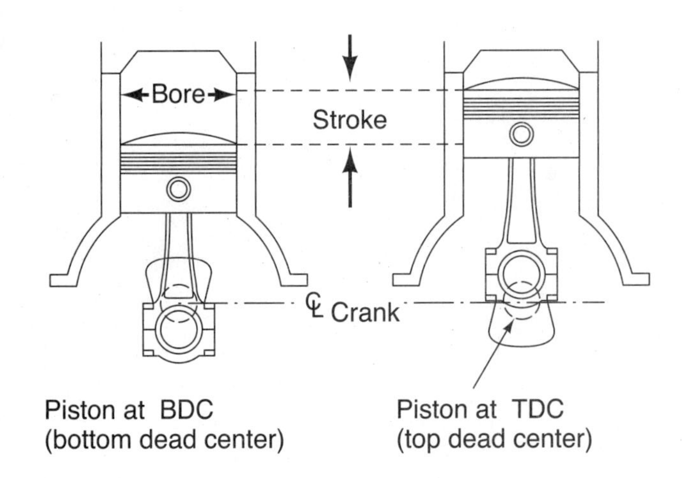

$$\frac{\text{Bore}}{\text{Stroke}} = \text{Bore stroke ratio}$$

Figure 1. The bore is the diameter of the cylinder. The stroke is the distance the piston travels from BDC to TDC.

Undersquare and Oversquare Engines

The size of the bore compared to the length of the stroke is described by the terms oversquare and undersquare. In an **oversquare engine** the bore is bigger than the stroke. An oversquare engine has the ability to spin (rev) faster because the shorter amount of ring travel reduces friction in the engine. An **undersquare engine** has more torque. Earth moving equipment uses undersquare engines to dig underground because the process requires huge amounts of torque; engine speed is not important. Some automotive engines are undersquare as well. An engine that has the same size bore and stroke is called a square engine.

ENGINE DISPLACEMENT

The size of the engine is determined by the swept volume. The swept volume refers to the area that the piston travels. That volume times the number of pistons defines the **engine displacement** or size **(Figure 2)**. Use the following formula to find the piston displacement:

$$\frac{\text{Bore}^2 \times \pi(3.14) \times \text{Stroke}}{4}$$

$^*\pi$ = roughly 3.14

You can also simplify this formula by dividing π by 4 and using the constant .785. Then the equation would read as follows:

$\text{Bore}^2 \times .785 \times \text{Stroke}$

For example, a V8 engine with a bore of 3.75 in. and a stroke of 3.62 in. would have a piston displacement of:

$$\frac{3.75" \times 3.75" \times 3.14 \times 3.62"}{4}$$

= 39.96 cubic inch displacement (cid)

or $$\frac{9.53\text{ cm} \times 9.53\text{ cm} \times 3.14 \times 9.2\text{ cm}}{4}$$

= 655.92 cubic centimeters

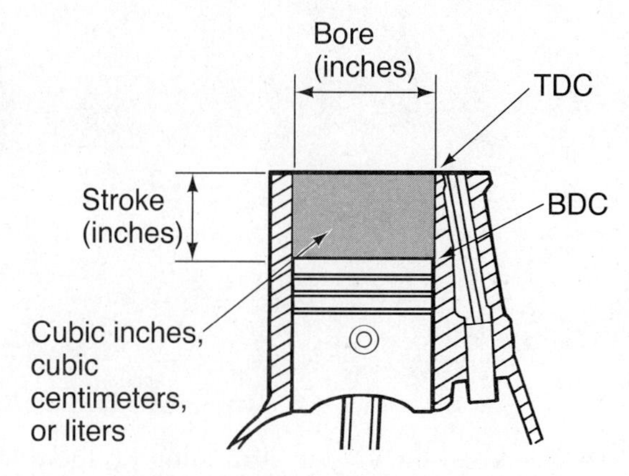

Figure 2. Displacement is the volume of the cylinder between TDC and BDC.

To find the engine displacement multiply the piston displacement times the number of cylinders:

$$\frac{\text{Bore}^2 \times \pi(3.14) \times \text{Stroke}}{4} \times \#\text{ of Cylinders}$$

To use the example of our engine again:

$$\frac{3.75" \times 3.75" \times 3.14 \times 3.62"}{4} \times 8 = 320\text{ cid}$$

or $$\frac{9.53\text{ cm} \times 9.53\text{ cm} \times 3.14 \times 9.2\text{ cm}}{4} \times 8$$

= 5247.36 cc

So we have found the engine size in cubic inches and in cubic centimeters. Most manufacturers are now describing the engine size in liters. To convert cubic inches to liters:

Cubic Inches ÷ 61.025 = Liters

or 320 cid ÷ 61.025 = 5.2 liters

To convert cubic centimeters to liters:

Cubic Centimeters ÷ 1000 = Liters

or 5247.36 cc ÷ 1000 = 5.2 liters

If we were to bore this engine .060 in. oversize:

$$\frac{3.81" \times 3.81" \times 3.14 \times 3.62"}{4} \times 8 = 330\text{ cid}$$

or 330 cid ÷ 61.025 = 5.4 liters

or $$\frac{9.68\text{ cm} \times 9.68\text{ cm} \times 3.14 \times 9.2\text{ cm}}{4} \times 8 = 5414\text{ cc}$$

or 5414 cc ÷ 1000 = 5.4 liters

** See Appendix B for a more complete listing of standard and metric conversions.

Increasing the engine size increases the engine power. Generally speaking, the bigger the bore and stroke, the more powerful the engine.

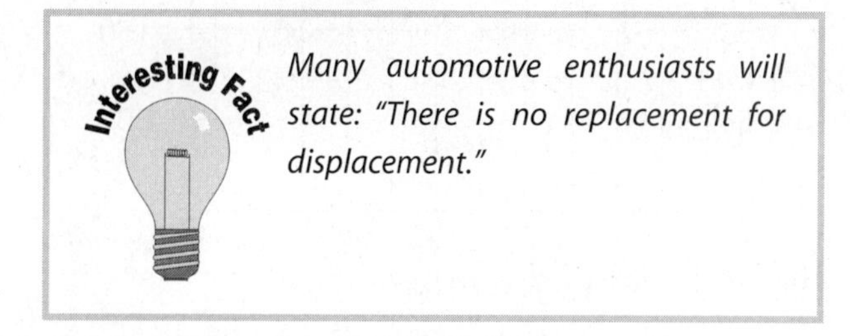

Many automotive enthusiasts will state: "There is no replacement for displacement."

TORQUE AND HORSEPOWER

An engine is widely advertised by its torque and **horsepower** ratings. Torque is the twisting force the piston exerts on the crankshaft. It is measured in foot pounds (ft.-lb) or newton-meters (N•m). The longer the stroke, the longer the lever working to spin the crank, so the greater the engine torque. The concept is the same as using a longer breaker bar to loosen a tight bolt; you can get more torque on it. The engine torque is transmitted to the wheels. You can feel engine torque as your head is pressed against the head rest when you accelerate. Torque changes throughout the engine speed (rpm—revolutions per minute) range. It builds as the engine gains speed and falls off when the engine is spinning so fast it can no longer get a full charge of air and fuel into the cylinders. The goal of the transmission is to keep the engine operating near the peak of the torque curve so that you have acceleration throughout the vehicle speed range.

Horsepower (hp) is the engine's ability to produce work. The work it is performing is applying torque at a particular rate. Horsepower is measured on a **dynamometer**. A combination of torque and horsepower allows an engine to gain speed. The more torque and horsepower an engine has, the faster the acceleration. Horsepower also changes throughout the engine's rpm range. The torque and horsepower curves show where the most useful power occurs. The flatter the curves, the longer the engine can use its peak power. **(Figure 3)** shows a very strong engine's torque and horsepower curves. Notice how flat the torque curve is and at what a low engine speed the torque is developed.

There are several different ways of measuring horsepower. Brake horsepower is a rating of the horsepower available at the crank. The **Society of Automotive Engineers (SAE)** has standardized two horsepower ratings commonly used by manufacturers. Gross horsepower is the power the engine can develop without any of the accessories being spun. Net horsepower is the horsepower available when the power steering pump, generator, exhaust system, and all other accessories are installed. Most manufacturers advertise the horsepower of an engine using this more realistic net horsepower rating. Some engine builders measure the torque and horsepower of an engine by mounting it on a dynamometer. The measuring equipment calculates the torque and horsepower at different rpms **(Figure 4 and Figure 5)**.

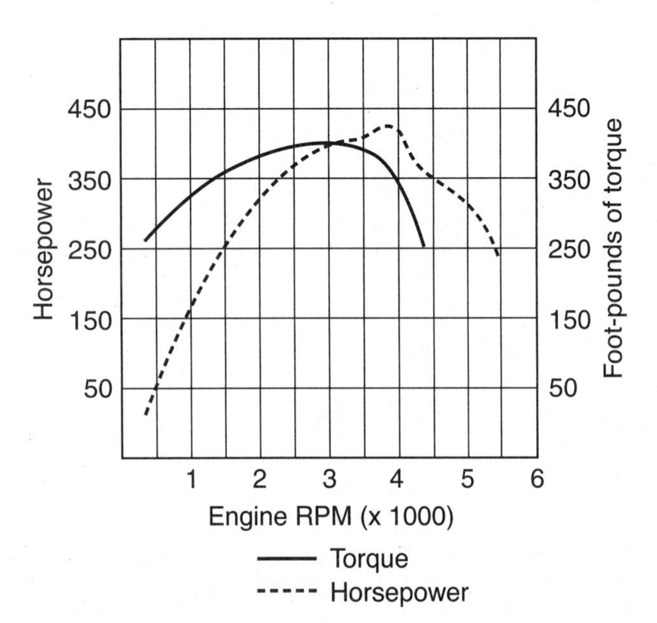

Figure 3. This engine develops almost 400 foot pounds of torque and over 400 horsepower. What a ride!

Figure 4. This engine is set up and ready to run on the dyno.

Figure 5. When the engine is running on the dyno these gauges display torque, horsepower, voltage, engine temperature, and rpm.

Some of today's vehicles can put out the same amount of horsepower as muscle cars of the 1960s and 1970s with an engine about half the size.

COMPRESSION RATIO

The **compression ratio** of an engine also helps determine its power. The compression ratio is a comparison of the volume of the combustion chamber with the piston at bottom dead center to the volume of the chamber with the piston at top dead center. An engine with a compression ratio of 8.5:1 means that the maximum cylinder volume is 8.5 times bigger than the minimum cylinder volume **(Figure 6)**. Most production vehicles have a compression ratio in the range of 8:1 to 11:1. We already know that a bigger maximum volume can provide more power. Compressing that charge makes the mixture more combustible. Then, when combustion does occur the pressure is greater because the expansion of gases takes place in a smaller area. That translates to more force on top of the piston. Engineers must ensure that the compression ratio is not so high that the air-fuel mixture self-ignites before the spark (preignition) or burns uncontrolled after the spark (detonation). These dangerous conditions can lead to serious engine damage. To some extent **preignition** and **detonation** can be minimized by using a higher-**octane** fuel. Some high-performance vehicles require the use of premium fuel, 92 octane. Most vehicles, however, are designed so that they will run safely on regular fuel, 87 octane.

COMPRESSION PRESSURE

Compression pressure is the amount of pressure the engine develops during the compression stroke. Most gas engines develop between 125 and 225 pounds per square inch (psi) while the engine is cranking **(Figure 7)**. The compression pressure while the engine is running is reduced because the combustion chamber does not have time to fully fill with air when the throttle is only partially open and the engine is spinning faster.

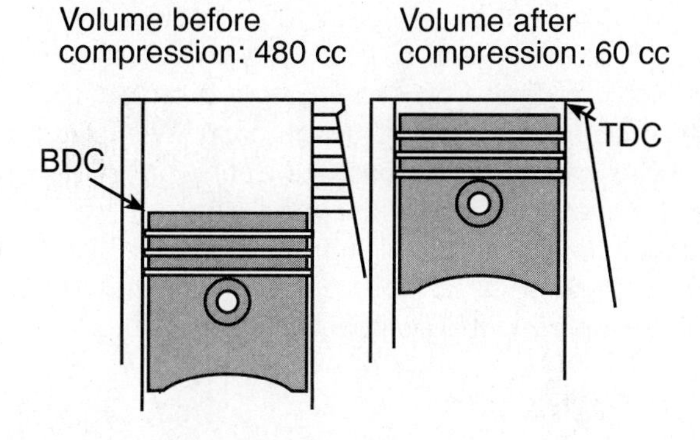

Figure 6. Compare the cylinder volume at BDC to the volume at TDC to find the compression ratio.

Figure 7. Measure the compression at the spark plug hole while cranking the engine over.

Compression is measured using a compression gauge as a method of diagnosing engine mechanical problems. You screw an adapter into the spark plug hole and attach a pressure gauge to the other end of the hose. Be careful when measuring a diesel engine's compression. Diesel engines have significantly higher compression ratios and compression pressures. Use a compression gauge designed for diesel engines; you could destroy an ordinary compression gauge using it on a diesel engine. We will discuss compression testing further in an upcoming chapter (Chapter 16).

VOLUMETRIC EFFICIENCY

Volumetric efficiency (VE) describes the engine's ability to breathe. The more air an engine can take in, the more power the engine can put out. Volumetric efficiency is determined by comparing the amount of air a cylinder can hold with the amount of air that actually gets into the chamber.

$$\text{Volumetric Efficiency} = \frac{\text{The volume of air that enters the cylinder}}{\text{The maximum volume of the cylinder}}$$

At lower engine speeds there is more clock time for the cylinders to fill; VE is higher. Once the engine starts revving higher the valves do not open long enough to allow a full charge to enter. Volumetric efficiency generally runs between 65 percent and 85 percent.

Turbochargers and superchargers increase the pressure of the air entering the cylinder. They can improve VE to 100 percent or higher. Many of the performance improvements made to engines are designed to allow the engine

Figure 8. Four valves per cylinder allow the engine to breathe more efficiently.

Figure 9. This engine uses roller lifters to reduce friction between the camshaft and lifters.

to breathe better. Engines with four or five valves per cylinder provide a larger opening for air to enter and exit the cylinder **(Figure 8)**. Freer flowing exhaust and headers allow the engine to exhaust the gases more fully. Variable valve timing is used to improve cylinder filling at higher engine rpms. Aftermarket camshafts typically open the valves further and hold them open longer to get more air in and more power out.

Many aftermarket modifications such as "hot" camshafts and modified computer chips are not legal for street use because they affect exhaust emissions. They are widely available for many vehicles, however.

THERMAL EFFICIENCY

Ideally, combustion in an internal combustion engine would use all the available **British thermal units (BTUs)**, a measure of potential energy, in the fuel. That would mean that all the heat energy available in the fuel would be extracted and develop power. All the heat of combustion would be used to cause expansion of gases to push on the pistons. Unfortunately, the engine components would melt without an adequate cooling system. The cooling system is required to prevent engine damage, but it significantly limits the thermal efficiency of the engine. The exhaust system also lowers thermal efficiency because hot exhaust gases are let out of the engine. Only about 35 percent of the potential heat energy in the fuel is actually harnessed to power the engine.

MECHANICAL EFFICIENCY

Modern engines are usually between 85 percent and 95 percent mechanically efficient. Friction between the piston rings and cylinder walls accounts for the biggest loss. Camshaft friction on the lifters also creates frictional losses **(Figure 9)**. Many manufacturers are using thinner rings, along with pistons with shorter skirts and a Teflon coating to minimize frictional losses. The use of lower viscosity oils, 5w-30 and 0w-30, also helps reduce engine friction.

FUEL EFFICIENCY

Fuel efficiency is very important to manufacturers and customers. The federal government has set **corporate average fuel economy (CAFE)** standards for passenger cars and light trucks. Enactment of the CAFE standards doubled vehicle fuel efficiency between 1975 and 1989. The current standards require that automobile manufacturers' fleets of vehicles have an average of 27.5 miles per gallon (mpg) for passenger cars and 20.7 for light trucks. This means that Ford, for example, must sell a huge number of fuel-efficient Focuses in order to average out the lower fuel economy of their Lincoln Navigators or face large fines. Much of the research and design of vehicles focuses on ways to improve fuel economy. Manufacturers use lightweight materials all over the vehicle to save fuel. Plastic intake manifolds and aluminum cylinder heads, blocks, and valve covers are all examples of newer measures aimed in part at improving fuel economy.

Summary

- Engine size is determined by the diameter of the bore and the length of the stroke.
- Oversquare engines are higher revving, while undersquare engines can produce more torque.
- Engine displacement is equal to:

$$\frac{\text{Bore}^2 \times \pi(3.14) \times \text{Stroke}}{4} \times \text{\# of Cylinders}$$

- Engine torque is the engine's twisting force on the crankshaft; it is measured in ft.-lb or N•m.
- Engine horsepower is the rate at which an engine performs work.
- The compression ratio of an engine is the maximum cylinder volume compared to the minimum cylinder volume.
- Gas engines generally produce between 125 psi and 225 psi of compression pressure while cranking the engine.
- Volumetric efficiency describes the ability of the engine to breathe. It is a comparison between how much air fills the chamber and how much air could fit in the chamber.
- The cooling and exhaust systems greatly reduce the thermal efficiency of the engine.
- Fuel economy standards and emissions drive much of current engine research and design.

Review Questions

1. The __________ is the diameter of a cylinder; the __________ is the distance the piston travels.
2. An engine with a bore of 4.0 in. and a stroke of 3.3 in. is __________square.
3. A V6 engine with a bore of 3.5 in. and a stroke of 3.8 in. has a displacement of __________.
4. A V6 engine with a bore of 3.95 in. and a stroke of 3.54 in. has a displacement of __________.
5. Technician A says that an engine's torque helps accelerate the vehicle. Technician B says that a flatter torque curve is desirable. Who is correct?
   A. Technician A only
   B. Technician B only
   C. Both Technician A and Technician B
   D. Neither Technician A nor Technician B
6. Technician A says that horsepower is measured in newtons. Technician B says that torque drops off at higher speeds. Who is correct?
   A. Technician A only
   B. Technician B only
   C. Both Technician A and Technician B
   D. Neither Technician A nor Technician B
7. Technician A says that an engine with a bigger bore has better volumetric efficiency. Technician B says that volumetric efficiency is greater at higher engine speeds. Who is correct?
   A. Technician A only
   B. Technician B only
   C. Both Technician A and Technician B
   D. Neither Technician A nor Technician B
8. Technician A says that variable valve timing can increase volumetric efficiency. Technician B says that the purpose of turbocharging is to increase VE. Who is correct?
   A. Technician A only
   B. Technician B only
   C. Both Technician A and Technician B
   D. Neither Technician A nor Technician B
9. Technician A says that gasoline engines are only about 35 percent thermally efficient. Technician B says that the cooling system reduces the thermal efficiency of the engine dramatically. Who is correct?
   A. Technician A only
   B. Technician B only
   C. Both Technician A and Technician B
   D. Neither Technician A nor Technician B
10. Technician A says that the CAFE standards dictate the manufacturers' average fuel economy. Technician B says that the standards regulate vehicle emissions. Who is correct?
    A. Technician A only
    B. Technician B only
    C. Both Technician A and Technician B
    D. Neither Technician A nor Technician B

Chapter 7 Engine Fuels and Combustion

Introduction

The fuel that customers use can have a significant impact on their engine's performance and durability **(Figure 1)**. Your customers may ask your advice about what type of fuel to use and why. You will also see drivability problems caused by fuel issues affecting combustion. Severe engine damage can result when combustion does not occur normally. It is important that you be aware of the causes of abnormal combustion. When you repair or replace an engine with a catastrophic failure you need to find the source of the problem so it does not happen again. Normal and abnormal combustion produce toxic emissions that are carefully regulated by the EPA. A lion's share of the research and design for future vehicle production is based on minimizing emissions and maximizing fuel economy. This chapter will explain the key attributes of fuel and help you better understand the combustion process.

Figure 1. Encourage your customers to use good quality fuel from a busy station.

GASOLINE

Gasoline is refined from petroleum. The oil is heated and pressurized to separate it into its different substances. Some of the lighter vapors condense to form gasoline, which is made of hydrocarbons (hydrogen and carbon). These hydrocarbons are then blended with additives to produce a gasoline fuel suitable for the automotive engine.

Octane Rating

The octane rating of a fuel describes its ability to resist spontaneous ignition, or engine knock. Engine knocking (detonation) or pinging (preignition) results when combustion occurs at the wrong time or at the wrong speed. Low-octane fuel can be a cause of preignition and detonation. If combustion begins before the spark, for example, combustion pressures may try to push the piston backwards at the end of the compression stroke. This results in a rattling noise from the piston. This is called preignition, and the sound is often described as pinging. Generally, automobile fuel is classified as regular, 87 octane; midrange, 89 octane; or premium, 92 or 93 octane **(Figure 2)**. The higher the octane number, the greater its resistance to knock. Octane is tested in two ways, by the research method and by the motor method. The advertised octane rating is the average of the two ratings. You will often see this described on the pumps as:

$$\text{Octane} = \frac{\text{RON} + \text{MON}}{2}$$

Figure 2. These typical octane choices are displayed showing the formula used to rate them. Notice that this fuel contains MTBE, a controversial additive designed to reduce carbon monoxide emissions. It is a proven carcinogen and is being phased out of gasoline production.

Customers should use fuel with the octane rating specified in their vehicle's owner's manual. Using a lower-octane fuel can cause premature and serious damage to the engine. Extended knocking or detonation can cause severe damage to pistons, rings, connecting rods, and rod bearings. Most engines specify the use of regular-grade gas, 87 octane. Some performance or luxury cars with higher-compression engines or forced induction systems require midgrade or premium fuel. It generally does not improve the performance of an engine to use a higher than specified octane fuel. Occasionally an engine will require a higher-octane fuel when pinging or knocking persists. It is usually engines that have problems such as carbon buildup in the combustion chamber, an inoperative **exhaust gas recirculation (EGR)** system, or clogged coolant passages that ping on the correct octane fuel.

Volatility

Fuel volatility is the ability of the fuel to vaporize (evaporate). The **Reid Vapor Pressure (RVP)** defines the volatility of the fuel. The RVP is the pressure of the vapor above the fuel in a sealed container heated to 100°F. The higher the pressure of the vapor, the greater the volatility of the fuel. This means that it will more readily vaporize.

Fuel volatility is adjusted seasonally in many parts of North America. A higher volatility fuel is allowed in the winter to help the engine start when it is cold. The fuel vaporizes more easily during compression rather than puddling along the cool walls of the combustion chamber. Too low a volatility fuel will cause hard starting and rough running at startup. A lower volatility fuel is used in the summer to reduce the amount of evaporative emissions and to prevent vapor lock in the fuel lines. The maximum volatility level is regulated by the Environmental Protection Agency (EPA) between June 1 and September 15. Depending on the state and month gasoline RVP may not exceed either 9.0 psi or 7.8 psi. This is controlled because if the volatility is too high excess vapors will form on top of the fuel in the tank and the excess pressure will have to be relieved through a fuel tank vent. These **hydrocarbon (HC)** or raw fuel emissions are a significant portion of overall vehicle emissions. Long cranking times, stalling, and rough running can also be caused by using fuel with an excessively high volatility in the summer. After the vehicle has been driven the fuel is already heated. Then, when a customer parks his car in a hot, sunny parking lot the fuel heats even further. The fuel in the lines may evaporate. Upon his return, when he attempts to start the engine, the fuel pump pushes air into the injectors. The engine has to crank and crank and will usually stall or run poorly for a minute or so after starting. These customer concerns are tricky to diagnose. Be aware that during the first cold snap or hot spell the fuel may not have been adjusted yet. Advise customers to use a high-volume gas station where the tanks are refilled frequently. This is good advice anyway to avoid getting stale or contaminated fuel.

Oxygenated Fuels

Oxygenated fuels are designed to reduce the **carbon monoxide (CO)** pollutant produced by the engine. These fuels have oxygen added by way of an oxygen-rich blend of ethanol or methanol. The Renewable Fuels Association reports that 30 percent of all fuel is blended with ethanol in 2004. Oxygenated fuels may be used in densely populated areas where pollution is a severe problem. They may also be used in high-altitude areas where the oxygen level in the air is already low.

Alcohol Blends

Some fuels have up to 10 percent ethyl alcohol mixed in with the gas. The **ethanol** increases the octane rating of the fuel. Most manufacturers approve of using ethanol as long as it does not comprise more than 10 percent of the fuel. Excessive alcohol is corrosive to fuel system components. It also increases the volatility of the fuel. This can increase emissions and the likelihood of vapor lock during the summer. Many people advise against the use of ethanol fuels during the summer **(Figure 3)**.

Methanol is another alcohol additive used to increase the oxygen level, octane rating, and volatility of fuel. Methanol is often specially blended with gasoline to create M85, which is 85 percent gasoline and 15 percent methanol. Only special flexible fuel vehicles can use M85. These vehicles use fuel system components resistant to the corrosive properties of methanol.

Figure 3. Use a fuel analysis kit to check RVP and alcohol content.

Sulfur Content

The sulfur content of fuel is carefully controlled. The EPA set corporate average sulfur content standards for gasoline manufacturers beginning in 2004. No fuel may contain greater than 300 parts per million (ppm) of sulfur, and its average sulfur levels must be 120 ppm maximum. These standards are phased down through 2007 when all refineries will have to produce fuel with an average of 30 ppm sulfur content and a maximum of 80 ppm.

You can smell a high-sulfur fuel; when it burns it smells like rotten eggs. (This same smell occurs when the catalytic converter is overloaded with fuel.) Fuel with high sulfur content will also form sulfuric acid, which contaminates the oil and is highly corrosive to engine components. Too much sulfur in the fuel can also form deposits in the fuel system and damage the catalytic converter. Reducing gasoline sulfur content will increase the efficiency and life span of the **catalytic converter**, a key emission control device.

The reduction in gasoline sulfur content will also reduce sulfuric acid emissions, which contribute to environmental pollution and corrosive acid rain. The EPA states that this reduction and a lower **oxides of nitrogen (NOx)** standard in the same ruling will result in passenger vehicles that are 77 to 95 percent cleaner than passenger vehicles on the road today.

Reformulated Gasoline

Reformulated gasoline (RFG) was designed for use in areas with high ground-level ozone problems. It is used in many cities throughout North America. Ground-level ozone is formed when hydrocarbon emissions react with oxides of nitrogen emissions in the presence of sunlight and heat. RFG is not terribly different from regular gasoline. It contains oxygenates to increase the oxygen content of fuel to at least 2.0 percent by weight. RFG also reduces volatility, as well as the levels of benzene, olefins, and sulfur.

COMBUSTION AND EMISSIONS

Combustion is the chemical reaction between fuel and oxygen that creates heat. It is a closely controlled burning of the air and fuel. Spark ignition occurs before top dead center on the compression stroke. The hot compressed air-fuel mixture is ignited, and a flame front develops. For normal combustion to occur the air-fuel mixture must be delivered in the proper proportions and mixed well, the spark must be timed precisely, and the temperatures inside the combustion chamber must be controlled. The flame can then move quickly and evenly (propagate) across the combustion chamber, harnessing the power of the fuel as heat. Pressure builds steadily as the gases expand from heat. The peak of this pressure develops around 10° ATDC to push the piston down on the power stroke **(Figure 4)**.

Gasoline is primarily made of hydrocarbons, carbon and hydrogen. Air contains roughly 21 percent oxygen and 78 percent nitrogen. If combustion were perfect, the hydrogen would combine with the oxygen to form water, the carbon would combine with oxygen to create **carbon dioxide (CO_2)**, and the nitrogen would go out the tailpipe unchanged. Unfortunately, combustion in our internal combustion engine is far from perfect. For one thing, the oxygen content is too low. Several other factors contribute to reduce the efficiency of combustion. The end result is that some of the hydrocarbons (HC) leave the tailpipe unburned, some of the carbon cannot find enough oxygen to combine with and leaves as carbon monoxide (CO), and part of the nitrogen is heated enough to combine with the oxygen to form oxides of nitrogen (NOx). These are the three major pollutants from the engine. These emissions levels are carefully regulated by the **Clean Air Act of 1990**. NOx levels have been further reduced by the EPA ruling mentioned earlier in the discussion of sulfur content **(Figure 5)**. Carbon dioxide is also emitted from our engines. This is considered a greenhouse gas, which contributes to global warming but is not currently regulated as a toxic emission.

HC emissions increase when the engine is cold and needs more fuel. They are excessive when there is an ignition fault. The combustion chamber is filled with air and raw fuel. If spark does not occur, that raw fuel (HC) is blown out the tailpipe. Anything that causes a misfire results in much higher HC emissions. CO is formed in high quantities when the air-fuel mixture is rich with fuel. The theoretically perfect ratio of air to fuel is 14.7 parts of air to one part of fuel, by weight. This ratio is called the **stoichiometric ratio**. There are many times within engine operation, startup, warmup, and acceleration for example, when the air-fuel ratio is much richer (below 14:1). Faulty engine sensors, fuel injection problems, and even a faulty thermostat or dirty air filter can cause extremely rich operating conditions. When

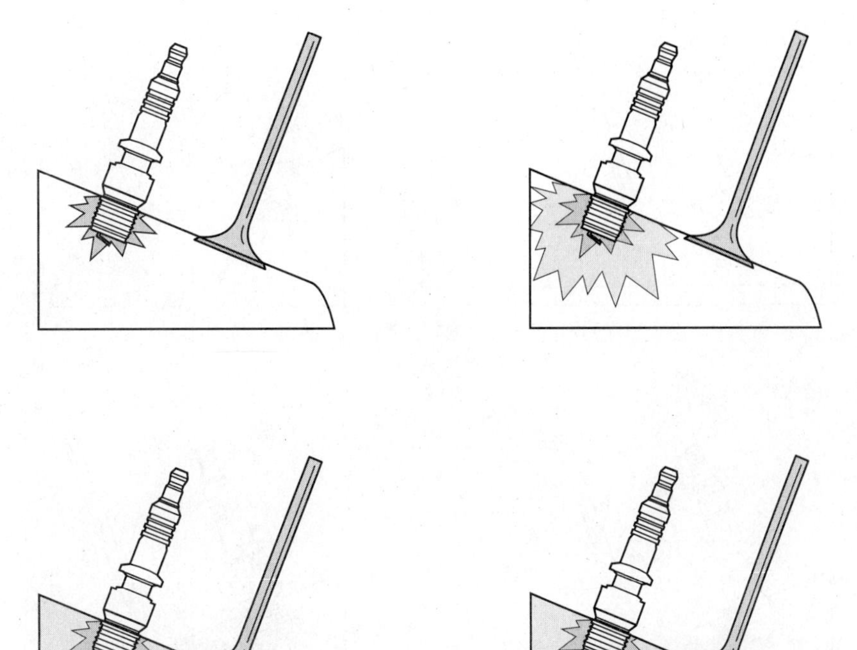

Figure 4. During proper combustion, the flame spreads steadily across the chamber and heats all of the gases.

Figure 5. Gasoline, oxygen, and nitrogen enter the engine; these five gases are measured at the tailpipe to assess combustion efficiency.

the air-fuel mixture is rich there is not enough oxygen to combine with the carbon, and too much CO is formed. NOx emissions increase when combustion is too hot. This generally happens when the vehicle is being driven under load. A faulty EGR system, pinging, cooling system deposits, or improper ignition timing are typical causes of high NOx emissions.

Engine mechanical problems can also cause a serious increase in vehicle emissions. If the engine is unable to produce adequate compression combustion will be inefficient and HC and CO emissions will increase substantially. Improper valve timing; worn rings, pistons, or cylinder walls; and burned or leaking valves can all increase HC and CO emissions. Carbon buildup within the combustion chamber caused by rich mixtures or oil consumption, or a cylinder head that has been resurfaced too much will increase the compression pressures and cause excessive NOx to form during combustion. When diagnosing a vehicle with emissions problems it is important to remember that the base engine condition plays a significant role in the control of the vehicle emissions.

ABNORMAL COMBUSTION

Normal combustion takes about three milliseconds (3/1000 of a second). Abnormal combustion, detonation, is more like an explosion, occurring as fast as two millionths of a second (2/1,000,000 of a second). The explosive nature of detonation can potentially cause engine damage. Pre-ignition is a form of abnormal combustion in which part of

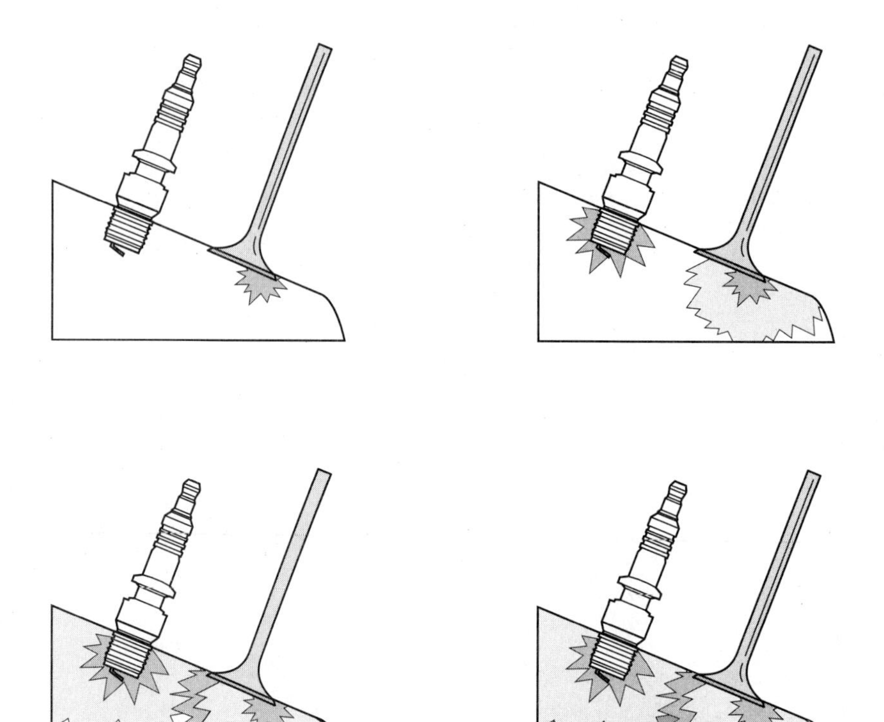

Figure 6. A hot spot starts a flame before the spark. After the spark the two flames collide causing spark knock or pinging.

the compressed air-fuel mixture ignites before the spark. Engine misfire is another form of abnormal combustion; it causes the engine to run poorly and lack power.

Preignition

Preignition means that a flame starts before the spark. This can happen when a hot spot in the combustion chamber auto-ignites the fuel. A flame front develops and starts moving across the chamber **(Figure 6)**. Then the spark occurs and the normal flame front develops. When these two flame fronts collide, a pinging or knocking is heard. Preignition causing pinging can overheat the cylinder and lead to the more damaging detonation. Minor and occasional pinging is normal in some vehicles, particularly those that are driven by a low-power engine, and does not cause engine damage.

Detonation

Detonation occurs when combustion pressures develop so fast that the heat and pressure "explode" the unburned fuel in the rest of the combustion chamber. Before the primary flame front can sweep across the cylinder the end gases ignite in an uncontrolled burst **(Figure 7)**. The dangerous knocking results from the violent explosion. Detonation causes piston and ring damage, bent connecting rods and worn bearings, top ring groove wear, blown head gaskets, and possibly complete engine failure.

Common causes of preignition and detonation are:

- Deposits in the cooling system around the combustion chamber
- Engine overheating
- Too hot a spark plug
- An edge of metal or gasket hanging into the combustion chamber
- Fuel with too low an octane rating
- A faulty EGR system
- Improper ignition timing
- Lean air-fuel mixtures (when there is too little fuel mixed with air)
- Carbon buildup in the combustion chamber
- A faulty knock sensor
- Excessive boost pressure from a turbocharger or supercharger

Knock Sensor

Most modern engines are equipped with a **knock sensor (KS)**. The knock sensor creates an electrical signal when it senses a particular frequency of knocking or detonation. This signal serves as an input to the engine computer, the **powertrain control module (PCM)**. When knocking is

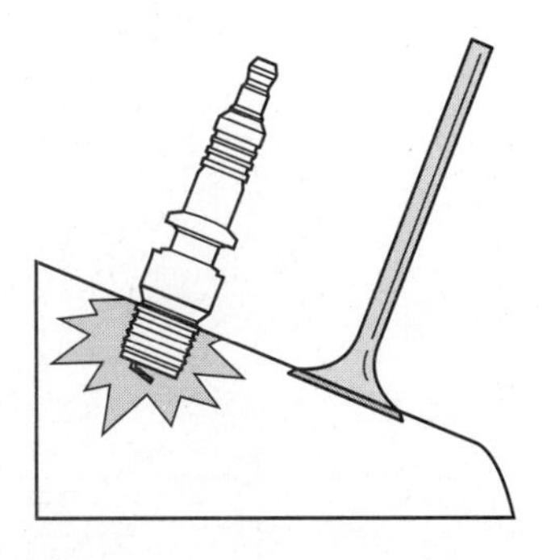
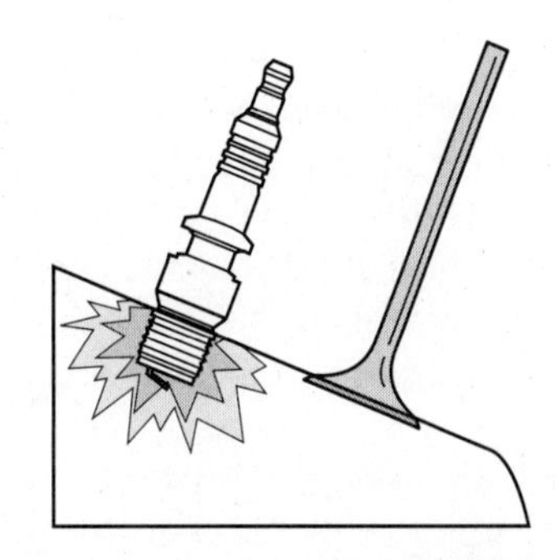
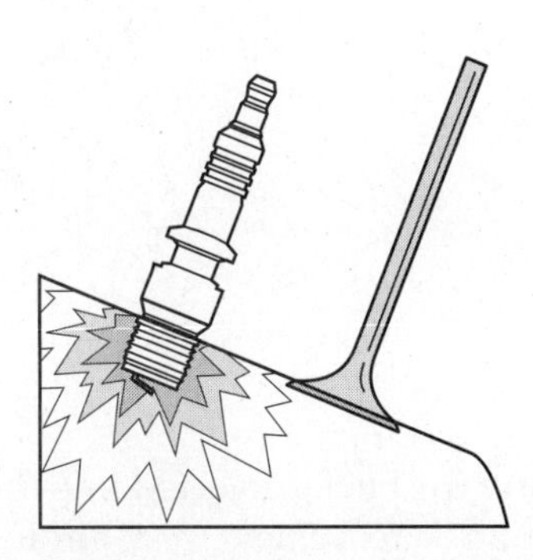
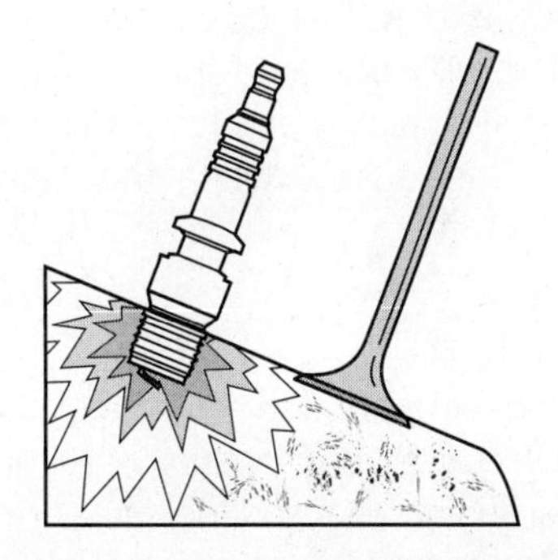

Figure 7. End gases explode before the flame can expand across the chamber. The hot pressure spike knocks the piston so hard it can blow a hole in it.

detected the PCM modifies the spark timing to reduce the potentially dangerous knocking.

Misfire

Misfire is another type of abnormal combustion. When an engine misfires it means that a cylinder (or cylinders) is not producing its normal amount of power. The cylinder is unable to burn the air-fuel mixture properly and extract adequate energy from the fuel. The misfire may be total, meaning that a flame never develops and the air and fuel are exhausted out of the cylinder unburned. Hydrocarbon emissions increase dramatically. Misfire may also be partial, meaning that a flame starts but sputters out before producing adequate power due to a lack of fuel or compression or good spark. When an engine is misfiring the engine bucks and hesitates; it is often more pronounced under acceleration. Technicians normally call this a miss or a skip. Extended misfire can destroy the catalytic converter by overloading it with fuel and causing it to overheat. Misfire can be caused by ignition or fuel system faults as well as engine mechanical problems. Typical engine mechanical faults that cause misfire are:

- Burned or leaking valves
- Valve stem or valve back carbon buildup
- Weak or broken valve springs
- Worn valve guides or lifters
- Broken, worn, or bent rocker arms or pushrods
- Worn camshaft(s)
- Improper valve timing
- Worn pistons or rings

DIESEL FUEL

Diesel power is commonly used for large commercial hauling trucks and for some light-duty trucks and passenger cars. Diesel engines can achieve better fuel economy than gasoline engines. The actual heat content available in **diesel fuel** is about twelve percent higher than that in gasoline. Diesel fuel is a heavier fraction of petroleum. Usually diesel engines use fuel grade No. 2-D. Diesel-powered engines require modifications to the engine structure and the fuel delivery system.

Viscosity

The viscosity of diesel fuel describes its ability to flow. The higher the viscosity is, the thicker the fuel and the harder it is for it to flow. The correct viscosity fuel ensures proper spray into the cylinder and good lubrication of the fuel system and engine components. If the viscosity of the fuel is too low it will run down the walls of the cylinders and dramatically increase ring friction. It will also take a toll on the injection pump, which operates at high pressures.

Cetane Rating

The cetane rating of diesel fuel is similar to the octane rating of gasoline. A higher-cetane-rating diesel fuel will reduce the tendency for a diesel engine to knock. The cetane rating of most No. 2-D fuel is between 42 and 48; a rating of 45 is generally recommended. The proper cetane rating helps diesel engines start when cold and reduces smoking during warm-up.

Cloud Point

The cloud point of diesel fuel describes the temperature at which the fuel will become cloudy. Cloudiness occurs when the paraffin wax in the fuel separates from the other parts of the fuel. This will cause a no-start condition if the wax clogs the fuel filter, pump, or injectors. Lower-temperature cloud point fuel is sold during the winter months to avoid these problems.

Sulfur Content

The sulfur content of diesel fuel is also being regulated. New low-sulfur diesel fuel reduces the formation of sulfuric acid. This reduces internal engine wear and the emission of sulfuric acid into the environment.

Diesel Fuel Contamination

Diesel fuel is very susceptible to contamination from water or sediment. The fuel systems are very sensitive; a small amount of water or sediment can keep an engine from starting. Water in diesel fuel commonly causes engines not to start when the temperature is below 32°F. Ice or sediment can restrict the fuel filter; it must be changed regularly. Contaminated fuel can significantly increase engine wear.

DIESEL COMBUSTION

Diesel engines go through the same four strokes as gasoline engines. They are very similar to gas engines in many respects. Combustion of the fuel, however, occurs through compression ignition; there is no spark plug. The extremely high compression ratio (17–25:1) and compression pressures (250–400 psi) of a diesel engine create a combustion so hot at the end of the compression stroke that the fuel auto-ignites. The temperature of the air in the combustion chamber is higher than the ignition point of the fuel. The fuel is injected directly into the cylinder gradually to prevent an explosion **(Figure 8)**. It takes a certain amount of time (lag time) to get hot enough to vaporize so the fuel burns at a controlled rate. If the lag time is too long the fuel can detonate as all the fuel in the combustion chamber ignites at once. A high cetane rating reduces the lag time. Diesel engines are built with heavier engine components than gasoline engines to withstand the higher temperatures and pressures of diesel combustion.

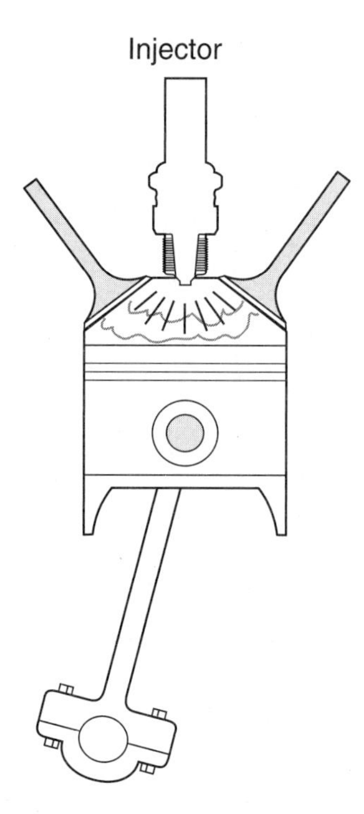

Figure 8. Fuel is injected into the cylinder over time to achieve a controlled expansion of the gases.

Diesel engines can operate with a much leaner air-fuel mixture than gasoline engines. Remember that the theoretically perfect mix of air to fuel in a gas engine is 14.7:1. This ratio changes to between 10:1 or lower under maximum load when cold and to about 22:1 when the engine is at a steady-state cruise. Diesel engines run an air-fuel mixture of over 100:1 at idle and about 20:1 under full load. This makes diesel engines significantly more fuel efficient than gasoline engines. Their exhaust emissions, however, are problematic. Diesel combustion produces significantly lower levels of hydrocarbons and carbon monoxide because of all the oxygen present during combustion. The increased temperatures of diesel combustion, however, contribute to much higher levels of oxides of nitrogen. Diesels also emit a lot of particulate matter from the exhaust pipes; you can see this in the thick black smoke emitted from diesels. Some states are already performing smog checks on heavy-duty diesel engines. Regulations for diesel emissions are becoming tighter, and regular emissions testing is part of the EPA's plan to reduce toxic air pollution.

Summary

- Gasoline is refined from petroleum and contains highly combustible hydrocarbons.
- The octane rating of gasoline describes its ability to resist knock; the higher the number, the greater the resistance to knocking.
- Customers should use fuel with the octane rating specified in their vehicle's owner's manual to ensure good performance and to prevent engine damage.
- A fuel's volatility describes its ability to vaporize.
- Higher-volatility fuel should be used in the winter to assist cold starts; lower-volatility fuel should be used in the summer to prevent excessive HC emissions and vapor lock.
- Oxygenated fuels contain alcohols and are designed to increase the oxygen content of the fuel and reduce CO emissions.
- The gasoline engine emits three regulated toxic pollutants: hydrocarbons (HC), carbon monoxide (CO), and oxides of nitrogen (NOx).
- Preignition and detonation are two types of abnormal combustion that can cause serious engine damage.
- Engine misfire causes bucking, hesitation, and increased emissions.
- Diesel fuel has twelve percent more potential energy than gasoline.
- Diesel engines operate with the same four strokes as gas engines but ignite the fuel from compression ignition.

Review Questions

1. A gasoline's octane rating describes __________.
2. How should a customer decide which octane fuel to use?
3. What problems can occur when fuel with an inappropriate volatility is used?
4. Oxygenated fuels are blended to reduce __________ emissions.
5. Technician A says that adding alcohol to gasoline increases the octane rating. Technician B says that adding alcohol to gasoline increases the fuel's volatility. Who is correct?
   A. Technician A only
   B. Technician B only
   C. Both Technician A and Technician B
   D. Neither Technician A nor Technician B
6. Technician A says that during normal combustion one flame front spreads rapidly across the combustion chamber. Technician B says that preignition occurs when another flame starts after the spark. Who is correct?
   A. Technician A only
   B. Technician B only
   C. Both Technician A and Technician B
   D. Neither Technician A nor Technician B
7. Technician A says that preignition is harmless. Technician B says that detonation can cause severe engine damage. Who is correct?
   A. Technician A only
   B. Technician B only
   C. Both Technician A and Technician B
   D. Neither Technician A nor Technician B
8. Each of the following is a likely cause of detonation *except:*
   A. Carbon buildup in the combustion chamber
   B. Engine overheating
   C. A lean air-fuel mixture
   D. Fuel with too high an octane rating
9. Technician A says that diesel engines go through the same four strokes as gasoline engines. Technician B says that diesels use two spark plugs to ignite the fuel. Who is correct?
   A. Technician A only
   B. Technician B only
   C. Both Technician A and Technician B
   D. Neither Technician A nor Technician B
10. Technician A says that diesel fuel with a lower cetane rating reduces diesel knocking. Technician B says that small amounts of water in diesel fuel can cause a no-start condition. Who is correct?
    A. Technician A only
    B. Technician B only
    C. Both Technician A and Technician B
    D. Neither Technician A nor Technician B

Lubrication System Operation

Introduction

A perfect engine will be reduced to scrap metal in a matter of minutes without a properly functioning lubrication system. The oil pump is as critical to the engine as your heart is to you. The pump pressurizes the oil, and the lubrication system delivers oil to all friction areas of the engine. The oil must be high quality and clean in order to perform its functions well. Careful evaluation of the lubrication system during engine repair is essential.

> **Interesting Fact**
>
> *Using engine oil with higher viscosity than is recommended by the manufacturer can cause premature engine wear. When the engine is first started, thicker oil takes longer to reach the critical oil areas such as the overhead camshaft bearings* **(Figure 1)**.

LUBRICATION SYSTEM PURPOSES

The lubrication system must perform several essential functions in order to minimize engine wear. The key functions are to:

- Reduce engine friction and shock
- Remove engine debris from components
- Help keep engine components cool

Reducing Engine Friction and Shock

Friction occurs whenever two components rub against each other. The contact between the camshaft and the lifter creates significant friction. The piston to cylinder wall scraping is another area of high friction. Engine oil lubricates these areas as well as the crankshaft main, connecting rod and camshaft **engine bearings**, the rocker arms, the timing chain and tensioner or timing gears, and the lifters. Adequate lubrication prevents rapid engine wear.

The camshaft, crankshaft, and connecting rod bearings, hydraulic valve lifters, rocker arms on pushrod engines, and timing chain tensioner are all fed oil under pressure. The bearings have a small clearance, usually .0005 in.–.002 in., between the bearing and the journal. Oil must take up this clearance to reduce friction and eliminate shock as the pistons force the crankshaft and connecting rods down during the power stroke. The thin film of oil actually prevents contact between the bearing and the journal. The oil clearances must be large enough to allow an adequate film

Figure 1. Use the specified oil to provide the best engine protection.

Figure 2. This crankshaft journal suffered severe damage from a lack of proper lubrication.

Figure 3. This aluminum oil pan with fins helps dissipate the heat of the engine oil.

of oil but not so large that too much oil can leak out from between the bearing and the journals. Low oil pressure and engine knocking will occur if the oil clearances are too great **(Figure 2)**. Most valve trains using pushrods deliver oil pressure through the lifters, up the pushrods, and to the rocker arms and valve tips to minimize wear and noise. Other rocker arm configurations may use a pressure line to splash the rocker arms or valve tips. Hydraulic valve lifters and timing chain tensioners require oil under pressure to function. A loss of oil pressure to the lifters will cause valve train clatter, a loud ticking noise from the top of the engine where the rocker arms contact the valves. A hydraulic timing chain tensioner may also create a metallic slapping noise on deceleration if it does not receive good oil supply.

The cylinder walls and timing chain or gears are splash fed with oil. The connecting rods throw oil on the cylinder walls as they move up and down the cylinders. This keeps a small amount of oil between the pistons and cylinders. It also helps the piston rings seal against the cylinder walls. Outlets in the oil passages throw oil at the timing chain or gears and other drive gears such as oil pump or distributor drive gears. Sometimes an oil slinger directs or sprays oil at a component.

Removing Engine Debris

Oil must pick up and hold engine debris in suspension to prevent component damage. During normal engine operation pieces of carbon from combustion and small filings from cylinder, gear, and bearing wear are picked up by the oil. This prevents them from causing abrasion between components. The oil filter is designed to filter most of these particles out of the oil as the lubrication cycle continues. Acids and fuel from combustion and the buildup of debris in the oil make frequent oil changes essential to long engine life.

Engine Cooling by Oil

The lubrication system helps cool engine components. Reducing friction certainly limits excess heat from developing. Further, the oil bath that friction components receive helps lower component temperatures. The oil picks up heat as it is forced between components. When the oil returns to the oil pan the oil dissipates much of this heat, and the lubrication system returns cooler oil back to the components. Some oil pans are made of aluminum and have cooling fins on them to help lower engine oil temperatures **(Figure 3)**.

LUBRICATION SYSTEM COMPONENTS

Each of the lubrication system components serves a vital role in maintaining proper oil pressure and flow. When performing engine repairs you must evaluate each component of the lubrication system to ensure that your repair work will last. The driver must also be alerted instantly if there is a loss of oil pressure to prevent serious engine damage.

Oil Pans

The oil pan is a reservoir for the oil. The oil in the pan is not under pressure. Oil leaks out the main and rod bearings back down into the pan. Return ports from the head allow oil to drain back into the pan after lubricating the valve train. Some oil pans are a simple stamped steel design. The ones made of aluminum often have fins to improve heat dissipation and help dissipate the heat in the oil.

Oil Pumps

The oil pump must consistently deliver oil pressure within a specified range to support engine operation. Most engines run between 10 and 75 psi depending on engine

Figure 4. The pick-up screen protects the pump from large pieces of debris.

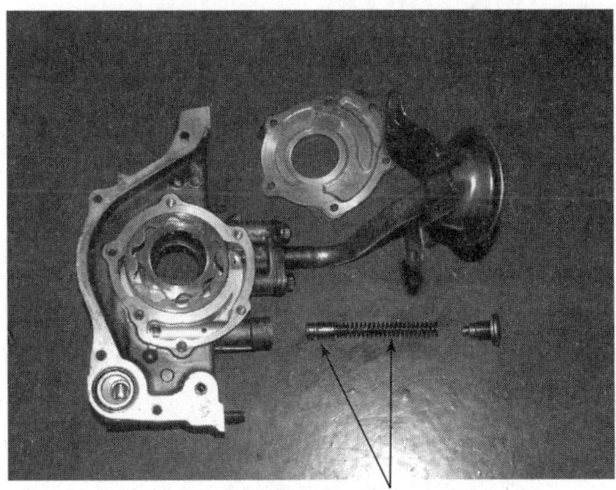

Figure 5. A disassembled gerotor oil pump shows the pressure relief valve spring and piston.

rpm. The oil pump picks up oil from the oil pan through a pick-up tube with a filtering screen on the end **(Figure 4)**. The oil pump then pressurizes the oil and sends it out the main oil gallery (passage). Engine oil pumps are **positive displacement pumps**. This means that the faster they spin, the more oil pressure they produce. The pressure level is controlled by an **oil pressure relief valve**. At higher engine rpms, or when the engine oil is cold and thick, the oil pressure developed could be high enough to damage engine components and seals. The pressure relief valve operates during high-pressure conditions to relieve excess oil pressure. It is simply a spring-loaded piston in a chamber **(Figure 5)**. When oil pressure is greater than the spring-calibrated spring tension, the oil pressure moves the piston, and a port is uncovered to return the oil to the pan.

Oil pumps all serve the same purpose and function similarly, but they are designed differently. Some oil pumps

Figure 6. This crescent style oil pump mounts on the front of the crankshaft.

are a gear-type or rotor-type pump. These pumps are driven by a gear off the camshaft or distributor. As the gear or rotor spins, oil is picked up into the larger-volume area. As the pump turns, the oil is forced into a smaller space, pressurizing the oil. A newer and increasingly common type of oil pump is the crescent-style pump **(Figure 6)**. This pump sits on the end of the crankshaft and is driven directly by the crank.

Oil Galleries

The engine has oil passages drilled through the engine block; these are called **oil galleries**. The oil galleries distribute oil to the necessary areas of the engine. When oil is forced out of the pump it is pushed through the oil filter and then out the main gallery or galleries. The main gallery feeds the camshaft bearings and hydraulic lifters. Oil is also pushed through the drilled galleries of the crankshaft to provide oil to the crankshaft main and connecting rod bearings. The cylinder walls are splash lubricated by the connecting rods **(Figure 7)**. The oil galleries are sealed by plugs in the end of the block. These passages must be cleaned during an engine overhaul.

Oil Filters

Oil filters trap carbon, metal, and other debris picked up by the engine oil. The filter element is made of specially calibrated paper designed to allow flow but block debris over a specified size, usually 10–20 microns. Not all oil filters are of equal quality. A heavier filter generally has more paper element in it and can do a better job. Many oil filters also contain a check valve to keep oil in the filter when the engine is turned off. This helps reduce engine wear during the damaging start-up period. Oil filters or their mounting housings contain a bypass valve **(Figure 8)**. The location depends on the manufacturer. The bypass valve opens to allow oil to circumvent the element when oil pressure in

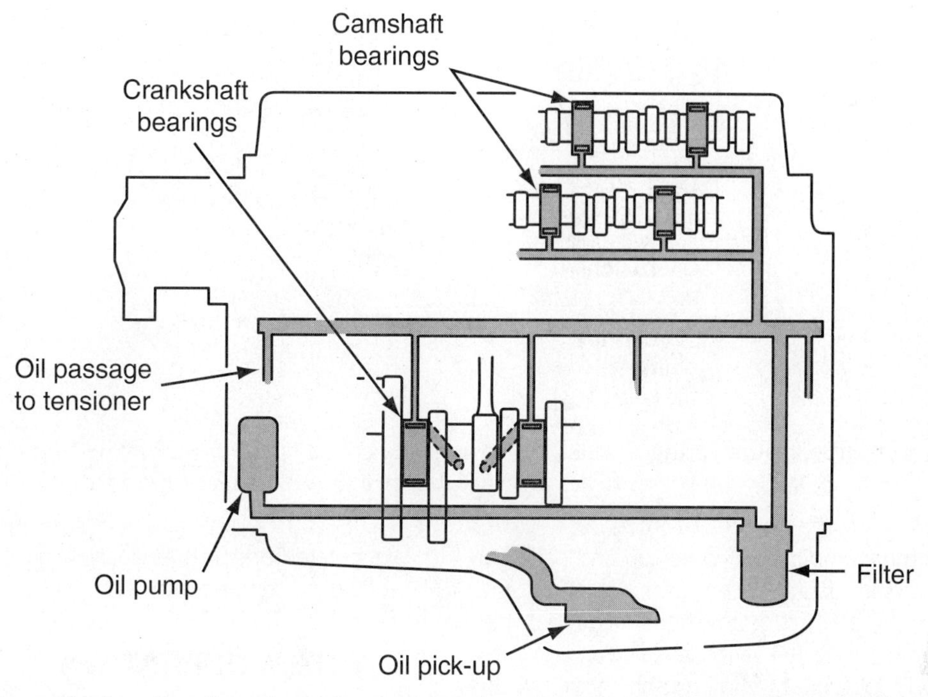

Figure 7. Follow the paths for oil to flow from the pump to the critical lubrication points.

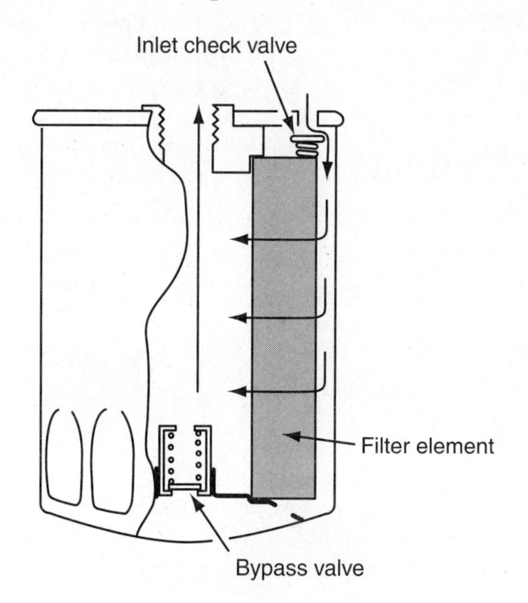

Figure 8. Oil normally flows through the paper filter element. If the element becomes clogged, the excess pressure will open the bypass valve to allow flow around the paper filter.

the filter is significantly higher (roughly 10 psi) than the pressure in the oil passage; this occurs when the oil filter is partially or fully restricted.

Oil Coolers

Turbocharged and some high-performance engines or engines in vehicles designed to tow or haul frequently use an **oil cooler** to help keep oil temperatures at a safe level. Oil temperature should generally run between 200°F and 250°F. Excess heat will prematurely break down the oil and allow premature engine wear. Oil coolers look like small radiators usually mounted near the side of the radiator. They exchange the heat in the oil with the air blowing across it.

Oil Warning Light and Pressure Switch

All vehicles have an oil pressure warning light to alert the driver if the oil pressure falls to a dangerous level. The oil warning light on the dashboard should come on with the key on and engine off and should go out immediately after the engine starts. The **oil pressure switch** or sending unit screws into a pressurized oil passage, often near the oil filter housing. The switch is normally closed, meaning that it provides an electrical ground path for the oil pressure warning light in its rest position when there is no oil pressure **(Figure 9)**. When the engine is started and oil pressure develops, the switch opens so the oil pressure warning light goes off. If the oil pressure ever falls below the pressure needed to open the switch, usually 3–5 psi, the oil warning light will receive a ground and come on. It is essential that customers know that they must immediately turn the engine off if the oil light comes on; serious engine damage will occur if they continue driving even for a few minutes!

Oil Pressure Gauge

Some vehicles provide an oil pressure gauge to give a more accurate reading of oil pressure. When oil pressure decreases the customer can see that the system should be

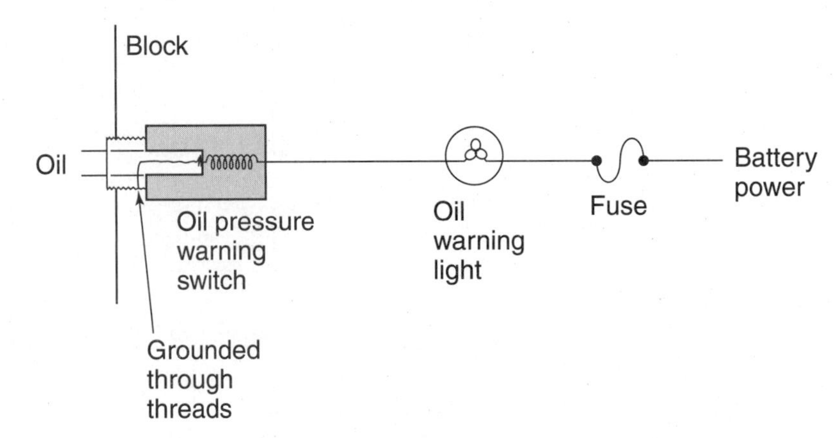

Figure 9. Without oil pressure acting against the spring, the contacts are closed providing a ground to the light turning it on. When oil pressure acts against the spring, the contacts open and the light turns off.

serviced before the warning light comes on and damage occurs. An oil pressure gauge system uses a sending unit that varies the resistance to ground for the gauge. This allows the gauge to reflect the varied levels of oil pressure. When diagnosing potential gauge problems or when checking engine oil pressure for engine diagnosis you should use a mechanical oil pressure gauge to get a precise reading of oil pressure.

Oil Level Indicators

An oil level indicator is generally an on/off switch similar to an oil pressure switch. It uses a floating arm in the oil sump (oil pan) that rides on the top of the oil. When the level falls below a calculated level, the float arm engages an electrical contact that provides a ground for the low oil level warning light.

Oil Life Indicators

Some newer vehicles are incorporating engine oil life data into their driver information systems. The display shows the percentage of engine oil life left or turns on a light alerting the driver that the oil should be changed. The powertrain control module (PCM) calculates the wear factors on the oil, such as engine load, fuel mpg, trip length, miles since the oil was changed, and engine temperature, to estimate when the oil should be changed. These systems can safely lengthen the interval between oil changes, saving the customer money and reducing oil use and the generation of toxic waste oil.

Oil Temperature Indicators

Some vehicles' instrument panels display engine oil temperature or use this information in calculating oil life. The information is inferred by an electrical signal from an oil temperature sending unit. The oil temperature sender is a resistor that changes its electrical resistance inversely proportional to temperature changes. The electrical signal generated operates a gauge or display panel to show the driver the temperature of the oil.

SYSTEM OPERATION

Oil is picked up through an oil pick-up tube and screen from the oil pan. The oil pump creates a low-pressure area to pull oil into the pump. The pump pressurizes the oil and sends it through the oil filter. When engine rpm is high or the oil is cold the pressure relief valve may be opened to allow some oil to drain back into the sump **(Figure 10)**. This reduces the oil pressure to a safe level. From the oil filter the oil is pumped through the main galleries and crankshaft to

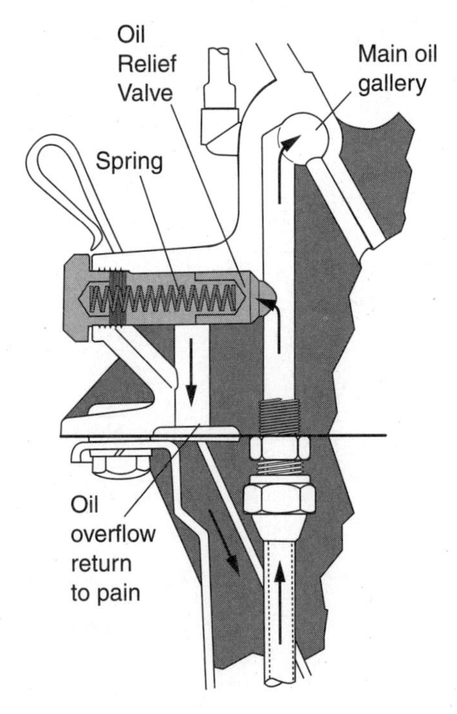

Figure 10. Excessive oil pressure will overcome spring tension and allow oil to drain back into the pan.

distribute oil to the engine bearings. The oil is sent through the bearings, where it slowly leaks back down into the sump. Oil that lubricates the valvetrain components, timing chain, and cylinder walls is also allowed to drain down into the crankcase. This process is repeated as the oil is continuously cycled through the engine.

POSITIVE CRANKCASE VENTILATION SYSTEM

The engine crankcase, the area below the crankshaft where the oil pan is, must be vented to prevent excess pressure from developing within it. As the combustion events occur some leakage of pressurized combustion gases past the rings, or **blowby**, is inevitable. If this pressure were not relieved the oil and gases in the crankcase could be blown out of the seals and back up past the rings into the combustion chamber. This would cause oil leakage and oil consumption (oil use through burning oil in the combustion chamber; this causes blue smoke out of the tailpipe). The **positive crankcase ventilation (PCV) system** prevents this by drawing the blowby gases back into the intake manifold. The PCV valve is typically mounted in the valve cover and allows a regulated amount of blowby gas to flow **(Figure 11)**. The PCV system is an emission system and must be properly maintained to avoid venting fumes into the atmosphere.

ENGINE OILS

You must use the proper oil for the engine to ensure maximum life. The manufacturer has dedicated hours and hours of research into which oil is best suited for the engine; generally you should use the oil specified by the manufacturer. Sometimes synthetic oil may be substituted for traditional mineral-based oil, but the same **viscosity**

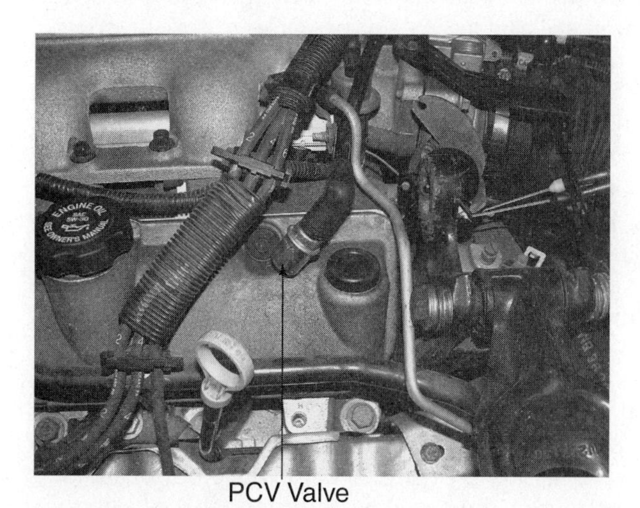

Figure 11. The PCV valve mounts in the rubber grommet in the valve cover just to its right. It is calibrated by an internal spring.

Figure 12. The starburst symbol shows that the oil is certified by the International Lubrication Standardization Approval Committee (ILSAC) for use in gas passenger cars and light duty trucks.

rating should be maintained. Thinner or lower viscosity oils are used in today's engines to reduce friction and improve fuel economy. Modern oils provide much better lubrication than earlier oils. Now, putting a thicker, higher-viscosity oil in an engine can cause premature engine wear. Oil standards are displayed on the oil container through the **American Petroleum Institute** service identification "doughnut" and by the starburst symbol **(Figure 12)**.

Oil Viscosity

Oil viscosity defines the oil's resistance to flow; the greater the viscosity, the thicker the oil. Oil viscosity needs have changed significantly over the past few decades. Better engine materials and tighter machining tolerances have made oils with lower viscosity essential for good engine lubrication, particularly at startup and through the warm-up period. Automotive engine oils are now multiple-viscosity-grade oils. Manufacturers commonly specify 5W-30 engine oil, for example. Some manufacturers are designing engines that use a 0W-30 or 0W-20 engine oil. The different ranges of viscosity are achieved by adding viscosity improvers to the oil. The 5W is the actual viscosity grade of the oil in cold temperatures; the W stands for winter. Then, viscosity index improvers, synthetic polymers, are added to allow the oil to flow like higher-viscosity oil, 30, when at normal operating temperature. The oil viscosity ratings are tested by the Society of Automotive Engineers (SAE). On the oil container, the viscosity will read SAE 5W-30. Using engine oil with higher viscosity may be recommended when towing, but generally the viscosity grade is no longer adjusted seasonally. If the oil viscosity is too high for the engine it will not lubricate the bearings as well when the engine is cold. The oil takes longer to reach the last areas of oil flow, causing premature wear on the engine. Using engine oil that has too low a viscosity index will not provide adequate protection when the engine is hot. The oil will leak out from between the bearings too quickly, reducing its effectiveness and the oil pressure.

Figure 13. The oil "donut" shows that the oil meets the current SL, GF-3, and energy conserving standards as tested by ILSAC and the American Petroleum Institute.

Oil Quality Rating

The oil is tested against quality standards by the American Petroleum Institute (API) and SAE. The API tests the oil and gives it the API certification **(Figure 13)**. The newest API ratings are "API Service SL." The letter S stands for service, but it can also be used as the designation for oil used in spark ignition engines. The letter L is the revision of oil. L is the highest-rated oil currently available; it was introduced in 2001. SJ was the highest-rated oil between 1997 and 2001. SH oil was used between 1993 and 1996. The next oil revision to designate the highest-quality oil available is SM oil. This means that API has tested the oil to even higher standards than the SL oil. You should use the newer oils on older engines, but you should not use an older revision on current engines. It quickly becomes difficult to find the obsolete oil. API uses a similar system to rate oil for commercial vehicles, CD-rated oil, for example. You can also remember the C as a designation for oil to be used in compression ignition (diesel) engines. Diesels currently use oil rated between CD and CH-4. These ratings apply to different types of diesel engines; it is not necessary or even recommended to just use the latest revision. Check the owner's manual or service information to determine the correct oil for use in a diesel engine.

Starburst Symbol

The **International Lubricant Standardization and Approval Committee (ILSAC)** created an oil standard that combines both the SAE and API ratings. When oil meets both the ILSAC standards, currently GF-3, the oil container displays a starburst symbol. Choose oil that shows the starburst symbol on the container.

Energy-Conserving Oil

Oil may display an energy-conserving designation in the lower portion of the doughnut. The American Society for Testing Materials (ASTM) certifies oil as energy conserving if it can produce a 1.5 percent increase in fuel economy over the standard test oil or as energy conserving II if it increases fuel economy by at least 2.7 percent.

Oil Additives

Engine oil has numerous additives in it to increase its performance in the automotive engine. The additives break down over time, which is part of the reason that frequent oil changes are required. Manufacturers may add:

- Detergents and dispersants to help keep engine components clean by suspending particles in the oil
- Antioxidants to minimize oil breakdown at high temperatures; they reduce varnish, bearing corrosion, and carbon buildup
- Antiwear additives to help the oil protect the high-friction areas of the engine; they help the oil prevent contact between the two parts
- Viscosity improvers to allow the oil to perform at all temperatures
- Pour point depressants to allow the oil to flow at lower temperatures
- Corrosion inhibitors to prevent rust and corrosion of engine components; they neutralize the acids formed during the combustion process
- Antifoam agents, friction modifiers, water repellants, dyes, and other additives to improve oil performance

Synthetic Oils

Synthetic oils are not made exclusively from a petroleum base. Some synthetic blends may use a portion of petroleum-based stock. Fully synthetic oils are made from a base stock processed from chemical compounds that do not exist naturally. More and more manufacturers specify synthetic oils for some engines. Other manufacturers warn against them or have specific procedures to follow when changing from petroleum-based oil to synthetic. Synthetic oil reduces engine friction, and its viscosity is more stable over a wider temperature range. It also theoretically lasts longer, although manufacturers' oil-change intervals must still be followed. Using synthetic oil may result in increased fuel economy because of the reduced internal engine friction. Many synthetic oil manufacturers claim that using their oil will lengthen engine life. While these sound like definitive reasons to use synthetic oils they also have some disadvantages. Some engine rebuilders claim that synthetic oil can reduce friction so much that the new rings used in rebuilding will not break in rapidly to the cylinder walls, causing excessive oil consumption. Synthetic oil is also much more expensive than traditional oil. Ultimately the choice, if approved by the manufacturer, is one that customers may make based on advertising and your advice.

Summary

- A properly functioning lubrication system is essential to prevent serious engine damage.
- The lubrication system must reduce engine friction and shock, remove engine debris from components, and help keep engine components cool.
- Oil forced into the small clearances between the engine bearings and journals prevents rapid bearing wear.
- The oil pump is the heart of the lubrication system; it pumps oil under the correct pressure throughout the engine.
- Oil galleries are drilled passages in the engine block to distribute oil.
- Oil filters must filter particulate matter out of the oil to prevent excessive engine wear.
- Oil viscosity grades are determined by the Society of Automotive Engineers (SAE); multiviscosity oils are currently used in automotive engines.
- Oil quality standards are tested by the American Petroleum Institute (API). These ratings are displayed in the API service doughnut.
- Oil meeting the SAE and API standards displays the ILSAC starburst symbol on the container.
- A variety of oil additives improve the performance of automotive engine oil.
- Synthetic engine oil may provide benefits to an engine, but be sure the manufacturer approves of its use in the particular engine.

Review Questions

1. What noises may occur if an engine loses oil pressure?
2. The __________ system can cause seal damage and oil consumption if faulty.
3. Describe three functions of the lubrication system.
4. Define each designation described in an API Service SL, SAE 5W-30 oil.
5. Technician A says that a thin film of oil is necessary between the bearings and journals to reduce the shock as parts move toward each other. Technician B says that engine oil reduces the friction in an engine. Who is correct?
   A. Technician A only
   B. Technician B only
   C. Both Technician A and Technician B
   D. Neither Technician A nor Technician B
6. Each of the following is a type of oil pump *except:*
   A. Worm
   B. Crescent
   C. Gear
   D. Rotor
7. Technician A says the oil filter commonly contains a pressure kick-down valve. Technician B says the oil pressure relief valve opens to maintain a safe oil pressure at higher rpms. Who is correct?
   A. Technician A only
   B. Technician B only
   C. Both Technician A and Technician B
   D. Neither Technician A nor Technician B
8. Technician A says the higher the viscosity number, the thicker the oil. Technician B says that using a higher-viscosity oil reduces engine friction. Who is correct?
   A. Technician A only
   B. Technician B only
   C. Both Technician A and Technician B
   D. Neither Technician A nor Technician B
9. Technician A says that an oil pressure sending unit can operate an oil pressure gauge. Technician B says that the oil pressure switch can turn on the low oil level warning light. Who is correct?
   A. Technician A only
   B. Technician B only
   C. Both Technician A and Technician B
   D. Neither Technician A nor Technician B
10. Technician A says that synthetic oil should be used in most modern engines. Technician B says that when using synthetic oil you can increase the length of the oil change interval. Who is correct?
    A. Technician A only
    B. Technician B only
    C. Both Technician A and Technician B
    D. Neither Technician A nor Technician B

Chapter 9 Lubrication System Service

Introduction

The last chapter explained the importance of the lubrication system in minimizing engine wear. This chapter describes the procedures used to maintain, diagnose, and repair the system. The oil and filter must be changed at appropriate intervals to prevent dirty, broken-down oil from allowing damage to engine components **(Figure 1)**. The system must be filled with the correct amount and type of oil at all times; leaks should be repaired promptly. Low oil pressure can dramatically reduce engine life; pressure should be checked whenever excessive engine noise is heard, when a gauge or warning light indicates low pressure, or before major engine repairs are undertaken.

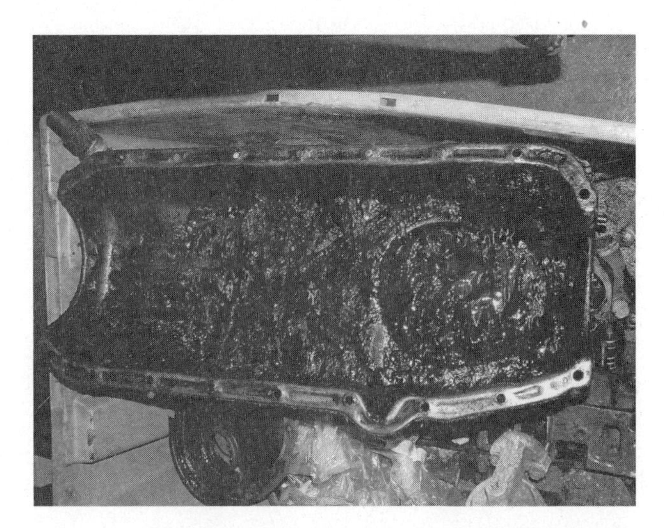

Figure 1. The sludge in the bottom of the pan reflects serious lubrication system neglect. The cost was a complete engine overhaul including oil pump, crankshaft, camshaft, and pistons.

Soon 0W-30 engine oil will be commonly used on new engines. Customers accustomed to using 10W-40 oil may require some reassurance that 0W-30 is best for the engine.

OIL AND FILTER CHANGE

Changing the oil and filter on an engine is the most important maintenance procedure to provide long engine life. Vehicle manufacturers typically recommend a maximum interval between oil changes; this is often every 7500 or 10,000 miles, or six months, whichever comes first. This is the maximum interval recommended; it is not necessarily the best interval for promoting long engine life. Manufacturers also specify that the oil should be changed at half the recommended interval if any of the following driving conditions occur:

- Driving on dirt roads or in dusty areas
- Routinely driving for short time periods or distances, under 15 minutes or 10 miles
- Operating the engine in temperatures below freezing
- Idling the vehicle for extended periods
- Driving for extended periods at high speeds
- Towing a trailer

It is easy to see that many people regularly encounter at least one of these conditions. Many technicians, repair facilities, and oil manufacturers recommend an interval of every 3000 miles or three months to customers. This is a relatively inexpensive way to help maximize engine life. It is important for you to communicate these reasons effectively to customers so that they can make an informed choice about how often they are going to change their oil. Some manufacturers produce vehicle models that use software to indicate oil life. In these cases an interval is not specified. When the service oil indicator is displayed customers should have their oil and filter changed promptly.

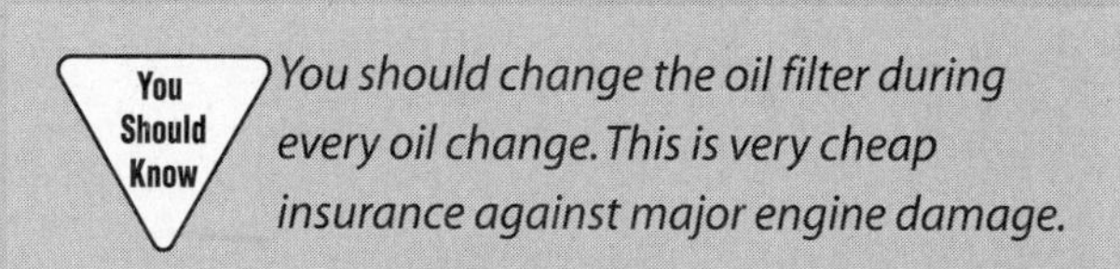

You Should Know

You should change the oil filter during every oil change. This is very cheap insurance against major engine damage.

Interesting Fact

When disassembling a low-mileage engine a technician found so much sludge in the oil pan that it had risen above the oil pump pickup and totally blocked it. The engine needed such extensive work that it was replaced with a rebuilt unit. The customer insisted that she had changed the oil every 5000 miles. Through further discussion the technician discovered that the owner had been adding a can of fuel injection cleaner at every fill-up. The excessive volume of this chemical contaminated the oil and turned it into a thick sludge. Thorough communication with the customer prevented another engine failure.

Before draining the engine oil be sure that it is at normal operating temperature. This ensures that the contaminants are suspended in the oil. Remove the oil drain plug, and allow oil to drain until it is dripping at a very slow rate out of the drain hole. Remove the oil filter with an oil filter wrench **(Figure 2)**. If the oil from the filter is likely to leak out onto the engine many technicians will use a plastic bag to catch the oil. This minimizes the amount of dripping after the oil change and prevents the customer from returning with a concern that they have an oil leak. If the oil filter is mounted at an angle above horizontal, poke a small hole in the end of the oil filter. This allows the oil in the filter to drain back into the engine rather than leak onto the engine during removal. Check the oil filter and the mounting base to be sure the oil filter seal is removed from the engine. If the old gasket is left on and the new filter is "double gasketed" the new filter will soon loosen up and create a dangerous oil leak. The old oil filter must be drained for twenty-four hours or crushed before disposal. It is hazardous waste and must be disposed of through an appropriate waste management service.

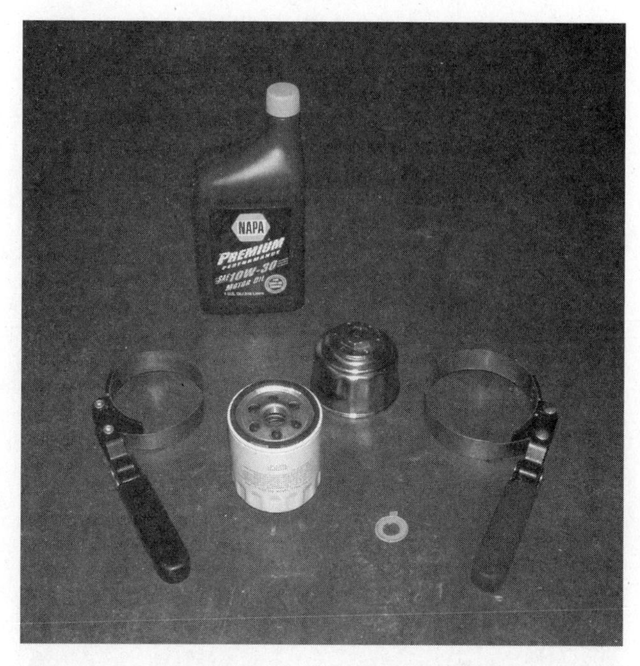

Figure 2. The tools for an oil and filter change: fresh oil and filter, a new drain plug gasket, and an assortment of oil filter wrenches.

You Should Know

Used engine oil is toxic. Use latex gloves when changing oil, or wash your hands thoroughly after contact to minimize any health hazards.

Run a thin layer of clean oil around the new oil filter seal. This helps it seal and allows for easy removal at the next oil change. Install the filter hand tight; there is no need to use a filter wrench. Check the drain plug gasket to be sure it can safely be reused. Some manufacturers recommend replacing the seal at every oil change. Put the drain plug back in the oil pan, and tighten it immediately. *Never* thread an oil drain plug in by hand and walk away from the engine for any reason. Carry the wrench or socket with you to install the drain plug. A loose drain plug can cost the repair shop an engine and even you your job.

You Should Know

Many fine technicians have lost an engine due to a loose drain plug. Develop a strict routine that you follow for every oil change that includes remembering when you tightened the oil drain plug before turning the key. Do whatever it takes to prevent serious engine damage from a simple job.

Refill the engine with the specified amount of the recommended oil. Turn the key on, and observe the oil warning light to be sure it is functional. Start the engine, and watch the light until it goes out **(Figure 3)**. If it does not turn off within about 15 seconds turn the engine off and check your work. Once the oil warning light goes out let the engine run for a few minutes to allow the oil to circulate throughout the engine and give you time to check for leaks. It is important to check the seal at the drain plug and the oil filter before returning the vehicle to the customer. An oil change is a simple procedure, but a mistake can be disastrous. Turn the engine off, wait a minute for the oil to drain back down into the pan, and check and adjust the oil level as needed. Install an oil sticker indicating the date or mileage at which the oil should be changed next **(Figure 4)**. It is important to be sure the customer knows when the oil is due to be changed again.

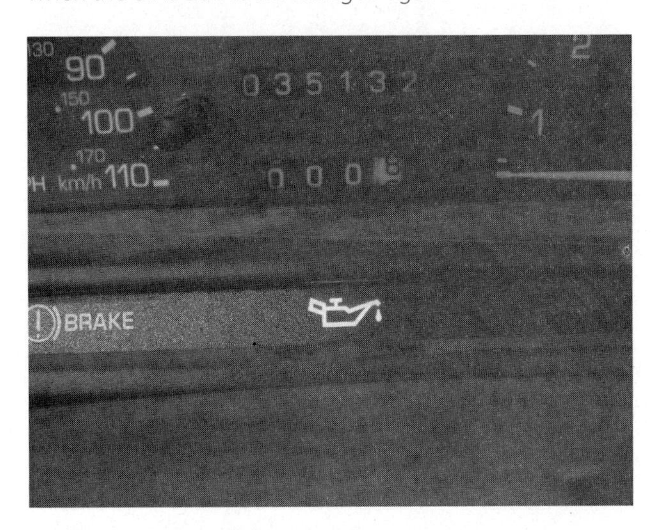

Figure 3. Make sure the oil warning light comes on with the key on and then goes out after the engine is started.

Figure 4. Install an oil change sticker showing when the next oil change is needed.

You Should Know *The oil change interval specified by the manufacturer may be deceiving. Be sure to read the recommendations for oil change intervals for vehicles experiencing severe service. The majority of drivers operate their vehicles within at least one of the criteria for severe service; this cuts the recommended oil change interval in half.*

Most repair facilities include other services along with an oil change. This may be the only time a customer has any sort of preventative maintenance check performed. You are often asked to check all other fluid levels, check and adjust tire pressures, and check the suspension, steering, brake, and exhaust systems. The customer usually appreciates your locating a problem before it becomes too serious. Carefully inspect the engine and transmission for leaks. If necessary, use a compatible dye in the leaking fluid and locate the source using a black light. It is important to inform the customer of any leaks and repair them right away to avoid further or serious damage. Follow your shop's full list of procedures to provide the customer with quality service at every visit.

You Should Know *Be careful about installing oil treatment additives to engine oil. Some manufacturers will void the engine warranty if unapproved additives are found in the oil.*

OIL LEAKS

Oil leaks are dangerous. They may start as small drips, but they can become serious at any time. Oil leaking on a hot exhaust manifold or pipe can start a fire. If a customer runs the engine with low oil levels he can quickly destroy an engine. Oil leaks are commonly found at the front or rear main seals, valve cover gaskets, camshaft seals, oil filter or oil filter housing, oil pan, drain plug, or occasionally head gasket. Often it is simple to detect the source of the leak. Sometimes when the oil has spread across the engine and leaked down from a barely visible area it can be hard to determine which gasket or seal needs replacing. In these cases use an engine dye and a black light to pinpoint the source of the problem. Put a small bottle of engine dye in the oil fill. Run the engine while looking over the engine with the black light. If the leak is slow the engine may have to run for quite a long time. In some cases you may have to ask the customer to return in a few days. The dyed oil will show up bright yellow or orange under the black light. Follow the trail to the source to definitively determine the cause of the leak. After repairing the leak, check the PCV system to be sure it is not restricted. A restriction in the PCV system can cause oil leaks by allowing pressure to build up behind seals and gaskets. Let us look at an example of a repair to correct an oil leak.

Front Main Seal Replacement

Before undertaking a significant job like replacing the front main seal, access the specific service information. Read through it first, and keep it handy to refer to throughout the job. To replace the front main seal you will have to remove all of the accessory drive belts. This is an excellent time to replace them if they show any signs of wear or have high mileage on them. Look closely for fraying, cracks, glazing, or missing ribs. Many newer vehicles use one serpentine belt to drive all the accessories. The manufacturer typically holds the belt taut with an automatic tensioner. To release the tension there are often flats for a wrench or a spot for a breaker bar so that you can rotate the tensioner. While the tension is relieved slip the belt off one of the pulleys. Then you can release the tensioner. On some models it may be necessary to lock the tensioner with a pin or special tool to prevent it from extending too far and breaking apart. On some vehicles multiple belts are used. Each belt is tensioned individually. Most adjusting mechanisms consist of a threaded screw that can be tightened or loosened. The pivot point bolts must be loosened before you can loosen or tighten the belt.

Once the drive belt or belts are removed you will have to remove the crankshaft pulley. Remove the bolt retaining the pulley. The pulley is usually press fit on the snout of the crankshaft. You will have to use a puller to properly remove the pulley. A few pulleys slide on and off the crankshaft and are held on only by the bolt. On front-wheel-drive vehicles where the engine is mounted transversely you will usually have to remove the inner fender splash shield to gain enough room to install a puller. On some longitudinally mounted engines you may have to remove the fan and/or the radiator to install a puller. Spray some penetrating oil on the seam between the pulley and the crankshaft, and then tighten the puller to slide the pulley off the crank.

With the pulley off, use a seal remover or prybar to remove the old seal. Be careful not to scratch the crankshaft. Check the crankshaft for wear or a ridge where the seal rides. Thin sleeves that fit over the crankshaft snout are available for some engines when the seal surface is damaged. Check that the key and keyway are in good condition and properly positioned. Thoroughly clean the crankshaft and the hole where the seal fits. Apply a little oil to the lip of the seal, and use sealant on the outer circumference only if the manufacturer specifies it.

You may be able to tap the pulley back onto the crankshaft if recommended. Use a tool that contacts the inner metal circumference of the pulley. In other cases you may need a crankshaft pulley installation tool. This is similar to a puller, but it presses the pulley onto the crankshaft. Be sure the pulley is fully seated, and tighten the crankshaft pulley retaining bolt to the proper specification.

Properly route the new drive belt(s), and allow the tensioner to tighten the belts. You should check the tension of the belt(s) with a belt tension gauge. This tool slides over the tightened belt and deflects a needle to read the tension. Compare this reading with the specification. If it is incorrect you will need to verify that the proper belt is installed in the correct routing and check that the tensioner is not damaged; you may have to readjust the tension if it is manually adjustable. If the belt is left too tight it could snap or damage accessory or even engine bearings. If the belt is left too loose it will quickly become glazed and may slip, preventing proper rotation of the components. Slipping could cause the water pump to spin too slowly and allow the vehicle to overheat, for example.

Finally, look over your work, and then start the engine. Listen for any belt noises, and look closely to be sure the new seal is not allowing any leakage. Correct any imperfections before returning the vehicle to the customer.

OIL PRESSURE

Test oil pressure when a gauge or warning light indicates low pressure. You may also check oil pressure to gauge the condition of the engine bearings and oil pump. This can be good preventative maintenance on a high-mileage engine. It is much less expensive and more convenient to repair or replace a worn engine than a failed engine.

Run the engine up to **normal operating temperature (NOT)** before performing an oil pressure test. The manufacturer will specify a pressure range when the engine is warm and at a particular rpm. Some systems are checked for a minimum pressure at idle, while others are checked at 2000 or 3000 rpm. To test oil pressure, remove the oil sending unit, and screw the correct adapter into the hole. Install the oil pressure gauge, and operate the engine at the specified rpm. Compare your results with the specifications **(Figure 5)**.

When the oil pressure is low it is important to determine the cause of the problem before further damage is incurred. First be sure that the correct viscosity oil is being used and that it is not diluted with fuel. You can smell excess fuel by sniffing the dipstick. If you are unsure, attempt to light the dipstick with a match or lighter. If the oil creates a flame it is contaminated with fuel and should be changed before further diagnosis. Other causes of low oil pressure are a worn oil pump, a loose or restricted or plugged oil pump pick-up tube, a stuck-open pressure relief valve, worn camshaft bearings or rocker arm shafts, and worn main or rod bearings **(Figure 6)**. Try using a stethoscope to listen to the different areas of the engine for

Figure 5. This engine has almost 40 psi of oil pressure at 2000 rpm. That is well within specifications for this older Jeep.

Figure 6. This oil pump pick-up screen is completely clogged with sludge. The engine had no oil pressure.

unusual noises to determine which area should be inspected first. Then disassemble as required to locate the source of the problem. You can usually visually inspect the components for excessive wear that would cause low oil pressure. A worn oil pump will show significant scoring on the gears or rotors as well as the housing **(Figure 7)**. Main and rod bearings will often be worn down to copper and scored.

The procedure to check a crankshaft-mounted oil pump is very similar to that for replacing a rear main seal. First remove anything necessary in order to access the pump. In some cases that will include removing the timing cover as well. On a camshaft- or distributor-driven oil pump you will have to remove the oil pan. Often this is quite straightforward. In other cases you may need to remove

Figure 7. The rotors of this oil pump are severely scored and worn, causing low oil pressure.

an exhaust pipe, drive axle(s), and/or suspension crossmembers or parts of the engine cradle. Refer to the manufacturer's service information, and use the specific procedure for the vehicle you are working on. With the oil pan off, remove the oil pump and disassemble it to evaluate its condition. You should also remove a main bearing cap and a connecting rod bearing cap to check the condition of the engine bearings. Look for excessive scoring or wear on each of the bearings; if you can see the copper backing in the bearing they are significantly worn. There is no point in installing a new oil pump if the engine bearings are too worn to hold oil pressure.

There may be times when you need to prove the cause of low oil pressure—for a warranty repair, for example. The oil pump can be evaluated through measurement as well as inspection. Clearance from rotor or gear tip to housing, gear endplay, and gear backlash can all be checked with feeler gauges and the clearances can be compared with specifications. Measure rotor diameter with a micrometer, and check cover flatness with a straightedge and feeler gauges **(Figure 8)**. Refer to the manufacturers' service information to locate the procedures and specifications needed to analyze the condition of the pump you are working on.

If the oil pressure is too high be sure that the engine was warm enough and that the correct viscosity oil is in the crankcase. Terribly dirty oil can also cause a high pressure reading; change dirty oil and retest before replacing components. Oil pressure that is too high is not good for the engine; it can cause oil consumption and damage to the bearing materials and valvetrain. If the oil is hot, clean, and of the correct viscosity, check the oil pressure relief valve. It may be sticking closed. Sometimes you can check it visually. Look for wear or ridges on the piston or in the bore that could cause sticking. Other times you will have to replace the PRV to determine whether it was the cause of the

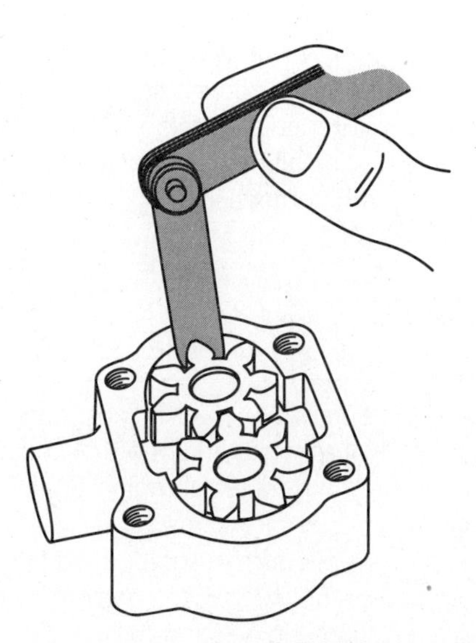

Figure 8. Check the gear-to-housing clearance as part of the evaluation of an oil pump.

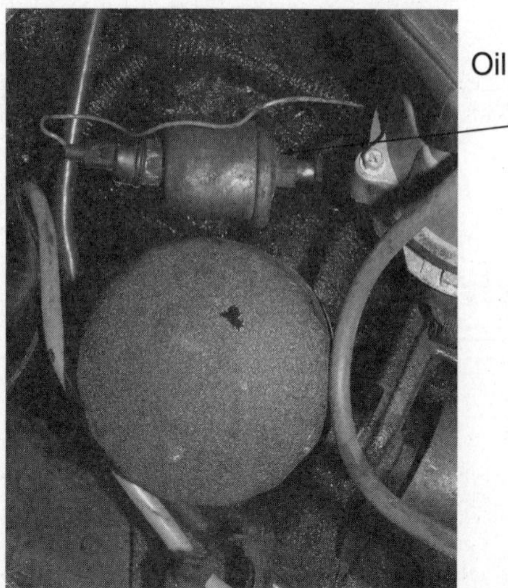

Figure 9. The oil pressure sending switch is typically located on the oil filter housing.

excessive pressure. If applicable, check the oil cooler for physical damage that could cause a restriction. It is often mounted in a vulnerable spot next to the radiator and is mounted low to pick up fresh air. Loosen the output line with the engine off, and be sure oil flows freely. A plugged oil gallery is the other cause of high oil pressure readings. Definitively eliminate all the other causes before disassembling the engine to inspect the oil passages.

EXCESSIVE OIL CONSUMPTION

When too much oil is allowed into the combustion chamber it burns and produces blue smoke out of the tailpipe. When **oil consumption** is too great customers will find their oil low between oil changes. A small amount of oil consumption is normal. An engine that uses more than a quart of oil in 2000 miles has an excessive oil consumption problem. The primary causes of oil consumption are worn valve seals or guides or worn pistons or rings. If the valve seals or guides are worn the blue smoke will be most evident when the engine is started first thing in the morning. It may also be noticeable when first accelerating after an idle period. When blue smoke occurs at all speeds whenever you accelerate the cause is typically worn piston rings. There is a tremendous difference in price to repair these two problems. We will discuss cylinder leakdown testing in an upcoming chapter; the procedure provides a way to check how well the rings are sealing.

CHECKING THE OIL PRESSURE WARNING LIGHT

If the oil pressure warning light is on and no engine noise is heard check the circuit electrically first. The oil pressure sending unit is typically located near the oil filter housing **(Figure 9)**. Most often you can remove the single wire from the unit and the light will go off; then ground the wire to clean, bare metal to make the light come on or the gauge rise. Be certain to check the manufacturers' wiring diagrams and service information first; this procedure could cause damage on some systems! If the circuit appears to function as designed, run an oil pressure test to check your findings.

To replace an oil pressure switch, sending unit, or temperature sensor, unscrew it from its bore. If it is physically damaged be certain you extract all the pieces from the engine block. If specified, use a small amount of thread tape or pipe sealant to prevent leakage. Tighten the unit to specification. Always check for proper operation of the system after replacing a component. Also check for and correct any leaks from the new part.

COMPONENT REPLACEMENT

You should replace individual components such as oil pan gaskets or valve cover gaskets following the manufacturers' specific service information. There are far too many engine designs to generalize about removal and installation procedures. The commonalities that do exist are:

- Lubricate an oil seal with a fine film of fresh oil before installation.
- Use the proper seal installer, not a hammer and punch, to install new seals.
- Use sealants on gaskets only if the manufacturer recommends it; then follow the procedure closely.
- Tighten all gasket covers using the correct torque specification and the proper tightening sequence.

Some manufacturers are using reusable gaskets and torque-to-yield attaching hardware. Replace fasteners whenever specified.

- Evaluate the condition of an engine before investing a lot of time and money into repairing expensive leaks. Excessive blowby on a worn engine can cause the seals to fail again in a relatively short period of time.

Remember, improper installation is the primary cause of seal and gasket failures; do it correctly the first time to prevent costly comebacks.

Summary

- The engine oil and filter should be replaced at an interval appropriate to the engine operating conditions or when the oil life indicator displays that it is necessary.
- Many technicians and repair shops recommend oil and filter changes every 3000 miles.
- Drain the engine oil when it is hot to ensure that contaminants are suspended in the oil.
- Always fully tighten the oil drain plug immediately after installing it.
- Lubricate the new oil filter seal with clean oil.
- Check your work carefully when performing oil changes; small mistakes can cause big problems.
- Use an oil dye and black light to positively pinpoint the cause of an oil leak.
- Check oil pressure when a warning light or gauge indicates low pressure, when an abnormal engine noise is heard, or on a high-mileage engine to check engine condition.
- Low oil pressure is often caused by a worn oil pump or worn engine bearings; repair the problem immediately to prevent further engine damage.
- High oil pressure is not good for the engine; check the oil and the pressure relief valve as common faults.
- Follow the manufacturers' procedures carefully when replacing gaskets and seals to prevent premature failure.

Review Questions

1. What factors should you consider when recommending an oil change interval to a customer?
2. What procedure can you use to pinpoint the source of an oil leak?
3. What are three things you should double-check when performing an oil change?
4. What are three causes of excessively high oil pressure?
5. After removing the oil pan to check an oil pump you should also inspect the __________.
6. Technician A says that frequent oil changes can prolong engine life. Technician B says that changing the oil and filter too often can damage the engine bearings. Who is correct?
   A. Technician A only
   B. Technician B only
   C. Both Technician A and Technician B
   D. Neither Technician A nor Technician B
7. Technician A says that the new oil filter seal should be lubricated with oil. Technician B says the oil filter should be installed with an oil filter wrench. Who is correct?
   A. Technician A only
   B. Technician B only
   C. Both Technician A and Technician B
   D. Neither Technician A nor Technician B
8. Technician A says that oil treatments should be added at every other oil change to help keep the engine clean. Technician B says that you should use a higher viscosity oil for the winter months. Who is correct?
   A. Technician A only
   B. Technician B only
   C. Both Technician A and Technician B
   D. Neither Technician A nor Technician B
9. Technician A says that a faulty oil pressure gauge can cause low oil pressure. Technician B says that a stuck-open pressure relief valve can cause low oil pressure. Who is correct?
   A. Technician A only
   B. Technician B only
   C. Both Technician A and Technician B
   D. Neither Technician A nor Technician B
10. Technician A says that low oil pressure can be caused by worn cam bearings. Technician B says that contaminated oil can cause low oil pressure. Who is correct?
    A. Technician A only
    B. Technician B only
    C. Both Technician A and Technician B
    D. Neither Technician A nor Technician B

Cooling System Operation

Introduction

The cooling system regulates engine temperature by allowing it to warm up quickly and maintain consistent temperatures under varying driving and ambient conditions. Only about 35 percent of the heat generated by the engine is used to produce power, so the cooling system must remove excess heat from the engine to prevent serious damage. It cycles coolant through the **radiator** and **heater core** to dissipate the excess heat energy into the air. Engine overheating is a common cause of engine mechanical failures. When repairing an engine you must also evaluate the cooling system to be sure that existing problems will not destroy your good work. Combustion temperatures may exceed 4500°F and normally average between 1200°F and 1400°F. This heat must be removed from the engine components quickly to prevent damage. Cooling passages allow coolant to flow around the cylinders and the valves to keep the block and head cool.

Removing the thermostat to "fix" an overheating problem can cause excessive engine wear, high emissions, and poor fuel economy. The increased rate of circulation may not even allow the engine to cool adequately.

COOLING SYSTEM OPERATION AND FUNCTIONS

A belt-driven water pump pushes fluid through passages in the block and head when the engine is started **(Figure 1)**. Until the engine is warmed up coolant is typically circulated from the water pump through the water jackets in the block, through coolant passages in the head, and around the thermostat back to the pump **(Figure 2)**. Once the engine coolant reaches the temperature at which the **thermostat** is rated to open, coolant flows through the open thermostat to the radiator. The radiator has a large surface area to absorb and then dissipate the heat as air passes across it. Coolant then flows into the return radiator hose, where suction from the water pump allows the cycle to continue **(Figure 3)**. When the coolant temperature starts to increase beyond the normal operating temperature, often at low vehicle speeds, a fan blows air through the radiator to help it cool more effectively. Most cooling systems operate

Figure 1. The water pump usually mounts to the front of the engine. This pump is driven by an accessory belt.

Figure 2. Coolant passages provide coolant flow into the water jackets that surround the cylinders.

Figure 4. Low coolant temperature will cause excessive engine wear. This is often caused by a stuck open thermostat.

at between 14 and 18 psi to increase the coolant's boiling point because systems regularly run above the boiling point of water (212°F). When the driver wants heat in the cabin coolant flows through a smaller radiator called the heater core. A fan blows the heat out of the vents.

To properly control the temperature of the engine and the passenger cabin the cooling system must:

- Allow the engine to reach its normal operating temperature quickly
- Remove excess heat
- Maintain the temperature within the designed range
- Provide heat to the passenger cabin

When an engine is running below NOT the engine is not working as efficiently as it can **(Figure 4)**. The pistons have not yet expanded to properly fit the cylinder walls. This allows excessive blowby into the crankcase, which contaminates the oil over time. The engine will also consume some oil. On a cold morning you can often see cold

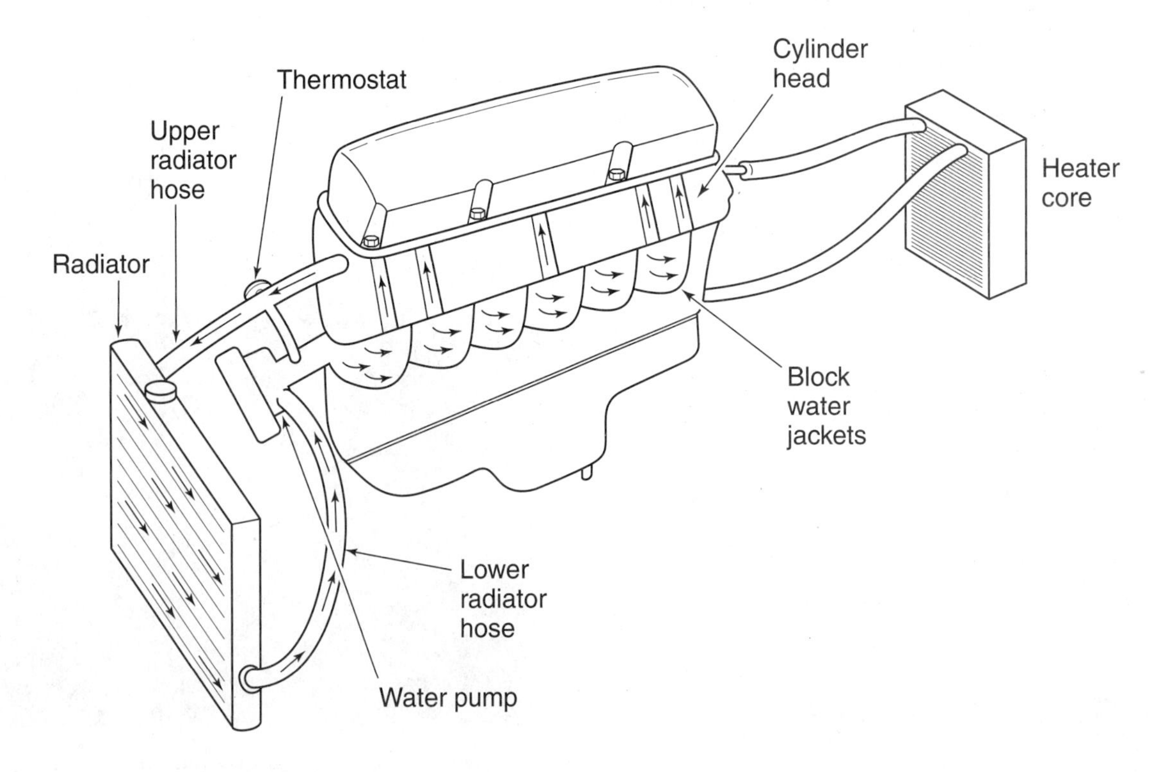

Figure 3. Coolant flows from the pump to the block, to the head, and through the radiator in a traditional cooling system.

vehicles with frost on their windshields pulling onto the highway with blue smoke coming out of their tailpipes. Running an engine below NOT also increases the vehicle emissions. The powertrain control module (PCM) delivers more fuel to the engine while it is cold, and some of that sticks to the relatively cool walls of the combustion chamber and is exhausted as hydrocarbons. When the engine runs too cool the oil becomes contaminated with blowby gases and carbon, forming sludge in the oil pan. The inadequately lubricated cylinders also wear more at the top, causing cylinder bore taper. Running an engine below NOT for extended periods significantly increases internal engine wear. The cooling system is designed to pick up the heat from combustion and circulate the hot coolant throughout the engine to heat it quickly and evenly.

The combustion flame temperature can exceed 4500°F. Some of the heat is used to expand the gases in the chamber and place pressure on the piston. Excess heat must be removed by the cooling system, or engine parts could literally melt. Coolant surrounds the cylinders to cool the combustion chamber. Valve temperatures rise above 1200°F. Cooling passages around the valve seat pick up heat during the time that the valves are closed.

The cooling system must regulate the engine temperature by maintaining a steady coolant temperature. Most modern engines are designed to run between 195°F and 220°F. Coolant temperatures can frequently reach 240°F when the vehicle is idling. We have discussed the problems caused by an engine running too cool. When an engine runs too hot serious damage can occur. Aluminum heads overexpand, warp, and cause head gasket failures. Excessive engine temperatures can also cause cylinder heads or blocks to crack, valves to burn, and pistons to melt. At the very least the coolant could boil over and the customer could find herself stuck on the side of the road.

Finally, the cooling system must deliver heat to the passenger cabin to defrost the windshield and provide warmth for the passengers. When the heat is turned on, coolant flows through the heater core, which is just like a small radiator **(Figure 5)**. The heater fan blows the heat across the heater core and into the correct vent: to the feet, the windshield, or the center of the cabin.

Figure 5. The heater core is located in the heat distribution box, which is usually under the passenger side dash.

COOLING SYSTEM COMPONENTS

Each of the cooling system components plays an important role in the overall function of the system. Understanding the purpose of each component will help you when you must diagnose problems with the system.

Water Pump

The water pump is usually belt driven off either the engine accessory belt or the timing belt. In a few applications it is gear driven off the camshaft at the back of the engine. The water pump circulates coolant throughout the engine and through the radiator **(Figure 6)**. It picks up coolant from the radiator outlet hose and sends it to the engine, often to the block first and then the head. The water pump has a seal around the impeller shaft that holds the pulley. When this seal leaks, coolant will begin to drip from the weep hole on the bottom of the water pump snout **(Figure 7)**. The water pump gasket seals the mounting end of the pump.

Interesting Fact

Some vehicles use a reverse cooling system, meaning that the coolant is circulated to the head first and then the block. This design can speed up the warm-up period, thereby reducing emissions and engine wear. The valves gain heat soonest, so coolant around the head gets hot faster. This allows the coolant to equalize the temperature of the engine areas more quickly. This design also cools the combustion chamber more efficiently so compression pressures can be higher without detonation.

Figure 6. The water pump impeller pushes coolant through the engine and radiator.

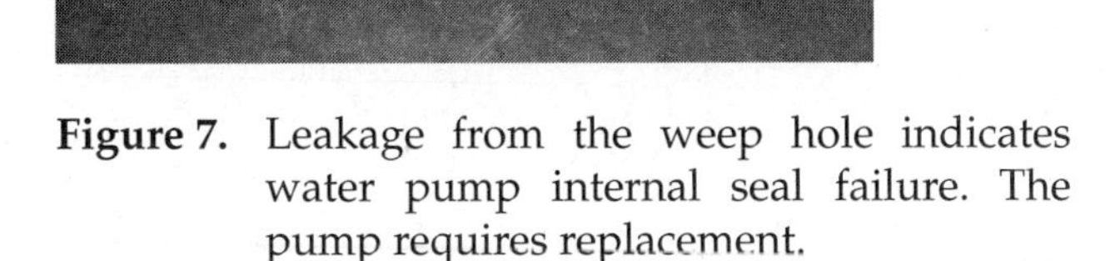

Figure 7. Leakage from the weep hole indicates water pump internal seal failure. The pump requires replacement.

Water Pump Belts and Hoses

The water pump belt may be a V-belt or a serpentine belt driven by the crankshaft pulley. The V-belt may drive the water pump only, or the pump and other accessories. The serpentine belt usually drives all the engine accessories **(Figure 8)**. The belt must be tensioned correctly to provide proper water pump operation. The water pump may be driven instead by the timing belt. When this is the case technicians usually replace the water pump when the timing belt is replaced as preventative maintenance.

The water pump hoses are made of two layers of reinforced rubber. Many hoses are molded to provide the appropriate bends to fit their application. The water pump inlet hose often contains a wire coil to prevent the hose from collapsing under the suction of the pumping action. Hoses are connected to their flanges with a variety of hose clamps. Replacement clamps are typically of the worm gear style shown in **Figure 9**.

Radiator

The radiator removes heat from the hot coolant. Coolant runs through small core tubes with fins around them. The fins provide a great deal of surface area to collect heat. They also allow cool air to pass across them, providing the cooling effect. Radiators are either cross-flow or down-flow. In the cross-flow radiator coolant flows through horizontal core tubes from one side of the radiator to the other. The core tubes are run vertically in a down-flow radiator; the coolant runs from the top to the bottom of the radiator **(Figure 10)**. The radiator housing may be made of

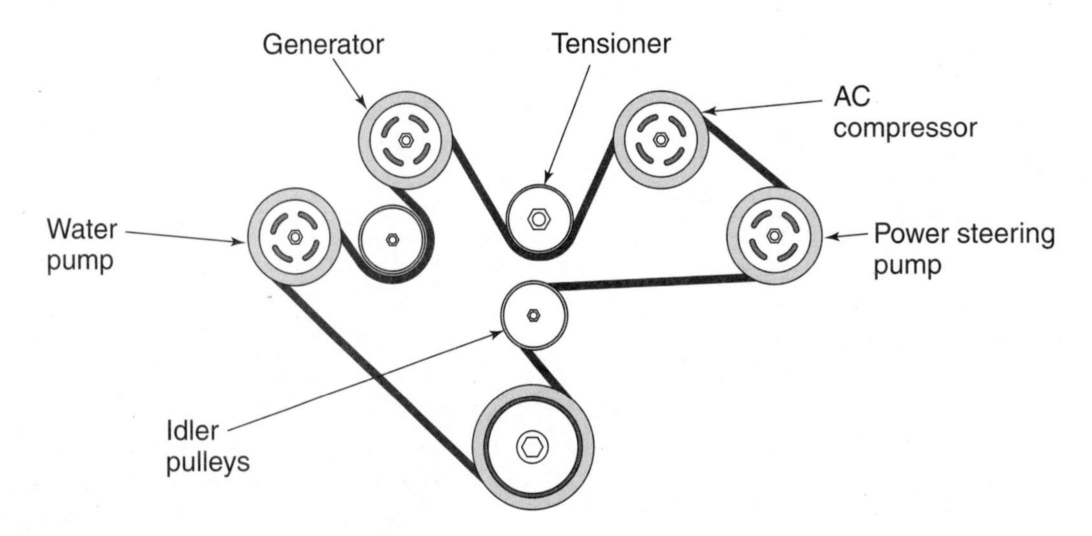

Figure 8. The automatically tensioned serpentine belt drives all the engine accessories.

Figure 9. An assortment of replacement hose clamps.

Figure 11. A typical multifunction radiator cap.

brass, aluminum, or plastic. The fins are made of brass or aluminum because those materials dissipate heat well and resist corrosion.

Pressure Cap

The **pressure cap** seals the system at the top of either the radiator or the reservoir **(Figure 11)**. The cap provides three important functions other than sealing. It must maintain 14–18 psi of pressure in the system. This increases the boiling point of the coolant from 212°F (the boiling point of plain water) to about 257°F to 265°F. Each 1 psi of pressure added to the system increases the boiling point of the coolant mixture roughly 3°F. As you know, cooling systems regularly operate at temperatures above 212°F. Without this added pressure the coolant would boil over. The 50/50 mixture of antifreeze and water also increases the boiling point slightly. The pressure cap must also relieve excess pressure. As the coolant heats up, the pressure in the system increases. The cap has a relief valve to maintain the system operating pressure. If the system pressure gets too high it

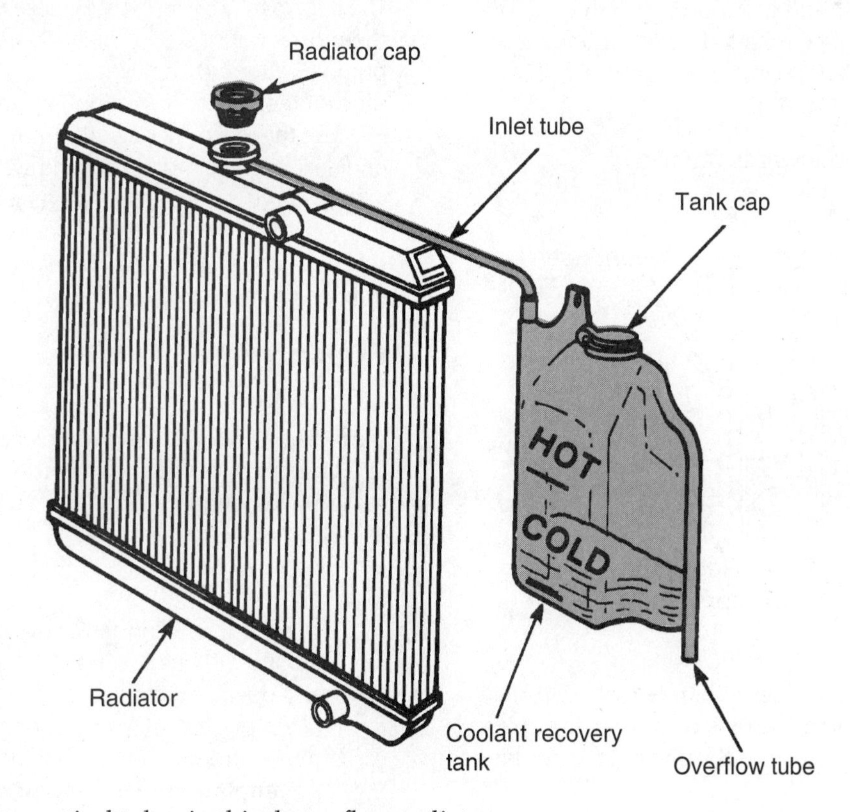

Figure 10. Notice the vertical tubes in this down-flow radiator.

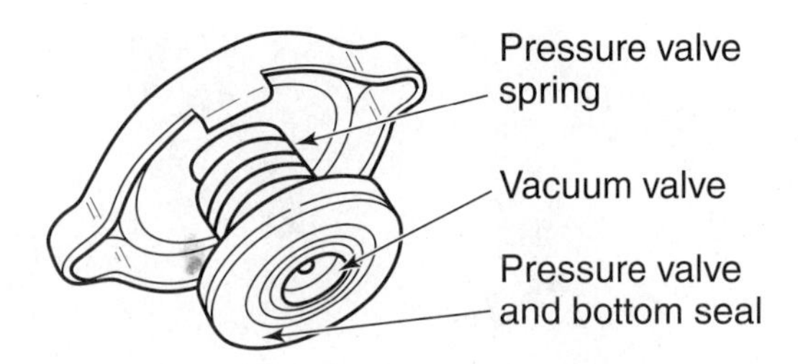

Figure 12. The springs and valves of the pressure cap provide sealing, pressure relief, and vacuum relief.

Figure 14. The bypass hose directs coolant from the head back to the water pump when the thermostat is closed.

can damage components and cause leaks. Finally, the cap must release vacuum from the system developed as the coolant contracts when the engine cools down after running. This prevents potential collapse or implosion of the hoses or the radiator **(Figure 12)**.

Thermostat

The thermostats currently used are mechanical devices that respond to the temperature of the coolant. An internal wax pellet expands and contracts with heat to open and close the thermostat (t-stat). Most modern engines use thermostats rated at 185°F to 195°F; this is the temperature at which they open. The purpose of the thermostat is to make sure the engine runs hot enough to support efficient combustion and minimize engine wear. When NOT is reached the t-stat opens, and coolant can flow through the radiator. When the engine is started cold the t-stat stays closed and blocks coolant flow to the radiator. This allows the coolant to circulate only through the engine and provide a quick warm-up. The thermostat opens only when the engine is approaching its normal operating temperature. The t-stat must stay closed any time the engine is operating at lower than ideal temperatures. Once the engine is warm the t-stat opens and closes partially to regulate coolant flow through the radiator; this maintains a consistent temperature **(Figure 13)**. The cooling fan's job is to prevent engine overheating.

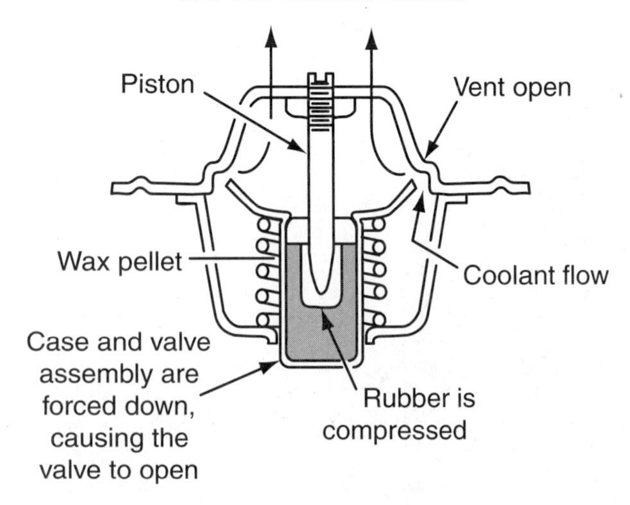

Figure 13. As the wax pellet expands from the heat of the coolant, it overcomes spring pressure and opens the thermostat for coolant flow to the radiator.

A thermostat bypass port is provided to allow coolant to circulate through the engine when the t-stat is closed. The bypass may be an external hose running from near the thermostat housing to a coolant port on or near the water pump **(Figure 14)**. The bypass may also be an internal passage off the thermostat mounting area.

Use the recommended thermostat in the vehicle. If the engine is overheating the cooling system needs more than a lower-temperature t-stat. The vehicle is likely to continue to overheat anyway; it is well after the t-stat is fully open that a cooling system fan and radiator have to keep engine temperatures at a safe level.

Cooling Fan

The cooling system relies on a fan to blow air across the radiator when the vehicle is idling or moving slowly. The fan is usually unnecessary when the vehicle is being driven at highway speeds. Three types of fans are used:

- Electric
- Clutch
- Flexible

The electric cooling fan is by far the most common fan system found on passenger cars today. It saves engine power by operating the fan only when it is needed. The fan is a simple electric motor with blades attached to catch air. Control of the fan is typically provided by the PCM. An **engine coolant temperature (ECT) sensor**

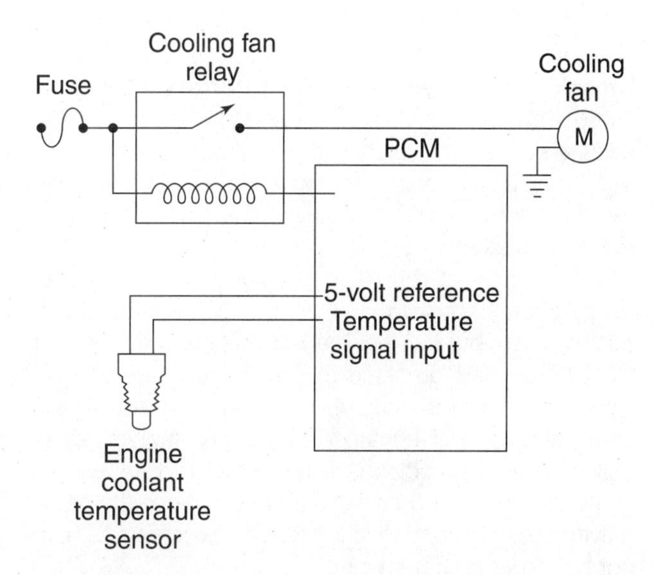

Figure 15. When the ECT provides information to the PCM that the coolant is getting too hot, the PCM turns on the cooling fan relay. The relay provides power to the fan motor.

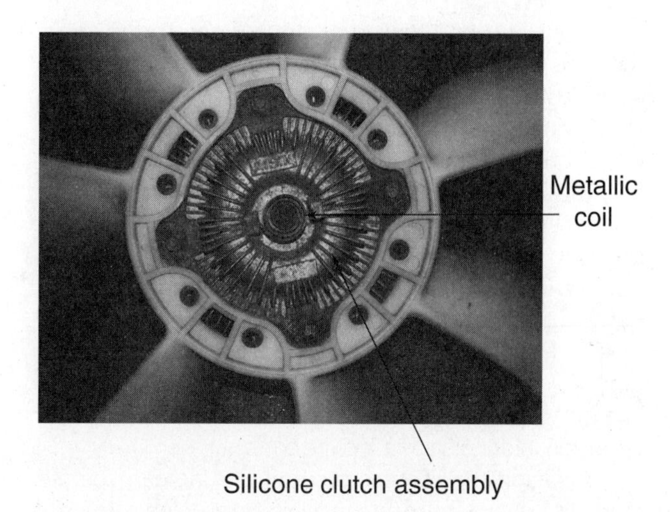

Figure 16. The metallic strip operates a valve that controls oil flow into and out of the clutch assembly.

informs the PCM about the temperature of the coolant through a voltage signal. When the PCM sees a voltage that corresponds to its preprogrammed temperature (usually around 220°F) for turning on the fan, it activates the fan. The PCM does not usually control the power or ground to the fan directly; it turns on a relay that provides power to the fan. **Figure 15** shows a typical fan electrical circuit.

Clutch fans are mounted on the snout of the water pump and belt driven off the crankshaft pulley. They may be either temperature sensitive or speed sensitive, though most original-equipment fans are temperature sensitive. The temperature-sensitive fan spins with the water pump when the coolant is hot and nearly freewheels when the engine is cool. It uses a bimetallic strip that expands and contracts with heat and cold, respectively. When the bimetallic strip has hot air blowing across it, it opens a passage in the clutch to allow silicone fluid into a clutch chamber **(Figure 16)**. This "locks" the fan to the water pump shaft, and the fan spins to cool the engine down. When cool air blows across the clutch, such as when the vehicle is being driven down the highway, the fluid is directed out of the clutch chamber. This reduces the drag on the fan and prevents the pump from driving it. This saves horsepower when the fan is not needed. A speed-sensitive viscous clutch fan works similarly; the clutch slips when the force required to turn the fan is high, as it would be when the vehicle is traveling at high speeds.

The flexible fan is solidly mounted to the snout of the water pump. The fan blades are angled steeply during low-speed operation to pull the most air across the radiator. As engine rpm increases the blades flatten out to reduce the drag and, therefore, the horsepower requirement.

Warning Lights and Gauges

Vehicles have a warning light to alert the driver when the engine is overheating. Many vehicles also have a coolant temperature gauge so the driver can monitor the engine temperature. The warning light may be operated by a temperature-sensitive switch that provides a ground for the light. The PCM may also send coolant temperature information from the ECT to the instrument panel (IP). Then the instrument panel turns the warning light on or operates the gauge. Some gauge systems use a separate temperature sensor to provide a signal to the gauge. It, like the ECT, is a thermistor, meaning it changes electrical resistance proportionally with temperature.

Heater Core

The heater core looks like and is constructed like a small radiator. It is located under or behind the dash inside the passenger compartment. When hot coolant flows through the heater core the fins give up heat. A fan blows this heat across the heater core and into the cabin.

COOLANT

There are now several different types of coolant in use on modern vehicles. Most coolants are made from a base of **ethylene glycol (EG)**, but the additive packages differ. Some environmentally friendlier coolants are made from a base of the less-toxic propylene glycol. Most coolants cannot be mixed, even if their colors are similar. When you are unsure of what coolant is in a cooling system you should fully drain the system before adding coolant. All coolants must:

- Resist freezing
- Reduce rust and corrosion inside the cooling system
- Lubricate the water pump and seals

The EG coolants can be broken down into two general categories. The older traditional green EG coolant is produced to be effective for about 2 years or 30,000 miles. The newer extended-life coolants come in many colors and are designed to function effectively for 5 years or 150,000 miles. All coolants must provide rust and corrosion resistance for the internal components of the cooling system. The difference in the life expectancy is due to the type of corrosion inhibitors added to the coolant.

The traditional EG-based coolant uses inorganic borate salts, silicate, and phosphate to minimize rust and corrosion. The inhibitors break down over time, and the coolant becomes acidic. The acidity also leads to electrolytic corrosion as a little battery is formed with acidic liquid in the presence of dissimilar metals (aluminum heads and cast blocks, for example) and excess electrical energy generated by the ignition system. The coolant must be replaced to prevent accelerated corrosion.

Many European manufacturers specify coolants with no phosphates to reduce sediment and scale buildup when mixed with hard water. Asian manufacturers, on the other hand, may specify a coolant with little or no silicates. These coolants may be dyed a rainbow of colors; blue, pink, yellow, orange, or red.

The newer extended-life EG coolants use a different type of corrosion inhibitor, which lasts longer. They use **Organic Additive Technology (OAT) corrosion inhibitors**. These corrosion inhibitors are effective for 5 years or 150,000 miles and are formulated without silicate or phosphate. Using traditional coolant to top off a system filled with extended-life coolant will effectively destroy the benefits of the OAT additives.

Propylene glycol (PG) coolant is manufactured because it is less toxic than EG coolants. It is certainly not safe to ingest, but it is less likely to kill pets and other animals. Animals are attracted to coolants because they are sweet. Less than half a cup of EG coolant is potentially fatal to most pets (and humans). EG coolant should not be mixed with PG coolant because the toxicity will be increased. EG and PG coolants also have different specific gravities so their combined antifreeze protection level cannot be accurately determined.

Coolants in production automotive engines should be mixed with water in a 60/40 to 50/50 ratio to provide good cooling and protection from freezing. Ideally, distilled water, which has no minerals, should be used, but it is common practice to use tap water. A 50/50 mix of coolant and water will not freeze above −34°F, and a 60/40 mix should protect the coolant from freezing at least down to −40°F. Straight water should not be run in any cooling system because of the added functions of the coolant. Pure coolant should not be used either. Coolant's ability to transfer heat is dangerously lower than water, and it freezes at 0°F.

Summary

- Combustion temperatures can reach over 4500°F; the cooling system must rapidly remove heat to prevent meltdown.
- The cooling system is designed to allow the engine to reach its normal operating temperature quickly, remove excess heat, maintain the coolant temperature within the designed range, and provide heat to the passenger cabin.
- Coolant is circulated by the water pump through the block, head, and radiator to maintain the proper operating temperature.
- The radiator removes excess heat from the coolant; its cooling fins are typically made of brass or aluminum.
- The cooling system pressure cap must seal the system, maintain adequate pressure, relieve excess pressure, and vent vacuum.
- The thermostat blocks flow to the radiator to allow quick engine warm-up and provides flow through the radiator when the coolant is hot to remove excess heat.
- The cooling fan may be electric, clutch, or flexible.
- Most passenger cars use an electric fan that is turned on and off by the PCM. This type of fan uses the least horsepower from the engine.
- Cooling systems provide a warning light, and often a gauge, so the driver is aware of an overheating condition.
- Coolants are most commonly produced from a base of ethylene glycol. Some less-toxic coolants use a base of propylene glycol.
- Coolant must protect the system from freezing, prevent rust and corrosion, and lubricate the water pump and seals.
- Traditional ethylene glycol should be changed every 2 years or 30,000 miles; extended-life coolant uses OAT additives to increase its effective life span to 5 years or 150,000 miles.

Review Questions

1. Most cooling systems run at temperatures between __________ and __________°F.
2. A thermostat should __________ when the coolant reaches NOT and __________ when the coolant temperature is below normal operating temperature.
3. Describe three functions of the cooling system.
4. List three reasons why straight water should not be used in cooling systems.
5. Technician A says that some cooling systems circulate coolant through the head first and then the block. Technician B says that this circulation route lowers emissions. Who is correct?
   A. Technician A only
   B. Technician B only
   C. Both Technician A and Technician B
   D. Neither Technician A nor Technician B
6. Technician A says that the water pump is usually belt driven. Technician B says that the water pump creates a little suction to help pull coolant from the radiator. Who is correct?
   A. Technician A only
   B. Technician B only
   C. Both Technician A and Technician B
   D. Neither Technician A nor Technician B
7. Technician A says the cooling system pressure cap should hold at least 25 psi of pressure on the system. Technician B says the cap should not hold vacuum. Who is correct?
   A. Technician A only
   B. Technician B only
   C. Both Technician A and Technician B
   D. Neither Technician A nor Technician B
8. Technician A says that the heater core is located next to the radiator. Technician B says the heater core is constructed like a radiator. Who is correct?
   A. Technician A only
   B. Technician B only
   C. Both Technician A and Technician B
   D. Neither Technician A nor Technician B
9. Technician A says a modern electric cooling fan is usually turned on at around 120°F. Technician B says that the PCM controls the cooling fan on most modern vehicles. Who is correct?
   A. Technician A only
   B. Technician B only
   C. Both Technician A and Technician B
   D. Neither Technician A nor Technician B
10. Technician A says that some coolant should be changed every 2 years or 24,000 miles. Technician B says that ethylene glycol coolant with OAT additives should not need to be changed for the life of the vehicle. Who is correct?
    A. Technician A only
    B. Technician B only
    C. Both Technician A and Technician B
    D. Neither Technician A nor Technician B

Cooling System Maintenance, Diagnosis, and Repair

Introduction

The cooling system must be properly maintained to provide trouble-free operation for the customer. Failures can be extremely inconvenient and expensive. A cooling system failure may be the root cause of an engine failure. When repairing or replacing an engine carefully inspect the cooling system, and replace worn components. Whenever a vehicle is brought to you for an oil change you should check the cooling system visually for potential problems. Many customers request a seasonal check of their cooling system. In the fall or spring check the belts, hoses, coolant condition and antifreeze protection, and fan operation, and check the system for leaks. During maintenance services complete a thorough evaluation of the system, and perform any recommended maintenance procedures. When diagnosing a cooling system failure be sure you locate the root cause of the concern.

Interesting Fact *One study of consumers indicates that over 70 percent of automobile owners do not perform preventative maintenance on their cooling system. Your communication with the customer may make the difference between a great vacation and a troublesome road trip.*

COOLING SYSTEM MAINTENANCE

Cooling system maintenance includes checking the belts for glazing, cracking, and proper tension; inspecting the hoses for signs of age, testing the system for leaks, and checking the coolant condition and antifreeze protection level. The coolant should be flushed and changed at the suggested intervals.

You Should Know *Never remove the pressure cap on a very hot engine. When the pressure is released the boiling point is lowered so coolant may spray out on you. Allow the engine to cool slightly and release the pressure from the cap slowly* **(Figure 1).**

Initial Inspections

Begin any diagnosis of cooling system concerns and all maintenance operations by looking closely at the belts, hoses, and coolant. Many checks of the cooling system can be performed using a thorough visual inspection. Look carefully at the cooling fan and water pump belts. Signs of wear include cracking, glazing, fraying, or contamination from a fluid, causing softness. Replace any belts that show signs of wear. When replacing or overhauling an engine the drive belts are typically replaced unless they are almost new. Check the belts for proper tension following the manufacturer's specification. A general measure is that a belt should deflect about 1/2 in. for each foot of belt running free between pulleys or tensioners. When installing a new belt use a belt tension gauge unless the vehicle uses an automatic tensioning system. An automatic tensioner is typically marked with a scale; be sure the belt is running in its normal range. A belt that is installed too tight can

Figure 1. Be careful not to open the system when it is hot. Releasing the pressure will allow the coolant to boil and spray on you.

damage the bearings in the water pump or generator. It can also damage the main bearings by placing too much upward tension on the crankshaft. If the belt is too loose it may slip. This can cause overheating, and it will glaze the belt and cause premature failure.

Inspect each of the coolant and heater hoses for cracking, brittleness, or softness. Make sure the radiator outlet hose does not easily compress with a light squeeze; it must resist collapse under suction from the water pump. Replace a hose that has any signs of wear; if it fails the damage will be much more expensive than the price of replacing the hose. Pay attention to the hose clamps; look for signs of leakage. Use the correct replacement hose for the application. Installing a universal hose and trying to bend it to fit can result in coolant restriction. It is a good idea to replace the cooling system hoses during a complete engine overhaul or replacement, especially on an older vehicle or one with high mileage.

To replace a cooling system or heater hose, drain or recover the coolant, depending on your shop's equipment. Under no circumstances should you allow the coolant to enter the shop drains. Used coolant is a hazardous waste and needs to be handled accordingly. Use a hose or seal removing tool to release the hose at both connecting ends. Sometimes the hose will come off in pieces. Thoroughly clean the sealing surfaces. You may have to use a light sandpaper or emery cloth to remove corrosion from the mating surface. Then place the new hose clamps on the hose, and slide the proper hose fully into position. Tighten the clamps securely. Refill the system as discussed later in this chapter. Make sure the system is full of coolant, and pressure test the system for leaks before returning the vehicle to the customer.

You Should Know *Inspect the fan blades carefully for cracks or damage before working around them. If part of a fan blade lets loose while you are hanging over it you could be killed or seriously injured.*

Check the date of the last coolant change in the service records, and calculate the mileage since the service. Check the coolant for the proper level. Do not mix antifreeze types. If you are unsure of what type of coolant is in the system, drain and refill it. Check the coolant for discoloration, contamination, rust, or scale. It is often apparent at a glance that the coolant system needs to be flushed and refilled with fresh coolant **(Figure 2)**. Blocked cooling system passages can cause overheating and serious engine damage. If the small passages around the valves become laden with deposits the valves may burn or the cylinder head can crack.

After an engine is overhauled or replaced and the coolant passages have been fully cleaned most technicians will flush the radiator or replace it if its condition is questionable.

The **pH level** of coolant mixture should be between 7.7 and 10.3. The manufacturer will specify the desired pH range. A pH level below 7.0 is dangerously acidic; the

Figure 2. This reservoir is full of sludge and contaminated coolant. This system needs to be flushed.

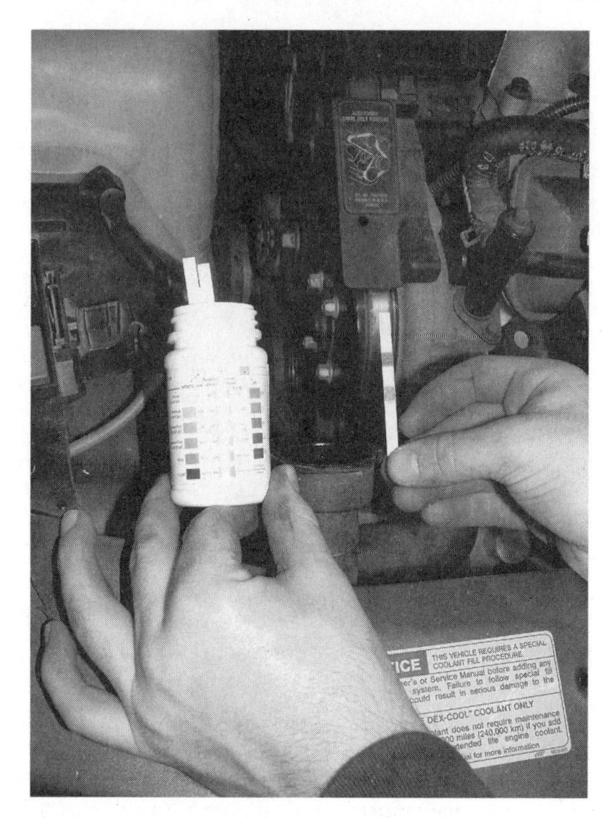

Figure 3. A coolant test strip can tell you the pH level of the coolant, its freeze-up protection, and its cleanliness.

coolant will eat away at the internal parts of cylinder heads, radiators, heater cores, and water pumps. The high acidity contributes to electrolysis, the electrochemical decay of metals caused by the presence of two dissimilar metals in an acidic or electrolytic fluid. Over time coolant becomes more acidic. You can test the pH level of coolant with a simple test strip **(Figure 3)**. Dip the test strip in the coolant, and compare the color with the pH chart on the bottle. Most test strips also indicate the coolant concentration and level of contamination.

Another way to test the condition of the coolant is to test its electrolytic action using a voltmeter. Stick one voltmeter lead in the coolant and the other lead to the battery negative post. The voltmeter should show less than .2 volt if the coolant is in good condition. Coolant that produces a voltage of .5 volt or more is electrolytic; flush and refill the coolant to prevent damage to cooling system components.

Test the antifreeze protection level of coolant using a hydrometer or refractometer. Pull coolant into the hydrometer, and read the scale as indicated by the arrow or floating discs. When testing PG coolant use a refractometer to get an accurate indication of antifreeze protection **(Figure 4)**. If the coolant is allowed to freeze inside an engine it can easily crack the cylinder head or block.

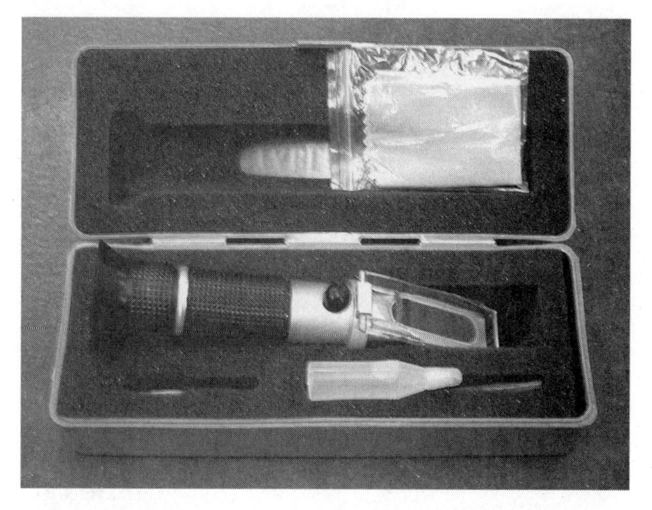

Figure 4. This refractometer can be used to test the concentration of any type of coolant as well as the specific gravity of battery electrolyte.

You Should Know *Engine block* ***core plugs*** *or "freeze" plugs are not designed to protect the engine in the event of freezing. They are drilled in the block to remove the sand after the casting process. Sometimes when an engine freezes these plugs will pop out and prevent the block from cracking. Other times the customer may not be so lucky!*

Look at the radiator fins for corrosion and to be sure they are all still there. As the fins corrode they will fall out from between the tubes. Gently sweep your fingers across the fins; if they fall away as dust under light pressure the radiator needs replacement. Inspect the radiator for leaks. You will often see coolant-colored stains on the tubes or tanks when a radiator is seeping or leaking. Also be sure that the fins are not plugged up by dirt, leaves, or bugs. Use low air pressure to blow the fins clear.

Always verify proper cooling fan operation during a cooling system inspection. We will discuss specific test procedures for the different types of fans later in this chapter. Also inspect the fan shroud and air dams. These plastic casings surrounding the fan(s) and radiator are critical to proper airflow to maintain safe cooling system temperatures. Look at the shrouding and dams for cracks, loose mounting, or missing components. These are not optional components; repair or replace any faulty components.

Coolant Flush and Refill

Many manufacturers recommend replacing the coolant every 2 years or 24,000 miles. Newer vehicles may have an extended service period of up to 7 years or

100,000 miles. Check the coolant condition as discussed above, and change coolant whenever it is indicated. Use a coolant exchanging and/or recycling machine to collect the used coolant and replace it with new or recycled coolant. Some of these machines simplify the process of refilling the system because the coolant is refilled under pressure, which prevents air from being trapped in the cooling system. Some manufacturers do not allow the use of recycled coolant. Follow their recommendations to ensure that no warranty is voided and to minimize the risk of cooling system troubles.

Most technicians replace the thermostat after flushing a cooling system. The low cost of the part versus the inconvenience and expense of problems caused by a failed thermostat justifies routine replacement.

If your shop does not have coolant recovery or recycling equipment you will need to drain the cooling system thoroughly and refill it with the correct coolant. Be sure to collect all of the old coolant, and place it into an appropriate hazardous waste container. It is illegal to drain coolant into shop drains. Used coolant should be recycled. A hazardous waste management service should collect your waste coolant. To drain the system fully remove the lower radiator hose, and allow the coolant to drain into a drain pan.

To refill the system mix coolant with water in the correct ratio. Follow the manufacturer's recommendations for the proper type of coolant and the correct ratio of coolant to water. The lubrication and anticorrosion effects of coolant are essential to proper system operation. Using the proper ratio will also prevent coolant from freezing, which can cause dangerous overheating or even a cracked engine block.

It is critical that you get all the air out of a cooling system during the refilling process. Air pockets in the system can cause overheating, erratic engine temperatures, hot spots in the engine, and a lack of heat. If the t-stat is sensing the temperature of air rather than coolant it will not open at the correct time. Refill the system, and run the vehicle with the pressure cap off. This allows any air bubbles to escape to the top of the system and out the reservoir or the radiator. Some manufacturers use a **bleeder valve** to assist in refilling a system **(Figure 5)**. Open the bleeder valve while filling the system to allow air bubbles to rise to this open high point of the system. There are also tools available that place the cooling system under a vacuum to refill it. Follow the instructions to use the tool effectively. Once you think the system is full, install the pressure cap, and let the vehicle run until the cooling fan cycles on and off. This ensures that the thermostat and fan are properly controlling the engine coolant temperature. Allow the engine to cool to a safe level, and remove the pressure cap one more time. The residual pressure in the system should force any remaining air out of the filler neck. Top the system off to the proper fill level. Perform a pressure test on the system to check the system for leaks before returning the vehicle to the customer.

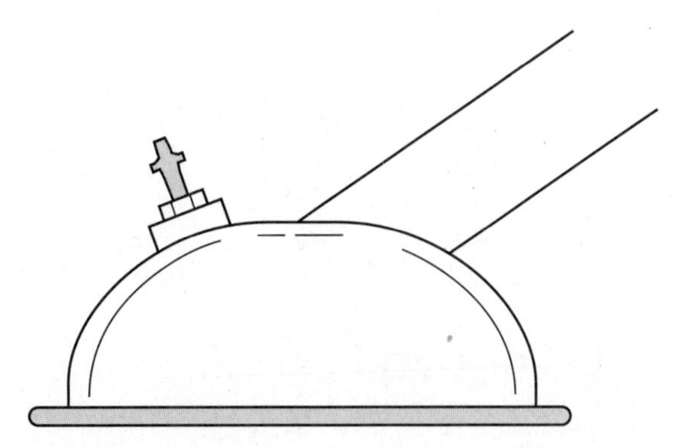

Figure 5. A bleeder valve is often placed on the thermostat housing to help get all the air out of the system when refilling.

COOLING SYSTEM PRESSURE TESTING

When a cooling system has a leak you may need to use a pressure test to locate the source of a leak. After an engine overhaul or replacement, pressure test the system before returning the vehicle to the customer. Do not open the system when it is hot! If the coolant is over 212°F it will boil and spray out when it is exposed to atmospheric pressure. When the system is cool enough, install a pressure tester in place of the cap, and pump it to the rated system pressure as indicated on the cap **(Figure 6)**. The tester should hold steady pressure for at least 5 minutes. Visually inspect the system for leaks. Sometimes it will be difficult to find the source of a leak around the back of the engine or near where the intake manifold bolts to the head. Put the appropriate dye in the cooling system, and use a black light to ensure that you can find the source of the leak. Use the correct adapter with the pressure tester to check the cap as well. It should hold the rated pressure and relieve excess pressure.

Figure 6. Pump the rated system pressure on the coolant and check for leaks.

To test the radiator or pressure cap attach the proper adapter from the cooling system tester onto the cap. Pump pressure onto the cap, and be sure it holds its rated pressure, 15 psi, for example. Attempt to pump a higher pressure to verify that it will release excess pressure. Finally, use a vacuum pump and attempt to pull a vacuum on the cap. If the cap fails any of these tests replace it. These failures can cause coolant loss, boiling over, and implosion of the radiator.

CAUSES OF OVERHEATING

Engine overheating can quickly cause severe engine damage. Customer concerns should be addressed and remedied immediately to prevent further failure. Diagnosis should always begin with a check of the basics. Make sure the system is full of clean coolant. Check the water pump belt for proper tension and signs of slippage. Other causes of engine overheating include:

- Thermostat failure
- Cooling fan faults
- Plugged radiator tubes
- Restricted internal coolant passages
- Excessive deposits in coolant passages
- Air in the system
- Defective radiator caps
- Defective water pumps or eroded impellers
- Collapsed radiator outlet hoses
- Preignition or detonation
- Fuel system faults causing lean air-fuel mixtures
- Dragging brakes or wheel bearings

Use your knowledge of component functions and system operation to guide your testing and diagnosis. The procedures below explain diagnosis of common cooling system problems.

THERMOSTAT DIAGNOSIS

When a thermostat fails it will usually cause predictable symptoms. If the vehicle is exhibiting signs of a faulty thermostat replace it. Do not take chances with a thermostat; the failure of this simple and inexpensive component can destroy an engine!

When a thermostat sticks open the customer will likely report that the temperature gauge stays low all the time or there is no heat. Sometimes a stuck-open thermostat can cause a customer concern of poor fuel economy. If the engine coolant is always running cool the PCM will deliver more fuel to provide proper combustion. To check whether the t-stat is sticking open let the engine cool down thoroughly if possible. Start the engine, and monitor the temperature of the radiator inlet hose. It should stay cool until the engine approaches normal operating temperature. If it starts to warm up as the engine warms up it means that the t-stat is open and allowing coolant flow to the radiator. You can also remove the t-stat and visually see if it is hanging open. If you take the time to remove it you should really replace it.

A common cause of overheating is a t-stat sticking closed. When it is stuck closed the engine will overheat rapidly and during all modes of engine operation. Failure of the cooling fan, in contrast, will generally produce the symptom of overheating only while in traffic, traveling slowly, or at idle. With a faulty t-stat the temperature gauge will rise steadily into the red zone after engine startup.

Often a thermostat sticks irregularly. The customer may notice that the temperature gauge goes up unusually high and then quickly drops back down to a normal level. This may happen occasionally or consistently. The thermostat is likely sticking and should be replaced now before the vehicle seriously overheats.

To replace a thermostat drain some coolant from the system, or drain the whole system if the coolant condition warrants it. Remove the thermostat housing and the faulty thermostat. Note which side of the thermostat faces up. On some vehicles the t-stat is integral to the housing and the whole assembly must be replaced **(Figure 7)**. Clean all the old gasket or o-ring material away from the housing and the cylinder head. Install the new thermostat in the proper direction. Most t-stats are installed with the jiggle valve up. Reinstall the thermostat housing with the correct gasket or o-ring, and tighten the housing down. Refill the system, being careful to get any air pockets out of the system. Run the vehicle to normal operating temperature while monitoring the temperature gauge and the radiator inlet hose. Allow the fan to cycle on and off, and then drive the vehicle to be sure the new thermostat is functioning flawlessly. It is possible to get a brand new but faulty thermostat; verify that yours is operating properly. Inspect carefully for leaks around the t-stat housing before returning the vehicle to the customer.

Figure 7. The thermostat on the left is integral to the housing.

COOLING FAN DIAGNOSIS

When the cooling fan is inoperative the vehicle will typically overheat while idling, in traffic, or traveling slowly. Once the vehicle is up to cruising speeds the natural flow of air across the radiator is generally enough to keep the coolant temperature below the danger zone. Always check for proper cooling fan operation as part of diagnosing an engine overheating concern.

The first check of a clutch-type fan should be for leakage by the clutch seal. Look for a greasy buildup near the seal, which indicates fluid leakage. Replace a leaky clutch; a clutch will not operate properly without the correct amount of silicone fluid. To test a temperature-sensitive clutch fan, turn the engine off and allow it to cool slightly. Rotate the fan while it is cool, and it should spin with just a little drag. Run the engine, and allow it to warm up fully. Turn the engine off, and immediately try to rotate the fan. You should notice significant resistance as you attempt to turn the fan. If it spins easily the clutch is not holding and should be replaced.

You Should Know *An electric fan can come on at any time, even after the engine is turned off. Keep your hands and tools safely away from an electric fan.*

Thorough diagnosis of an electric cooling fan circuit requires a solid understanding of electrical principles and test procedures. For practical purposes a few simple tests are explained here, but for complicated problems the service manual will provide you with the correct troubleshooting procedure. The electric cooling fan should come on when a predetermined temperature is reached, usually somewhere near 220°F. Use an infrared pyrometer or a pyrometer on a digital multimeter to check the cooling system temperature. Do not rely on the vehicle's gauge. Some older vehicles use a switch to turn the cooling fan on, but newer vehicles rely on the engine coolant temperature (ECT) sensor, the PCM, and a **relay** to turn the fan on. Check the circuit fuse(s) first. To find the cause of an inoperative fan you should split the circuit in half by determining whether it is the control side of the circuit at fault or the fan and wiring itself. Use a scan tool when possible to activate the fan. On many modern vehicles disconnecting the ECT will cause the PCM to turn the fan on. If the fan itself will not come on during these tests you can supply power and ground directly to the motor harness to check the wiring and be sure that the motor is actually faulty. If the fan does not come on replace it. If the fan operates during any of the preceding tests the control side of the circuit is at fault. This includes the ECT sensor, the coil side of the relay, the PCM, and the wiring. Check the circuit wiring and components using the manufacturer's diagnostic procedures. **Figure 8** shows a typical PCM-controlled cooling fan circuit.

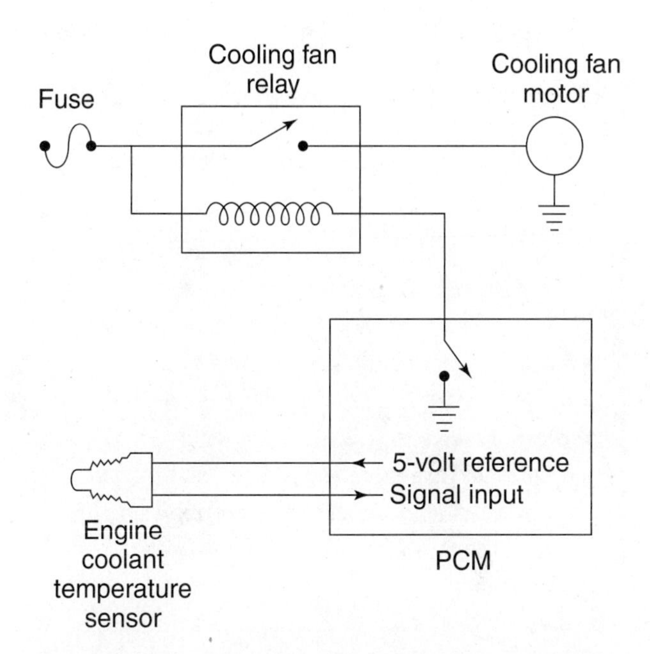

Figure 8. The PCM energizes a relay to turn the fan on when the coolant temperature reaches a preprogrammed value.

INTERNAL COOLANT LEAK DIAGNOSIS AND REPAIR

A faulty head gasket or a cracked cylinder head or block can cause a number of different symptoms. The most obvious sign of one of these problems is excessive white sweet-smelling smoke coming from the tailpipe. Other times the symptoms may include erratic heat and engine temperature; consistent or intermittent overheating; rough running; misfiring, especially at startup; excessive pressure in the cooling system; coolant overflow; coolant loss without an external leak; low compression on two adjacent cylinders; and coolant mixed in the oil, creating a fluid in the lubrication system that closely resembles a coffee milkshake. Prompt repair of a vehicle exhibiting any of these symptoms is essential to prevent more severe engine damage. Overheating typically causes failures of the head gasket, head, or block. Do not simply replace the faulty part. Thoroughly evaluate the cooling system to find the cause of overheating before calling the job complete.

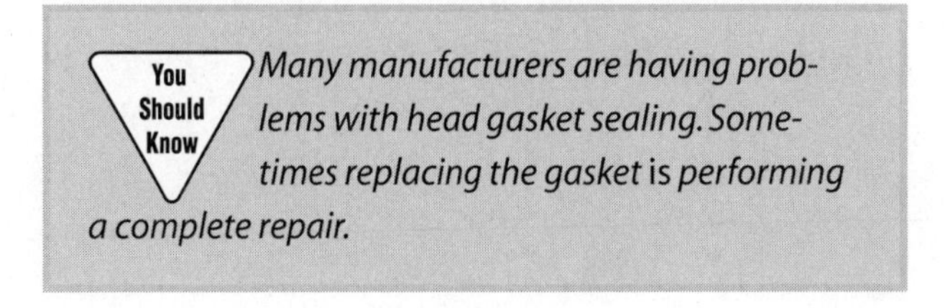

You Should Know *Many manufacturers are having problems with head gasket sealing. Sometimes replacing the gasket is performing a complete repair.*

These failures will very often cause the telltale symptom of white smoke blowing out of the tailpipe as coolant leaks into the combustion chamber and burns. You may have witnessed a head gasket failing while driving; all of a sudden a cloud of white smoke comes from the tailpipe of the unlucky person's vehicle. When excessive white smoke is coming from the tailpipe replace the head gasket as described below, and inspect or test the cylinder head and block.

> **You Should Know** *Check the manufacturer's service information, including technical service bulletins (TSBs). TSBs are bulletins that a manufacturer publishes when it has a common problem and has identified an effective correction. Some manufacturers are recommending that a liquid "stop leak" chemical be added to the coolant to try to seal the head gasket rather than replacing the gasket. Use only the chemical recommended by the manufacturer.*

Occasionally, the crack in the head gasket, head, or block is very small and will cause less dramatic symptoms. When the engine is cool the metals in either of the suspect parts contract and enlarge the crack. Coolant may leak into the affected cylinder and foul the spark plug or cause small amounts of white smoke to exit the tailpipe. Symptoms may occur for a short time after startup; then the engine heat can cause enough expansion in the metal to seal the gap. To confirm your diagnosis remove the spark plugs, allow the engine to cool *thoroughly*, and place rated system pressure on the coolant using a pressure tester. Let the vehicle sit for as long as is practically possible but for at least one-half hour. Disable the fuel system, have an assistant crank the engine over, and watch closely for coolant spray out of a cylinder. Sometimes it is helpful to place a paper towel in front of each plug hole so you are able to check each cylinder. Any coolant leaking from a cylinder indicates a sealing failure. If you are uncertain of the result from a cylinder you can perform a cylinder leakage test (as described in Section Three, "Engine Evaluation") to confirm your suspicion.

> **You Should Know** *If the engine does not crank over at all or at a normal rate of speed stop your testing because the engine could be hydrolocked. This means that a cylinder(s) is so full of coolant the engine can neither displace the coolant nor compress it. Putting a battery charger on the system and continuing to crank could crack a piston, a cylinder head, or even the block. Removing all the spark plugs will minimize the risk of this happening during testing.*

Combustion gases leaking into the cooling system may cause erratic temperatures, overheating issues, and problems of excess pressure and boiling over. The gases can form air pockets in the engine, preventing the thermostat from operating properly. When the leak is larger the gases produce enough pressure in the system to force the cap to relieve pressure, and leak coolant with it. One way to check for the presence of exhaust gases in the cooling system is to install a pressure tester on the filler neck. Do not pump any pressure on the tester. If compression pressure is leaking into the coolant the pressure on the tester gauge will rise rapidly. Sometimes the leak is large enough to diagnose by observation. Be certain that the engine is cool enough to run without the pressure cap on for a few minutes. Remove the cap and start the engine. If engine coolant bubbles violently and sprays out the top of the filler then the head gasket, the head, or the block has failed. When the problem is not this severe use a **chemical block tester** to confirm your suspicions of leakage **(Figure 9)**. This tester uses a chemical solution that changes color

Figure 9. Place the block tester on the coolant filler neck. Pump the bubble to pull air across the fluid. It will turn yellow if combustion gases are present in the cooling system.

when exposed to combustion gases. Remove the pressure cap, and start the engine. Place the tester on the top of the reservoir or radiator. Use the bulb to pull air (not coolant) from the top of the fluid. Pump the bulb several times with the engine running while watching the color of the liquid in the chamber. If it changes color, usually from blue to yellow, you have diagnosed a head gasket, head, or block leak.

> **You Should Know** *Today's vehicles use at least two, often several,* **oxygen sensors** *in the exhaust system to inform the PCM about the level of oxygen in the exhaust stream. These sensors help the PCM determine how much fuel should be injected to maximize fuel efficiency and performance and also to determine the efficiency of the catalytic converter(s). These sensors are quickly damaged when exposed to coolant-laced exhaust. When quoting a head gasket replacement after the engine has been burning coolant add in the cost of replacing the oxygen sensors; they are often each worth well over one hundred dollars.*

HEAD GASKET REPLACEMENT

To replace the head gasket you must have access to appropriate service information. No general procedure can adequately guide you through cylinder head removal and installation on a modern vehicle. The suggestions provided here are meant to reinforce and supplement the manufacturer's service information.

Carefully mark the location of fasteners, brackets, wires, and hoses as you prepare to remove the cylinder head. Be sure the head is cool before loosening the head bolts or the head may warp. Follow the proper sequence for loosening the bolts. Remove the cylinder head, and carefully inspect the gasket for signs of leakage. Even when a clear problem is seen in the head gasket, carefully inspect the cylinder head and combustion chamber for cracks. Aluminum cylinder heads should be sent out to a machine shop to be checked for cracks whenever they have been overheated. Cracks are often too small or too well hidden in a passage to see by eye. Check the head and deck for warpage, and have the machine shop plane (machine flat) either the head or deck if warpage exceeds the manufacturer's limits **(Figure 10)**. Clean all gasket surfaces thoroughly. Be careful to use the proper abrasive disc if using a die grinder to remove gasket material. Using too harsh an abrasive on an aluminum head can damage the sealing surface and cause another head gasket failure. Many manufacturers specifically prohibit the use of any abrasive discs on their cylinder heads. They may allow only a chemical gasket remover and a plastic scraper to remove all of the old gasket material. Also be meticulous in cleaning the gasket and metal particles out of the combustion chamber and valvetrain components to prevent component damage. Clean the threads in the head bolt bores. Use new head bolts when directed by the service information. Torque-to-yield head bolts should be replaced; they will not provide adequate clamping force if reused. Follow the manufacturer's torquing procedure and sequence carefully. This may include using a thread sealant on certain head bolts that run through a coolant passage.

Overheating is the most common cause of head gasket failures and cracked cylinder heads or blocks. Replace the thermostat. Many technicians will flush the engine and radiator before restarting the engine. Pressure test the system for leaks. Then run the engine to normal operating temperature, and let the engine run long enough to cycle the fan on and off. Be sure there are no air pockets in the system. Road test the vehicle to be certain that the engine and its cooling system are performing properly.

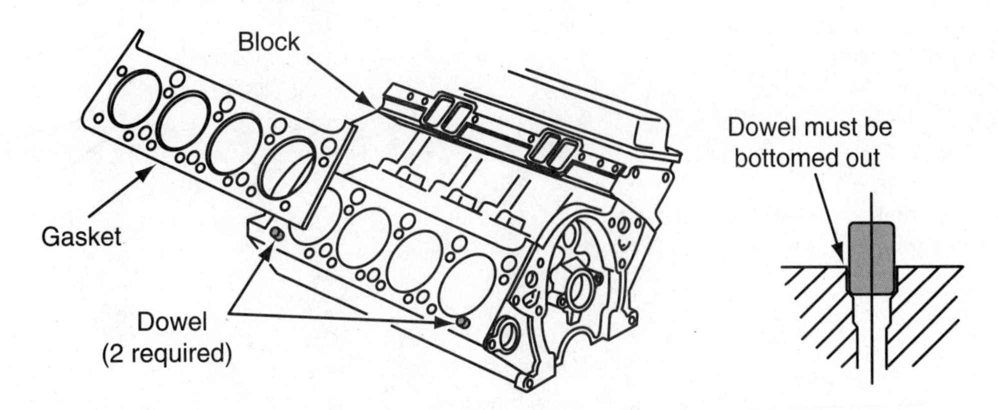

Figure 10. Many blocks have dowel pins to align the head gasket during installation.

Summary

- Cooling system failures can cause severe engine damage.
- Proper cooling system maintenance can prevent costly and inconvenient failures.
- Check the belt, hoses, and coolant condition regularly and as the first step in problem diagnosis.
- Check the coolant for contamination, pH level, electrolytic action, and antifreeze protection.
- Replace or flush coolant as recommended by the vehicle or coolant manufacturer.
- Pressure test a cooling system to check for leaks and locate the source.
- There are many possible causes of engine overheating; use your knowledge of the system to help you diagnose the fault.
- A thermostat that is sticking open can cause low engine operating temperatures, premature engine wear, a lack of heat, and poor fuel economy.
- A thermostat that is sticking closed will cause overheating during any type of driving.
- An inoperative cooling fan will cause overheating while driving at low speeds, in traffic, or at idle.
- A blown head gasket will often cause clouds of white smoke to pour out of the tailpipe.
- Diagnose a blown head gasket or a cracked cylinder head or block using visual inspection of system operation, chemical tests, pressure testing, or cylinder leakage testing.
- As part of repairing a head gasket, head, or block failure be certain that the cooling system is operating properly to prevent repeated failure.

Review Questions

1. What is the range of acceptable coolant pH level?
2. Explain the procedure to test coolant for electrolytic action.
3. List four likely causes of overheating.
4. Describe four symptoms of a blown head gasket.
5. Describe the procedure to fully evaluate a radiator cap.
6. Technician A says to pressurize the system to 25 psi to check for leaks. Technician B says a dye may be put in the coolant to help locate the source of a leak. Who is correct?
   A. Technician A only
   B. Technician B only
   C. Both Technician A and Technician B
   D. Neither Technician A nor Technician B
7. An engine overheats only while idling or in traffic. Technician A says the thermostat may be sticking closed. Technician B says the cooling fan is probably inoperative. Who is correct?
   A. Technician A only
   B. Technician B only
   C. Both Technician A and Technician B
   D. Neither Technician A nor Technician B
8. Technician A says that a stuck-closed thermostat will cause low engine temperatures. Technician B says that a stuck-closed thermostat can cause head gasket failure. Who is correct?
   A. Technician A only
   B. Technician B only
   C. Both Technician A and Technician B
   D. Neither Technician A nor Technician B
9. Technician A says to replace the thermostat when replacing a head gasket. Technician B says that torque-to-yield bolts can be used only once. Who is correct?
   A. Technician A only
   B. Technician B only
   C. Both Technician A and Technician B
   D. Neither Technician A nor Technician B
10. Technician A says that a voltmeter test can confirm a blown head gasket. Technician B says that a chemical test can detect some head gasket failures. Who is correct?
    A. Technician A only
    B. Technician B only
    C. Both Technician A and Technician B
    D. Neither Technician A nor Technician B

Chapter 12 Intake and Exhaust System Operation and Diagnosis

Introduction

The better an engine can breathe, the more power it can produce. The intake and exhaust system provide the breathing tubes apparatus for the engine. To maximize volumetric efficiency the intake and exhaust manifolds must be carefully designed. Both must allow adequate airflow to provide strong engine operation. A plugged air filter can seriously restrict airflow, causing the engine to run with reduced power or even not start **(Figure 1)**. A restricted exhaust system can cause the same symptoms; the engine must be able to freely take air in and easily push air out to breathe well. The intake system must be well sealed so that a strong vacuum is formed. This ensures that air will flow into the cylinder when the throttle and intake valve open. A vacuum leak can cause rough running, stalling, and misfiring.

Figure 1. A dirty air filter can prevent the engine from starting. Check the basics first.

Rodents often plug up air filters and air filter housings by storing food in them. They are attracted to the warmth of the engine. Check underneath a filter for dog food, nuts, or bedding.

INTAKE SYSTEM

The intake system consists of the air inlet tube, an air filter, ducting, a throttle bore, and the intake manifold **(Figure 2)**. The opening and closing of the throttle plate in the throttle bore controls airflow into the engine. Some systems use a throttle cable to open and close the throttle plate, while many newer systems employ "drive by wire" technology. In these systems the PCM responds to an input on the throttle pedal and controls a motor at the throttle plate to open it to the correct angle. The intake system should provide airflow at a high velocity and low volume when the engine is operating at low rpms to allow good fuel atomization and mixing when the injector sprays into the air stream in the manifold near the intake valve. It should also provide a high volume of air when the engine is running at high rpms and fill the combustion chamber as much as possible in a short time. This keeps power high even at higher engine speeds. Volumetric efficiency will naturally fall at higher rpms, but the design of the intake and exhaust systems can reduce the losses.

Intake Manifolds

The intake manifold design is largely responsible for meeting the requirements of the intake system. Many

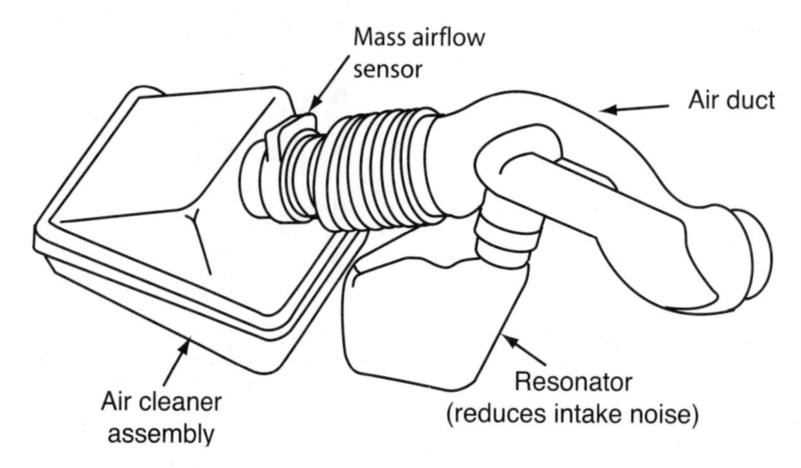

Figure 2. The air intake duct work connects to the throttle housing.

intake manifolds provide a compromise between low- and high-rpm operation; ideal operation of the intake system occurs only within a particular rpm range. The manifold is carefully designed to flow properly at the widest range of rpms. Tuned intake manifolds are designed so that natural pressure waves form in the manifold just as the intake valve opens. The added pressure forces more air into the cylinder. A lot of newer vehicles are using a variable intake system to make both low- and high-rpm operation more efficient **(Figure 3)**. In these systems two sets of runners are used. One set of runners is used at low-rpm operation. When the engine reaches a predetermined rpm an additional set of runners is opened to increase airflow volume when the valves are open for a shorter time. The runners are typically opened with a butterfly valve by an intake manifold runner control valve operated by the PCM **(Figure 4)**.

Modern intake manifolds are made of lightweight materials, either cast aluminum or plastics. Plastic intake manifolds have an advantage; they do not transfer as much heat to the intake charge. This allows for a denser, more powerful air charge. They are also much lighter, which contributes to improved fuel economy. Plastic manifolds can crack more easily or explode during backfire into the manifold, however. It is essential to use the proper torque sequence and specification when installing a plastic intake manifold.

Figure 3. This variable intake manifold has a short runner and a long runner for each cylinder.

You Should Know *Every ounce that a manufacturer can reduce from the weight of a vehicle is critical. Plastics and composite materials are used more and more to help engineers meet the CAFE standards. A plastic intake manifold may not seem like much of a weight savings, but multiply this advantage and other similar ones by the millions of vehicles produced, and the effect is significant.*

ENGINE VACUUM

The intake manifold should contain engine vacuum. Engine vacuum is created as the piston goes down on the intake stroke, increasing the volume of the chamber. As the area in the chamber increases the air pressure decreases. The pressure in the cylinder and in the intake manifold is lower than atmospheric pressure; this is called vacuum. Vacuum is typically measured in inches of mercury vacuum, in. Hg vacuum. An engine should produce between 18 and 22 in. Hg vacuum at idle at sea level. This is called intake manifold vacuum and can be measured by installing a vacuum gauge on a port on the manifold. Sometimes the vacuum is measured in pounds per square inch of manifold absolute pressure, MAP psi. As vacuum increases pressure decreases. MAP psi is typically between 7 and 9 at idle.

Adequate intake manifold vacuum is essential to a properly running vehicle. The engine must be able to

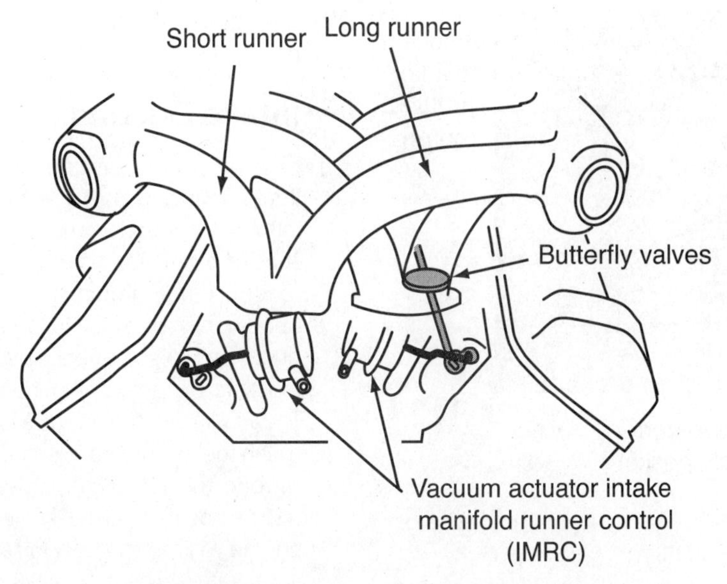

Figure 4. The airflow is controlled by butterfly valves in the short runners. The vacuum actuator pulls the linkage open at higher rpms.

develop adequate vacuum to quickly pull air into the cylinder when the intake valve opens. Air flows into the cylinders due to the pressure difference between the atmosphere and the intake. If the rings, piston, and valves do not properly seal the combustion chamber low vacuum will result. If the intake manifold or its gasket leaks vacuum will be lost. Powertrain control systems measure either the mass of airflow (by the **mass airflow (MAF) sensor**) into the intake or the manifold absolute pressure (via the **manifold absolute pressure (MAP) sensor**) in the intake manifold to help determine how much fuel must be mixed with the air to produce adequate power, maximum efficiency, and low toxic emissions. Leaks in the intake system throw these measurements off and cause drivability problems. Be sure to treat these electronic components carefully when removing them during engine service procedures, and reconnect the electrical connectors and vacuum lines if applicable.

Check for proper idle vacuum of 18 to 22 in. Hg vacuum using a vacuum gauge connected to the manifold **(Figure 5)**. If vacuum is low, listen for leaks. Often a light whistling noise can be heard where the engine is pulling fresh air in through a leaky seal or gasket. Check the intake manifold gasket, vacuum lines, and injector o-rings. You can use metered propane to help locate the source of a vacuum leak. Let a small amount of propane out of the end of a narrow hose. Run the hose around the gasket sealing surfaces and the injectors. If the propane can leak into the manifold the engine idle will smooth out and usually run more steadily as it receives more fuel. Even if it runs worse because it is getting too much fuel the change indicates that you have located a leak.

You Should Know *When using propane to diagnose vacuum leaks it is important that there be no possible ignition sources. A spark plug wire with faulty insulation that is allowing a spark to arc to a spot on the engine could create a dangerous, potentially explosive situation.*

Another safer and more efficient method of testing for intake leaks is to use a smoke leak detector. This machine uses shop air and a chemical to generate nontoxic smoke.

Figure 5. This gauge shows a steady reading of 18 in. Hg vacuum.

Figure 6. The smoke leak tester found the cause of this problem quickly.

Place the adapter into the intake ducting, and watch for smoke leaking out of the manifold. **Figure 6** shows smoke pouring out the manifold from a bad intake manifold gasket. This leak was so severe that the engine stalled consistently and would barely run off idle.

A typical engine breathes approximately 9000 gallons of air for every gallon of gasoline used. The air filter works awfully hard and deserves regular replacement, often once a year.

Air Filter

The air filter is designed to catch particulate matter in the air before it gets into the engine. Particles of dirt, rock, or sand are abrasive to the pistons, rings, and cylinder walls. When contaminants get into the oil they cause rapid wear of the oil pump and engine bearings. A vehicle should never be run without an adequate air filter. The air filter should be replaced when it is dirty or when specified by the manufacturer. The service information may not recommend replacement of the air filter until 30,000 or 60,000 miles. It does, however, call for more frequent inspection of the air filter. If the air filter is dirty it should be replaced promptly to be sure that engine performance is not affected. Do not attempt to clean an air filter with shop air. This can tear the filter element and allow larger pieces of debris into the engine. Remember, an engine that cannot breathe cannot produce power; a clean air filter is essential for proper performance. When an air filter is restricted the engine will generally lose power at high rpms and cause poor fuel economy. If it becomes really plugged the engine may stall, idle poorly, or fail to start. Do not neglect the basics of engine diagnosis and maintenance.

Intake Ducting

The tubes that connect the air filter housing to the intake manifold must be properly sealed. Many air ducts include a resonator chamber that looks like an irregularly shaped protrusion on the side of an intake duct. This is used to reduce the rumble of intake air as it rushes through the duct work. If outside air can sneak in through a leak in the duct work it is not filtered. Vehicles that use a MAF sensor measure the air coming into the engine from the air filter. The PCM uses this input as the primary indicator of how much fuel should be injected. If air is drawn into the engine beyond the MAF sensor through a leak in the duct it will not be measured. The PCM will deliver less fuel, and the engine will run too lean. In severe cases the engine will not accelerate, and it can backfire as the throttle is depressed. Carefully check the connecting clamps and the seams of the ducts for leakage.

EXHAUST SYSTEM

The exhaust system must provide a low-restriction outlet for the burned gases emitted at the end of the power stroke. The loud noise of combustion must be muffled as the exhaust valve opens to allow the spent gases out. The muffler(s) and resonators quiet and tune the noise from the exhaust. The exhaust must also be treated to reduce the amount of toxic pollutants emitted from the tailpipe. A catalytic converter works to reduce the HC, CO, and NOx emissions. The exhaust system is designed with a small amount of backpressure. Without it too much of the fresh air and fuel charge would escape out the exhaust during the valve overlap period. Some emission control devices use exhaust backpressure as a control for the exhaust gas recirculation system. With too much restriction in the exhaust system the engine will not be able to breathe properly. Excess pressure will remain in the cylinder and block the flow of fresh air into the engine. An exhaust system does not require maintenance, just periodic checks for leaks and testing for restriction when symptoms suggest a problem.

Exhaust Manifold

The exhaust manifold bolts around the exhaust ports on the cylinder head. It provides an extension of the exhaust ports and a means to connect the engine to the exhaust system. Most exhaust manifolds are made of cast iron, which can withstand the tremendous temperature changes and heat of the exhaust. They are designed to optimize flow out of the cylinder. Often there are short runners that quickly flow into a shared manifold. Many exhaust manifolds use a **header** design **(Figure 7)**. This design employs longer runners to keep the exhaust charges from

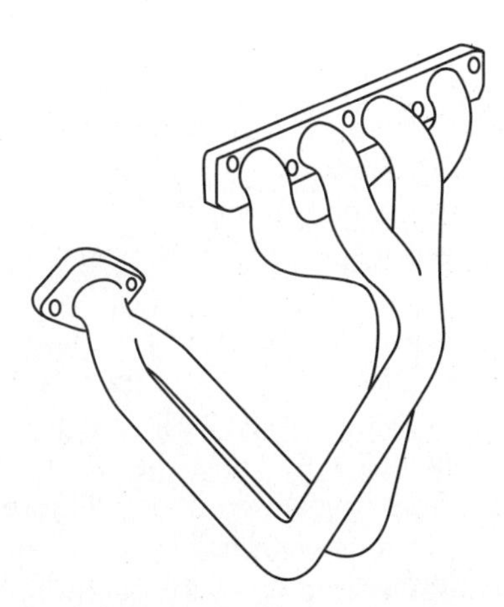

Figure 7. A header helps the engine breathe better by helping to clean out the spent gases more efficiently.

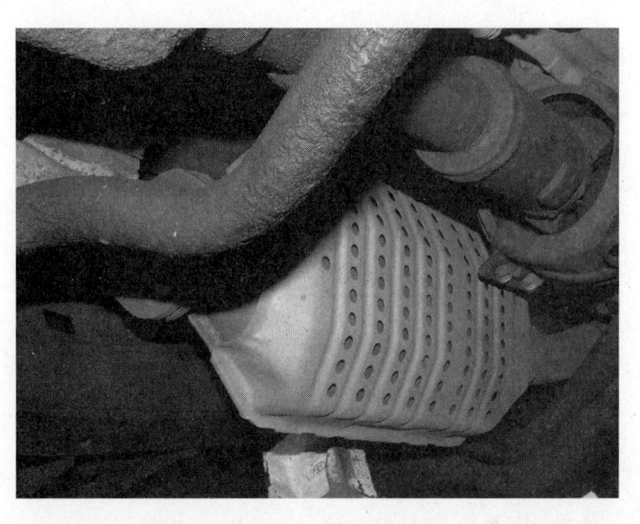

Figure 8. This catalytic converter is close to the manifold so it heats up quickly, which makes it function sooner.

each cylinder separate. The benefit is that backpressure from one cylinder will not impede the initial flow of exhaust out of another cylinder.

Catalytic Converter

After passing through the exhaust manifold the exhaust goes through a front pipe into the catalytic converter. Some vehicles eliminate the front pipe and bolt the main converter or a precatalytic converter directly to the manifold. The catalytic converter has catalyst material sprayed onto the huge surface area of a ceramic honeycomb. A catalyst is something that creates a chemical reaction. Modern converters are three-way catalysts (TWCs). They help oxidize hydrocarbons and carbon monoxide to form water and less-harmful carbon dioxide. They also chemically reduce oxides of nitrogen to form nitrogen and oxygen. Catalytic converters are found on all modern vehicles and greatly reduce polluting emissions. They regularly operate at temperatures between 1200 and 1600°F, and temperatures can exceed 2000°F. For this reason they use heat shields to prevent combustible materials close to them from catching on fire **(Figure 8)**.

Converters normally present some restriction to exhaust flow. Often they can fail and cause serious blockage. When a catalyst is overheated, usually from trying to oxidize too much fuel, the honeycomb can melt and become a mass through which exhaust cannot readily pass. A fuel injection or ignition system fault that is allowing raw fuel to exit the combustion chamber is a classic cause of converter meltdown. An engine with an ignition misfire should be repaired promptly to prevent permanent catalyst damage. Road shocks, temperature shocks, or technicians' blows can also break the honeycomb and cause restrictions. Any excessive exhaust restriction will reduce engine power, particularly at higher speeds when a greater volume of air must pass through the system.

All passenger cars and light trucks sold in the United States since 1996 must meet OBDII requirements and monitor the efficiency of the catalyst. If it is inefficient the PCM must turn on the **malfunction indicator light (MIL)** and set a **diagnostic trouble code (DTC)**. Always replace the converter with one that meets original equipment from the manufacturer (OEM) specifications. It is illegal to remove a converter and not replace it with another one. Used converters can be fitted only if they have been certified.

Mufflers

Mufflers are used to quiet the loud sound of the exhaust. The exhaust enters the muffler and flows around multiple baffles that break down the noise. Mufflers also naturally create some backpressure. Some enthusiasts and race car designers reduce or eliminate the restriction mufflers to improve engine breathing. Problems can occur in production vehicles when the baffles break and block flow.

Exhaust System Leaks

Exhaust system leaks are a safety concern and an annoyance. An exhaust leak can allow deadly carbon monoxide to enter the passenger compartment. You can usually locate a leak easily by sight and sound **(Figure 9)**. The cracked or rusted component is replaced, and the new one is installed. Occasionally an exhaust leak near an **oxygen sensor** can create drivability problems. The oxygen sensor (O_2S) informs the PCM about the level of oxygen in the exhaust stream. The PCM uses this to trim the fuel delivery to the correct proportion. If the exhaust system has a small leak it is common for the exhaust to leak out during the pressure pulses and for air to leak into the exhaust system during the low-pressure pulses. The rapid opening and closing of the valves create these waves of pressure. When

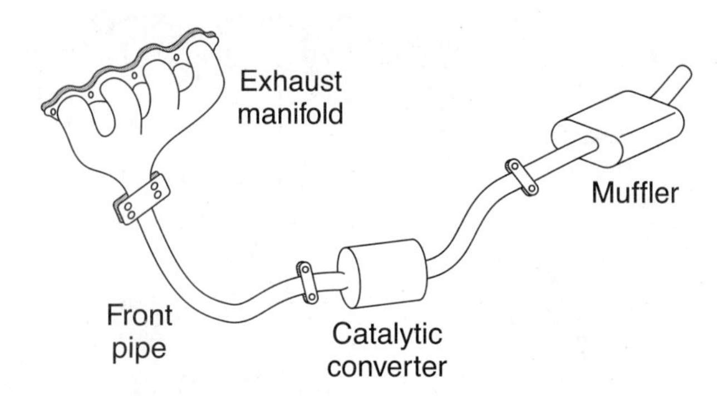

Figure 9. Check the whole exhaust system carefully for cracks or rusted holes.

outside air is pulled through a leak past the O_2 sensor it tricks the PCM into thinking there is excess oxygen in the exhaust. This can drive the PCM to consistently add more fuel until the engine runs poorly or the fuel economy decreases significantly. The PCM may also determine that the O_2 sensor is not operating properly and turn on the MIL. Check carefully at flanges and connections for leakage to diagnose or prevent this problem.

Exhaust System Backpressure

Too much exhaust backpressure can cause the same symptoms as a plugged air filter: a loss of power, particularly at high rpms; stalling; poor fuel economy; or a no-start condition. If you suspect a plugged exhaust tap lightly on the mufflers and converter to check for internal damage. Be gentle when tapping on the converter; you do not want to create a problem. Rattles in either component indicate problems. A broken muffler baffle can plug the outlet, as can pieces of broken converter honeycomb. Replace faulty components, and road test the vehicle again.

If nothing turns up during a rattle test you can perform a vacuum test to check for excessive exhaust backpressure. Install a vacuum gauge on the intake manifold. Rev the engine up to 2500 rpm, and hold it there for a minute. The vacuum gauge should show high and steady vacuum, 18 to 22 in. Hg. If the vacuum slowly falls toward zero the exhaust is plugged. Another way to check is to snap the throttle hard and watch the vacuum. It will fall to 0 in. Hg when the throttle is first opened, but it should return to 18–22 in. Hg almost instantly when the throttle closes. If the vacuum only gradually increases back toward a normal reading the exhaust is plugged. In both these tests you are checking to see that pressure is not building up in the chamber as a higher volume of air tries to move through the cylinder.

You can also use a backpressure gauge to test the exhaust system. Install the gauge into the exhaust ahead of the converter. The gauge will typically read from 0 to 10 psi in half-psi increments. Often the gauge has an adapter that screws into the front oxygen sensor port **(Figure 10)**. Other setups may require that you drill or punch a small hole in the front exhaust pipe and screw the gauge in. Always seal that hole before returning the vehicle to the customer. With the gauge installed check the pressure at idle, 2500 rpm, and under a hard, quick snap of the throttle. The pressure at idle should be near 0 psi, and it should stay below 1.5 psi at 2500 rpm. During a snap of the throttle the pressure should not exceed 4 psi. If the pressure is too high remove the converter, and inspect it for damage. It is more common for a converter to fail than a muffler, but you may also have to remove and check the muffler or mufflers. Converters are expensive, so be sure it will fix the problem!

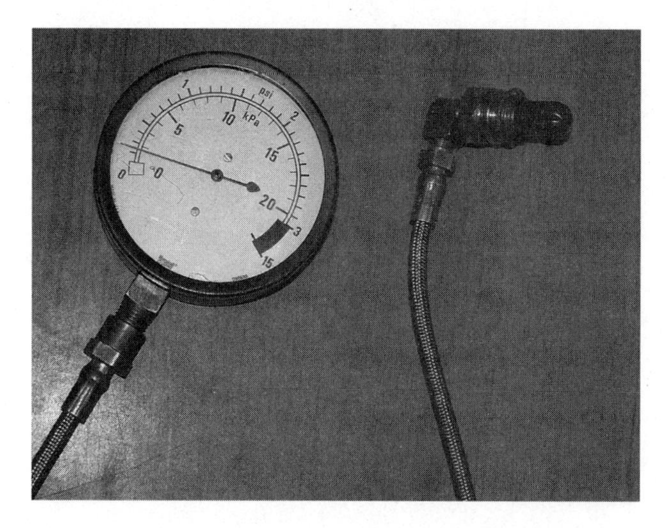

Figure 10. You can use an exhaust backpressure gauge to check for a plugged exhaust.

Summary

- The intake and exhaust systems are the breathing tubes for the engine; any restrictions will reduce engine power.
- The intake manifold is carefully designed to deliver air to the intake ports in adequate volume and at an appropriate speed.
- Variable intake manifold systems operate different runners at low and high rpms to maximize intake performance.
- Engine vacuum at idle should be 18 to 22 in. Hg.
- Vacuum leaks can cause stalling, rough running, and misfiring.

- A plugged air filter or air filter housing can cause low power, stalling, or a no-start condition.
- A cast-iron exhaust manifold is bolted to the exhaust ports on the head to direct exhaust gases out of the engine.
- A catalytic converter reduces the emissions of hydrocarbons, carbon monoxide, and oxides of nitrogen.
- Mufflers temper the loud roar of exhaust.
- Damaged mufflers and converters can create excess exhaust backpressure that reduces engine power and in severe cases can prevent the engine from starting.

Review Questions

1. Describe the functions of the intake manifold.
2. What problems can a dirty air filter cause?
3. A typical vacuum reading at idle is __________; a typical manifold absolute pressure reading at idle is __________.
4. What are two benefits of a plastic intake manifold?
5. The maximum allowable exhaust backpressure at 2500 rpm is __________.
6. Technician A says variable intake manifold systems are designed to improve midrange torque. Technician B says they operate two sets of runners to perform better at both low and high rpms. Who is correct?
   A. Technician A only
   B. Technician B only
   C. Both Technician A and Technician B
   D. Neither Technician A nor Technician B
7. Technician A says you can use propane to check for a vacuum leak. Technician B says you can use a smoke machine to check for a vacuum leak. Who is correct?
   A. Technician A only
   B. Technician B only
   C. Both Technician A and Technician B
   D. Neither Technician A nor Technician B
8. Technician A says that most air filters should be replaced every 10,000 miles. Technician B says that the air filter should be checked at every maintenance service. Who is correct?
   A. Technician A only
   B. Technician B only
   C. Both Technician A and Technician B
   D. Neither Technician A nor Technician B
9. Technician A says that a catalytic converter oxidizes hydrocarbons and carbon monoxide. Technician B says that a converter reduces carbon dioxide. Who is correct?
   A. Technician A only
   B. Technician B only
   C. Both Technician A and Technician B
   D. Neither Technician A nor Technician B
10. An engine has good vacuum at idle. It drops to 0 in. Hg when the throttle is snapped. Technician A says the exhaust is plugged. Technician B says that further testing is required. Who is correct?
    A. Technician A only
    B. Technician B only
    C. Both Technician A and Technician B
    D. Neither Technician A nor Technician B

Chapter 13 Turbochargers and Superchargers

Introduction

Turbochargers and superchargers are used to increase the volumetric efficiency of an engine **(Figure 1)**. They add pressure to the intake charge, often 8–14 psi above atmospheric pressure, to force extra air into the cylinders. Naturally aspirated engines use only the pressure difference between atmospheric pressure and engine vacuum to fill the cylinders with air. Forced induction systems, those with a turbocharger or supercharger, add significant power to an engine and can dramatically improve the horsepower to weight ratio. They are used to boost performance on vehicles. Manufacturers often use turbochargers or superchargers to maintain fuel economy standards; adding forced induction allows them to reduce engine size without sacrificing power. Aftermarket turbocharger or supercharger kits make popular, if expensive, add-on performance packages.

Figure 1. A turbocharger can make a small engine feel big!

In 1999 almost 4 percent of domestic vehicles and over 5 percent of imported vehicles were built with forced induction systems.

FORCED INDUCTION PRINCIPLES

On naturally aspirated engines, those without a forced induction system, volumetric efficiency falls as engine speed increases. Intake air at atmospheric pressure does not move fast enough to fill the chambers when the valves are opening and closing so quickly. This reduces the potential horsepower and torque of the engine. Forced induction systems pressurize the air in the intake manifold to force more air in when the valves open. This **boost pressure** dramatically increases volumetric efficiency at higher rpms. This can increase horsepower by up to 50 percent, though most production engines realize less than half that gain. Still, a turbocharger or supercharger can make a huge and very noticeable difference in performance. Both turbochargers and superchargers compress air to add pressure to the intake charge; the methods used differ.

TURBOCHARGING

Turbocharging an engine is the most efficient way of adding power. The turbocharger uses wasted exhaust gas pressure to operate; no power is lost in the process of creating boost pressure. Turbochargers spin at 100,000 to 150,000 rpm when producing maximum boost! It takes time to develop this speed as the throttle is opened, creating more exhaust pressure. The time it takes for the turbo to "spool up" is noticeable; this is called turbo lag. The engine is slower to get off the line, and the turbo feels ineffective until higher engine rpms are reached. Then the turbocharging effect kicks in, and you can feel the vehicle accelerate hard. Smaller, lighter turbochargers result in less lag time but lower boost pressures. Some engines use two sequential turbochargers: a very small, lightweight one to help acceleration at low rpms and a bigger-volume one to create plenty of boost to maximize power at higher rpms. Low-pressure turbochargers may provide boost of only 4–8 psi, but they reduce the lag time until the 8–14 psi from the full-size turbo kicks in.

Turbocharger

The turbocharger is made up of two impellers, one on the exhaust side and one on the intake side. The turbine, on the exhaust side, is spun by the pressure of the exhaust gases. On the same shaft, but on the intake side, is the compressor, which spins with the turbine to compress the intake charge. The exhaust gases never mix with the fresh intake charge **(Figure 2)**. The spinning compressor wheel forces the pressurized air into the intake ducting. The turbocharger gets extremely hot. Turbochargers have an oil feed to lubricate the bearings. Oil is delivered into the bearing area and returned out the other side. This increases the temperature of the oil considerably. Some turbocharged engines use an oil cooler to reduce oil temperatures. The oil change interval on turbocharged engines should be shortened, usually to every 2000 miles.

Intercooler

By adding pressure to the intake charge the turbo also adds heat. Uncooled intake air on a turbocharged engine can exceed 200°F. Heated air is less dense than cooled air. Thinner, heated air contains less oxygen and will not produce as much power as a cooler, denser cylinder full of air. An **intercooler** may be added to cool the intake air and thus make it denser. An intercooler looks very much like a radiator and may sit in front of it **(Figure 3)**. While the intercooler slightly restricts the airflow and reduces maximum boost pressures the net gain is positive, about 10 percent more power, when the engine is running hard. The intercooler may be air cooled or water cooled.

Figure 3. This large intercooler keeps the air charge denser by cooling the pressurized air.

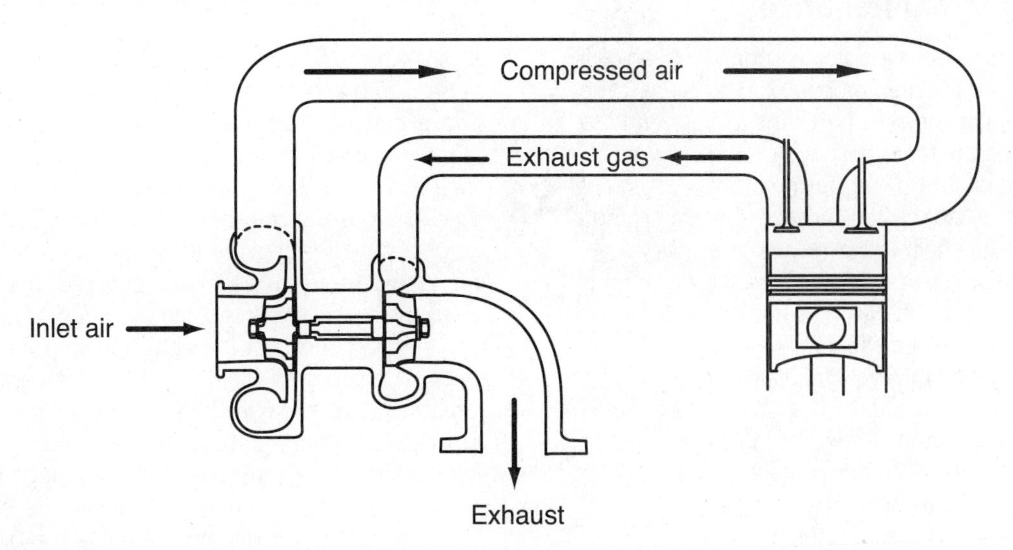

Figure 2. Exhaust gas spins the turbo and then pressurizes clean air for the intake.

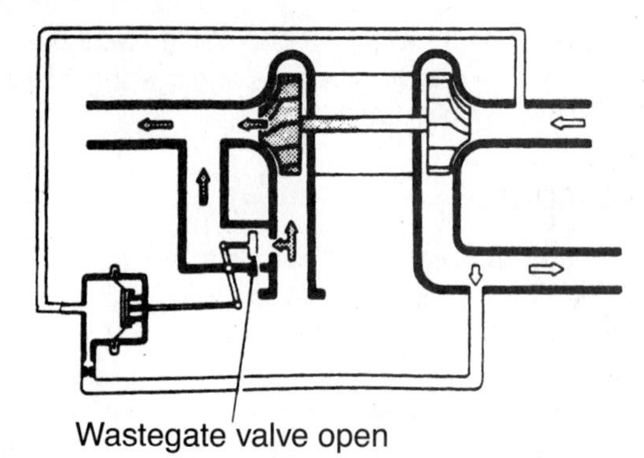

Figure 4. With the wastegate open, much of the exhaust gas is diverted through the bypass to slow the compressor down.

Wastegate

The **wastegate** is the control valve for the turbocharger. It prevents excess boost pressure that could cause detonation and severe engine damage. The wastegate opens to allow the gases in the exhaust manifold to bypass the turbine. This reduces compressor speeds and turbo boost pressures **(Figure 4)**. When the wastegate is closed, all the exhaust gases are used to spin the turbine. Some older systems used a mechanical spring-actuated wastegate; when boost pressure exceeded 6 psi it forced the wastegate open. Newer systems use the PCM to control wastegate operation. In this way the PCM can look at information about the intake air temperature, coolant temperature, engine load, throttle position, and engine knocking to determine when it is essential to open the wastegate. These systems allow for higher boost pressures whenever it is safe for the engine. A vacuum solenoid is used to operate the wastegate. The PCM can pulse the solenoid on and off to control boost pressure precisely.

Turbocharger Maintenance

Turbochargers must have a steady supply of clean oil if they are to last. When treated properly turbos can last the life of an engine. The oil must be changed frequently to provide long life to the turbo bearings. Oil breakdown and restrictions in the oil return line are the primary causes of turbo failure. The oil exiting the turbocharger is very hot. When an engine has been traveling under a heavy load, comes to a quick stop, and is shut off, the turbocharger is still spinning at high speeds. Just like it takes time for the turbo to spool up it takes time for it to wind down. As soon as the engine is shut off oil stops circulating, and the turbo bearings no longer receive fresh oil to cool them. The oil stuck in the turbo and in the turbo return line can bake at temperatures well over 400°F. This process is called coking and produces a white powdery substance that can restrict the oil passages. To prevent coking it is important to let the turbo bearings cool down for thirty seconds to a minute after hard operation before shutting the engine off. This is enough time for the turbo to cool down and the bearing temperatures to decrease. Not all customers know this about their vehicles. After a customer suffers a turbo failure explain to her the importance of letting the turbo cool down. Some manufacturers and aftermarket companies offer an add-on electric oil pump to circulate oil for a few minutes after engine shutdown.

Interesting Fact

One automotive technician drove a vehicle 250,000 miles with the original turbocharger still providing good boost. She drove it hard but consistently changed the oil at about 2000 miles and let the turbo cool down before shutting the engine off.

TURBOCHARGER DIAGNOSIS AND REPAIR

Bearing failure from a lack of good-quality and -quantity lubrication is the most common problem with the turbocharger itself. When bearings begin to fail the turbo whines as it spools up. It sounds quite loud, almost as though a police car were pulling up next to you with its siren blaring. When severe it can break the impeller blades and reduce boost pressure. Remove the intake ducting from the compressor, and see if the blades have been hitting the housing. If they have the turbocharger needs to be replaced. Whenever you replace a turbocharger for a bearing failure carefully check the oil lines for blockage. Many turbocharger manufacturers require that you replace the oil lines to adhere to the stipulations of the warranty **(Figure 5)**. Discuss proper turbocharger maintenance with your customer when you return the vehicle.

A worn turbocharger or a plugged oil return line can cause a turbo to consume a lot of oil, as shown by blue smoke coming from the tailpipe. When diagnosing the cause of oil consumption it is wise to check the base engine function first. If the engine is operating properly check the positive crankcase ventilation (PCV) system. This system vents the pressure that blows past the rings from the crankcase. If it becomes plugged the oil in the crankcase becomes pressurized and is forced into the combustion chambers past the rings and valve seals, and into the turbocharger oil drain. This causes excessive oil consumption and clouds of blue smoke. Finally, remove the intake duct from the turbocharger, and look for signs of oil in the duct. Often, if the oil return line is plugged, it will force oil pressure past the turbo seals and into the intake. Before condemning the turbocharger, remove the oil return line and check for blockage. After replacing a turbocharger be

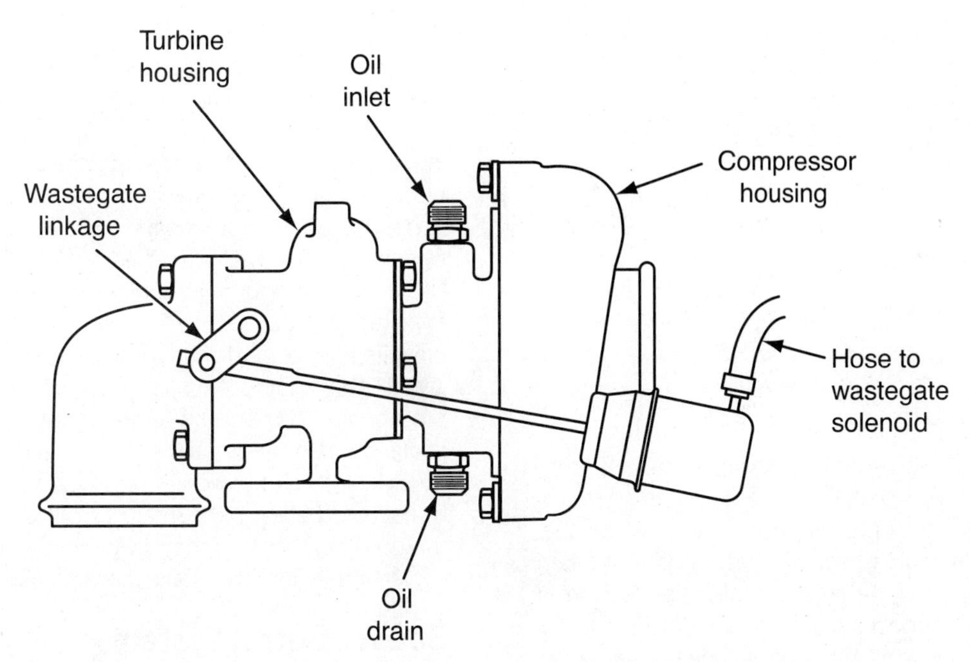

Figure 5. The turbo receives pressurized oil to its bearings. Check the inlet and outlet ports for clogging if the turbocharger is blowing blue smoke.

sure the lines are clean (or replace them if the manufacturer or turbo supplier recommends it) and the oil is changed.

Interesting Fact

One day, on that same high-mileage turbocharged engine, the technician thought she had finally blown the turbo. She started the engine up and soon saw a huge cloud of blue smoke billowing out her tailpipe. She had had no symptoms the previous day. Just as soon as she started plotting a way to find an inexpensive turbocharger the smoke stopped. It was a frigid Vermont morning, and sludge in the PCV line had frozen. She cleaned the line and continued to enjoy the power of that turbocharger.

There are many causes of low boost pressure; the least common is a worn turbocharger. Anything that limits the amount of airflow into or out of the engine will reduce turbo output. A worn engine, for example, will not develop adequate vacuum to fill the chambers with fresh air, nor will it compress the air fully enough to produce maximum combustion and pressure out of the exhaust. Check engine compression, as described in Section Three, as a first step in diagnosis. Something as simple as a plugged air filter may prompt a customer concern that the turbocharger is not working properly. You must also make sure that the wastegate system is functioning properly. Sometimes the wastegate can become stuck open. Many wastegates are now made so they are open in their rest position. The PCM must actively command the wastegate closed. Other times the PCM can command the wastegate open if it receives erroneous inputs or if an engine condition such as overheating or engine knocking exists. Refer to the manufacturer's service information for complete troubleshooting procedures for the electronically controlled wastegate system.

A leak in the intake ducting on a turbocharged engine can cause a very serious hesitation, bucking on acceleration, and lack of power. If boost pressure leaks out of an intake duct, at a cracked seam for example, the engine will hesitate horribly and lack power. Similarly, if the air taken into the engine is drawn after the mass air flow (MAF) sensor the engine can run so lean that it will buck and sputter as the throttle is opened.

You Should Know

One vehicle came to a shop with a new turbocharger, a new MAF sensor, and a new PCM. It still hesitated and bucked severely under heavy acceleration. Careful inspection revealed a split intake duct. Always check the simple things first!

Overboosting from the turbocharger feels great until the engine blows. Overboost is a very dangerous condition and should be repaired immediately. If the wastegate diaphragm rips, the vacuum from the solenoid will be ineffective; boost pressure will become excessive under high loads and at high rpms. Install a vacuum pump on the wastegate vacuum port. As you pump up a vacuum the wastegate arm should move. If it does not, replace the wastegate. Check to be sure that the vacuum line from the solenoid to the wastegate is intact. Verify the wiring to the solenoid. Many turbocharger wastegates are adjustable, and an overzealous customer could cause serious damage to the engine from detonation. Be sure the wastegate has not been tampered with.

SUPERCHARGING

Supercharging serves the same purpose as turbocharging, boosting the intake air charge, but the systems differ significantly. The supercharger is belt driven off the crankshaft **(Figure 6)**. It is parasitic, meaning that it takes horsepower away from the engine at all times, making it less efficient than a turbocharger. When it is blowing a lot of air the supercharger can take over 50 horsepower away from the engine. Luckily, blowers (as superchargers are commonly called) add plenty of power to the engine so the power gain is still noticeable. Superchargers spin at roughly 8000 to 17,000 rpm and are always spinning, so they do not cause the lag experienced in turbocharged vehicles. The power is almost instantaneous, boosting the vehicle quickly off the line.

Supercharger

There are several different types of superchargers, though they produce similar results. Most blowers are the positive displacement type, meaning they produce the same amount of compression with every rotation; the faster the supercharger spins, the more air it blows. This is the attribute that increases low-speed torque and eliminates lag time. The blowers can be either a rotor or screw design. This refers to the shape of the internal moving components. A Roots supercharger is a rotor-type blower. In both types the rotating parts come very close to each other but do not touch. The bearings that support the two shafts are bathed in a sealed lubricant from the factory. No external lubrication is required. This eliminates the need for extra oil changes on a supercharged engine. Some blowers have a plug that you can open to check the oil level. Higher engine temperatures from the increased compression and combustion temperatures still warrant an oil cooler on many applications. The supercharger picks up air from the intake duct work. It pressurizes the air charge as it is squeezed between the small gap between the rotors and the housing. The air is then pushed out through an intercooler and on into the intake manifold **(Figure 7)**.

Figure 6. The belt-driven supercharger sits on the top of the engine.

Boost Control System

Modern supercharging systems employ a PCM and a vacuum controlled **bypass valve** to control boost pressure. At idle or during rapid deceleration boost is not needed. The high vacuum in the manifold works on the bypass valve and holds it open. This allows air to vent back into the inlet duct rather than be forced into the intake. As the throttle is opened and load increases boost is desirable; the reduced engine vacuum allows the bypass valve to close. Full boost pressure is delivered to the intake manifold. The PCM can also control the bypass valve through a **boost control solenoid**. The solenoid opens or closes a passage supplied with boost pressure to the bypass valve. Normally the PCM energizes the solenoid to keep the boost pressure from acting on the bypass valve; this lets engine vacuum control the valve. When the PCM sees engine conditions that warrant reduced boost pressure it turns the solenoid off. This causes boost pressure to act on the bypass valve and open it up. The PCM can quickly cycle the solenoid on and off to control boost.

SUPERCHARGER MAINTENANCE AND REPAIR

Superchargers do not have a high failure rate. They are dependable as long as they receive clean air. Dirt particles can cause friction between the rotors and bearing wear. Dirt can be sucked into a blower through a vacuum leak. If the engine has a high or rough idle, stalling, or a telltale whistling, check for a vacuum leak and repair it promptly. Additional oil changes are not required on supercharged engines. Maintenance checks include driving the vehicle to check for proper performance and checking the supercharger oil if possible.

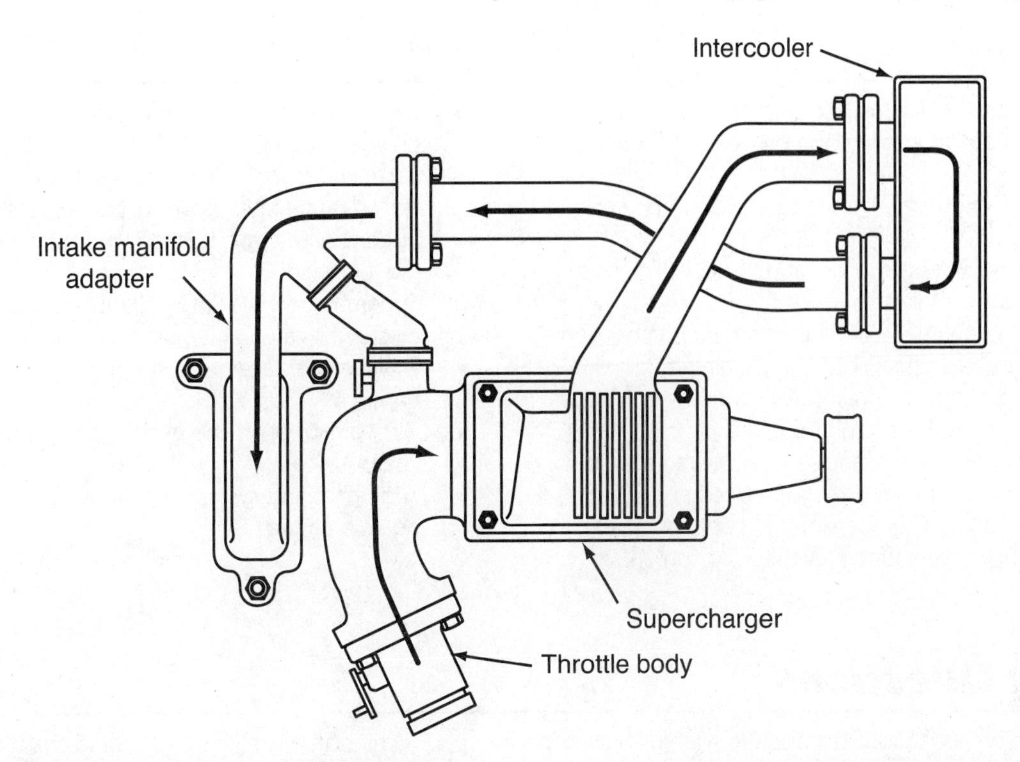

Figure 7. Airflow on a supercharged intake system.

If the supercharger itself fails it will cause reduced power and produce plenty of noise as the rotors scrape against each other. More likely the intake ducting or pressure ducting will develop leaks or the boost control system will develop a fault. Like a turbocharged engine a vacuum leak or a pressure leak will cause rough running, severe hesitation, and reduced power. Check the duct seams and clamped joints closely for leaks.

A fault in the boost control system can cause low power from low boost pressures or dangerous overboosting. Check the vacuum hose to the bypass valve, the boost hoses to and from the solenoid, and the electrical connections at the solenoid **(Figure 8)**. If the bypass valve sticks open boost pressure and power will be reduced significantly. If the boost control solenoid is electrically open it will allow boost pressure to force the bypass valve open constantly. Look for the bypass valve, and watch the actuator move as you move the throttle from the idle position to a quick snap open. If it does not move check the mechanical operation of the bypass valve. Remove the outlet line from the boost control solenoid to see if it is open and allowing boost pressure through to the bypass valve. If pressure is found check the wiring to be sure that it is receiving power and ground. If the solenoid has power and ground available but is leaking boost pressure through its port it is faulty and requires replacement.

Figure 8. Check the boost valve vacuum lines and connections.

Overboosting can occur if the boost control solenoid or bypass valve sticks closed. Also check the boost supply into the solenoid and the line coming out that goes to the bypass valve. Leaks in either line can prevent enough boost pressure from getting to the bypass valve to open it.

Summary

- Forced induction systems use either a turbocharger or a supercharger to increase the pressure of the intake charge.
- Forced induction systems increase engine power by improving volumetric efficiency.
- Some modern systems may add 8–14 psi of boost pressure to the intake.
- Turbocharged engines should have their oil changed at shorter intervals than naturally aspirated engines.
- Turbochargers take no power away from the engine as they increase torque and horsepower.
- Turbo lag is felt as the engine is accelerated from a stop until the turbo spools up to its operating speed.
- You should allow a turbocharger to cool down before shutting the engine off to prevent oil coking.
- Lubrication faults are the primary cause of turbocharger failures.
- Turbochargers typically use an electronically controlled wastegate to control boost pressure.
- Superchargers are belt driven and consume engine power to work.
- Superchargers are internally lubricated so additional oil changes are not required.
- Positive displacement superchargers eliminate lag time.
- Dirty air supply is the most common cause of supercharger wear.
- Supercharging systems use a bypass valve to control boost pressure.

Review Questions

1. An engine without a forced induction system is called __________.
2. What are typical boost pressures on today's forced induction systems?
3. How fast does a supercharger spin?
4. A turbocharger is driven by __________, while a supercharger is driven by __________.
5. When a turbocharger is replaced what else is often replaced?
6. Technician A says that a turbocharger blows pressurized exhaust gases into the intake. Technician B says that a supercharger makes more power by increasing the engine's volumetric efficiency. Who is correct?
   A. Technician A only
   B. Technician B only
   C. Both Technician A and Technician B
   D. Neither Technician A nor Technician B
7. Technician A says that if the wastegate sticks open engine power will be reduced. Technician B says that if the wastegate sticks closed severe engine damage may result. Who is correct?
   A. Technician A only
   B. Technician B only
   C. Both Technician A and Technician B
   D. Neither Technician A nor Technician B
8. Technician A says that you should change the oil every 2000 miles on a supercharged engine. Technician B says that supercharging systems often use an intercooler to keep the air charge dense. Who is correct?
   A. Technician A only
   B. Technician B only
   C. Both Technician A and Technician B
   D. Neither Technician A nor Technician B
9. Technician A says that a supercharger is an electric motor. Technician B says that a supercharger makes power by increasing the vacuum in the manifold. Who is correct?
   A. Technician A only
   B. Technician B only
   C. Both Technician A and Technician B
   D. Neither Technician A nor Technician B
10. Technician A says that a bypass valve opens to lower supercharger boost. Technician B says that the PCM can typically control the bypass valve through an electrically operated solenoid. Who is correct?
    A. Technician A only
    B. Technician B only
    C. Both Technician A and Technician B
    D. Neither Technician A nor Technician B

Engine Evaluation

SECTION OBJECTIVES

After you have read, studied, and practiced the contents of this section you should be able to:

- Thoroughly evaluate engine condition before mechanical repairs are made.
- Gather information about the source and extent of engine damage prior to teardown.
- Provide the customer with a reasonable understanding of the work that will be required for proper engine repair.
- Interpret vacuum gauge readings to diagnose several engine mechanical and drivability faults.
- Read spark plugs to gather information about the conditions within the combustion chamber.
- Identify which cylinder(s) in an engine is producing low power.
- Test cranking compression to identify combustion chamber sealing problems.
- Identify valvetrain problems through analysis of running tests.
- Pinpoint the source of combustion chamber leaks using a cylinder leakage test.
- Explain the causes of excessive blue smoke emitted from the tailpipe.
- Explain the causes of excessive white smoke emitted from the tailpipe.
- Understand some of the causes of excessive black smoke emitted from the tailpipe.
- Recognize irregular bottom-end noises and identify the causes by symptoms and tests.
- Recognize and identify the causes of irregular top-end noises.

Thorough testing before undertaking major engine repairs can guide your work and allow you to offer a realistic estimate to the customer.

Chapter 14 Vacuum Gauge Diagnosis

Introduction

The tests described in this chapter and in the following chapters in this section are used to fully evaluate the condition of the engine. You should perform a thorough analysis of the engine condition before recommending repairs. In some cases a relatively simple replacement of the valve seals will cure a problem of excessive use of oil. In other cases your evaluation will pick up several major engine weaknesses that may warrant installing a rebuilt engine. On a low-mileage engine with a distinct fault you and the customer may choose to repair the engine. Use these tests to gather enough information to make the best repair choice and to be sure that your work corrects all the problems in the engine. Using a vacuum gauge is a simple way to get initial information about the condition of an engine. You can use the tests explained here to begin diagnosis of a major engine failure or to pick out some smaller specific problems. Some of these tests provide the best way to confirm your suspicions about a problem that is otherwise difficult to diagnose. A vacuum gauge is a simple and inexpensive tool; it is a valuable addition to your toolbox.

Simple tests with a vacuum gauge can help you diagnose serious engine problems such as burned valves and snapped timing belts.

ENGINE RUNNING VACUUM

When an engine is running properly it should develop 18–22 in. Hg of vacuum at idle. This reading will drop about 1 in. Hg for each 1000 feet above sea level. To test for proper vacuum install a vacuum gauge on a port that supplies manifold vacuum. Usually there are a few small lines coming off the manifold **plenum**, the shared area of the intake manifold before it branches out to the runners. The reading should be steady. The vacuum will drop to 0 in. Hg when you snap the throttle open, but the vacuum should return quickly to its normal reading. If the vacuum is normal the engine is developing adequate compression. The compression rings and valves are in satisfactory condition. To gain a more precise measure of their effectiveness perform a compression test as explained in this section. If the vacuum reading is low or unsteady continue with further vacuum tests.

There are several possible reasons for a low but steady vacuum reading. Each possible cause should be investigated until the source of the problem is found. Some likely problems are:

- A vacuum leak from the intake manifold or manifold gasket, vacuum lines, brake booster, solenoids, or vacuum motors **(Figure 1)**
- Leaking valves
- Worn compression rings
- Incorrect or defective PCV valve
- Retarded cam timing
- Retarded ignition timing

The engine may also produce adequate but unsteady vacuum. Probable causes are:

- Erratic fuel mixture from a bad MAF or MAP sensor
- Leaking injectors
- Faulty ignition components

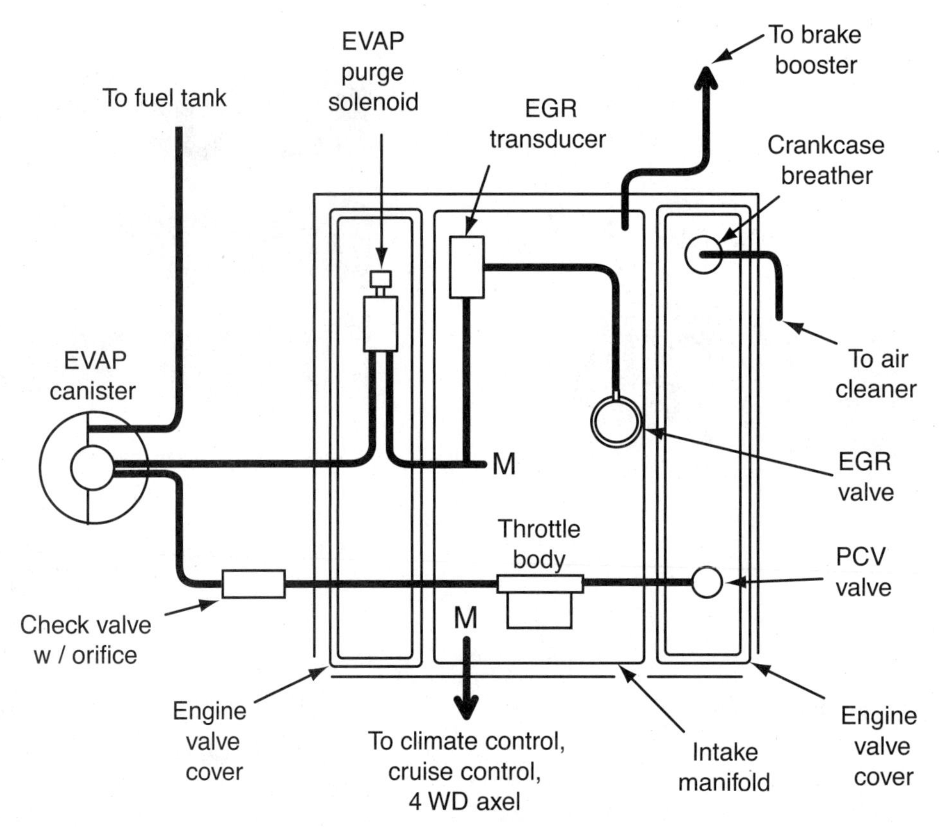

Figure 1. Look at all the vacuum lines for possible leaks.

- Sticking-open or leaking EGR valve
- One or more burned or leaking valves **(Figure 2)**
- Worn valve guides
- Weak or broken valve springs

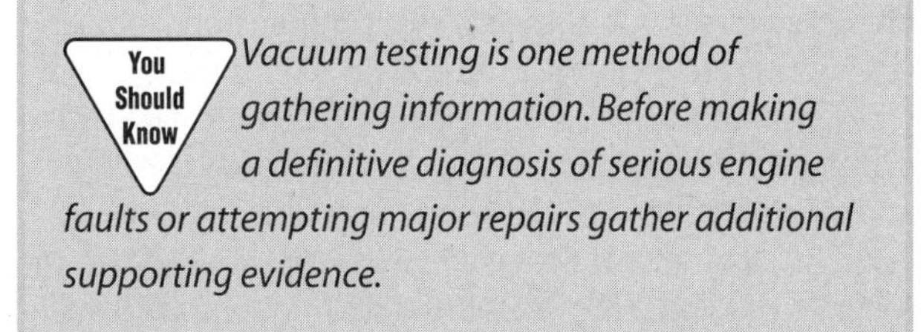

Vacuum testing is one method of gathering information. Before making a definitive diagnosis of serious engine faults or attempting major repairs gather additional supporting evidence.

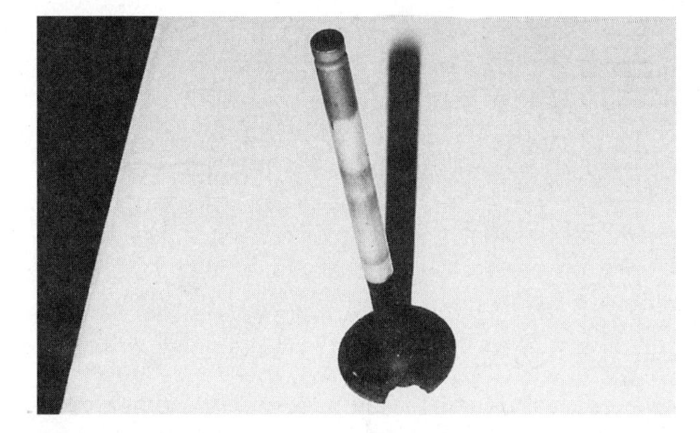

Figure 2. This burned valve caused the vehicle to run poorly.

ENGINE CRANKING VACUUM

To get an indication of the mechanical condition of an engine that will not start perform a cranking vacuum test. Install a vacuum gauge to a vacuum port, and crank the engine over while observing the gauge. Normal readings should show 3–6 in. Hg vacuum **(Figure 3)**. The colder the engine is, the lower the reading; the pistons have not yet expanded to form an ideal seal between the rings and pistons. A reading below 3 in. Hg vacuum indicates a loss of compression. There are several likely causes of low cranking compression:

- Low engine cranking speed
- Burned or leaking valves

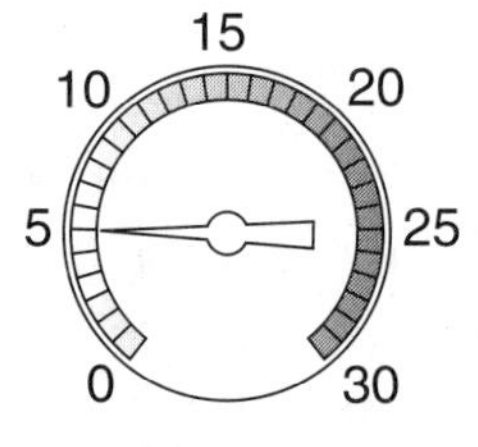

Figure 3. Normal cranking vacuum proves that the engine has enough compression to at least start.

- Worn compression rings
- A performance camshaft
- Improper camshaft timing

When a timing belt or chain snaps, the cranking vacuum will be near 0 in. Hg vacuum. The customer is likely to report that the engine was operating fine until it died. When you perform the cranking vacuum test on a car with a snapped timing chain or belt you will also notice that the engine cranks over extremely quickly. The starter is turning the engine against only mechanical friction; there is no pressure in the cylinders to slow the starter. To confirm this diagnosis you can often remove the oil fill cap and watch to see if the camshaft or valves are operating. If there is no movement of the valvetrain you have found a broken timing belt or chain.

DIAGNOSING WEAK VALVE SPRINGS

When the valve springs become weak the customer is likely to notice rough running and misfire only at high rpms. This is not a common problem; it is most likely to occur when a vehicle is frequently driven near or beyond the upper limits of rpm, indicated by the red line on the tachometer. The load on the engine is less of a factor. When the valves are opening and closing very fast at high rpms it requires all the strength of the valve springs to overcome the momentum of the valves and hold them on the profile of the cam. When the springs cannot achieve this the valves "float" over the nose of the cam and do not close as quickly as they should. This **valve float** causes the engine to miss and run poorly. Weak springs can be difficult to diagnose. A vacuum test can help confirm this problem.

It is best if you can install a vacuum gauge on the intake manifold and drive the vehicle at high rpms. Alternately, rev the engine over 2500 rpm in the shop while monitoring the gauge. If the vacuum reading is normal and steady through the lower rpm range but fluctuates significantly as engine speed is increased it is likely that the valve springs are weak **(Figure 4)**. If you are driving the vehicle while testing, your assistant should notice the fluctuation on the gauge as you begin to notice the symptoms of valve float. To positively confirm your diagnosis remove the valve springs, and test their tension as described in Chapter 26.

DIAGNOSING WORN VALVE GUIDES

When valve guides wear they can cause rough running, especially at lower engine rpms, because the valves do not seat squarely. This causes compression leakage, reduced vacuum, and a rough idle and low-rpm operation. They can also allow oil through the guide and valve seal. Worn guides will typically wear the valve seals as well because they allow the valve to rock side to side, stretching the seals. The result is oil consumption and excessive blue smoke after startup or a prolonged idle or deceleration.

To test for worn valve guides using your vacuum gauge, let the engine idle, and watch the gauge. Usually if the guides are badly worn the vacuum will fluctuate 3–7 in. Hg vacuum. Then when you increase the rpm to 2500 the needle will become much steadier if not stable **(Figure 5)**. This happens because as the valves move up and down faster in the guides they have less time to rock back and forth; momentum keeps them steadier. Usually, when guides are worn badly enough to cause valve sealing problems and rough running the valve seals have also been damaged. The combination of blue smoke from the tailpipe at startup and unsteady vacuum readings only at lower rpms should confirm your suspicion of worn valve guides. Always replace the valve seals when you replace valve guides.

DIAGNOSING STICKING VALVES

Many newer vehicles with tighter clearances are having a problem with valves sticking in the guides. Intermittently, especially when the engine is hot, the guide will stick in the valve and keep the valve either open or closed. Either state will cause misfiring on that cylinder. This can be tricky

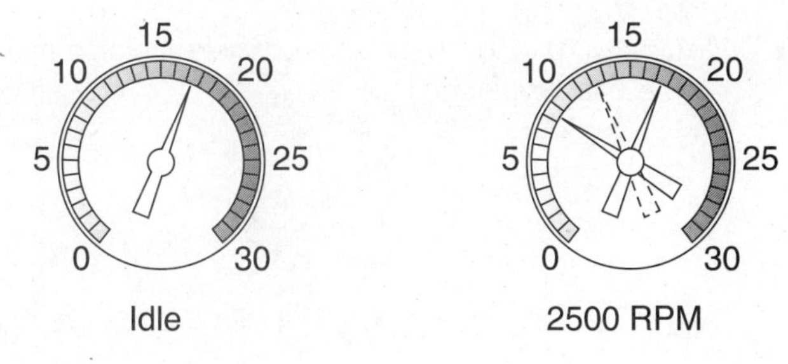

Figure 4. Look for fluctuating vacuum readings only at high rpm to diagnose weak valve springs.

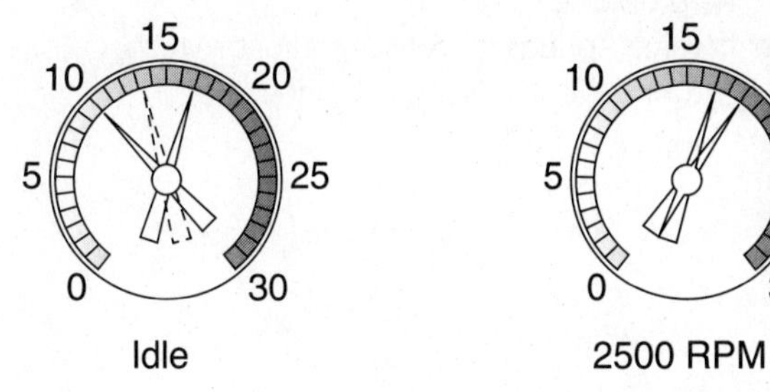

Figure 5. The unsteady readings at idle stabilize at 2500 rpm; this points to worn valve guides.

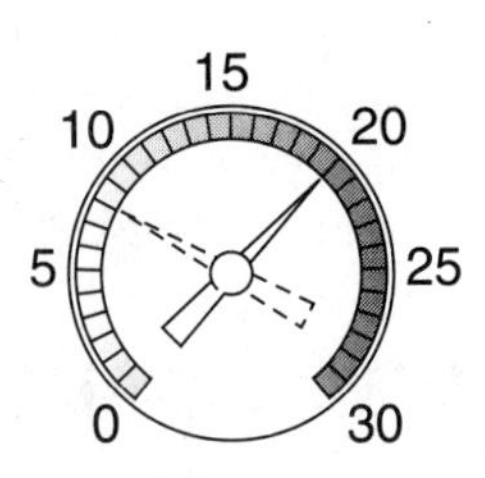

Figure 6. A sticking valve will cause an occasional drop in vacuum. You will notice the engine misfire when this occurs.

to diagnose because it may only happen from time to time. Your goal must be to get the engine to act up. Drive the vehicle with a vacuum gauge installed, and have an assistant monitor the vacuum reading while you feel for misfire. You may also notice a backfire through the intake or exhaust. When the valve sticks the vacuum will drop and return in regular cycles each time that valve is supposed to open or close **(Figure 6)**.

If the engine does not misfire on a road test, bring the vehicle back into the shop and let it sit for about 5 minutes for a hot soak period. Then start the engine up, and monitor the vacuum gauge and engine operation for sticking valves. If the valve is sticking the idle will be rough, and the exhaust will pop as the cylinder misfires. If the engine is misfiring squirt a little automatic transmission fluid (ATF) into an intake vacuum port with the engine running. Often this lighter oil will penetrate between the valve and guide and free it up. Observe the engine and vacuum gauge for smoother operation.

DIAGNOSING A RESTRICTED EXHAUST

An engine with an exhaust restriction will lack power at higher rpms and heavier loads. During these driving conditions the large volume of exhaust gases may start to back up in the system. The excessive pressure in the combustion chamber will not allow an adequate charge of fresh air in. This will cause the engine to lose volumetric efficiency and, therefore, power. To test for this problem install a vacuum gauge, and drive the vehicle or hold the engine rpm above 2500. If the exhaust is restricted the vacuum gauge will begin to drop the longer the throttle is held open. Ordinarily, the vacuum at 2500 should be near its maximum or close to 22 in. Hg vacuum. You can also snap the throttle open and watch the vacuum fall to zero as the throttle opens. If the exhaust system is operating normally the vacuum should rise right back up to normal almost immediately after the throttle closes. If the exhaust is restricted the vacuum will very slowly return to a normal or slightly lower reading. Usually it is the catalytic converter that gets plugged, but check the muffler(s) for loose baffles, too.

Summary

- A modern engine should produce a steady 18- to 22-in. Hg vacuum at idle.
- A common cause of low but steady readings is a vacuum leak.
- Fluctuating vacuum readings may be caused by improper fuel, improper spark, or engine mechanical problems.
- Cranking vacuum should be between 3 and 6 in. Hg vacuum.
- Low cranking vacuum typically indicates mechanical engine problems.
- Weak valve springs allow valve float, which causes rough running and misfiring at high rpms.
- Steady and normal vacuum at idle but fluctuating vacuum at higher rpms indicates weak valve springs.
- Worn valve guides can cause a rough idle and oil consumption.
- Fluctuating vacuum at idle that steadies as the rpm increases indicates worn valve guides.
- Sticking valves cause intermittent cylinder misfiring.
- Occasional drops from normal in the vacuum reading may indicate sticking valves.
- A restricted exhaust system causes a significant loss of power; the engine cannot breathe.
- Vacuum will drop off as the engine is held at high rpms if the exhaust is plugged.

Review Questions

1. A modern engine should develop between __________ and __________ in. Hg vacuum at idle.
2. What is valve float, and what symptoms will the driver notice when it occurs?
3. What combination of conditions indicates that the valve guides are worn?
4. How can you try to verify that a valve is sticking when an engine is misfiring in the service bay?
5. List three probable causes of low and unsteady vacuum readings.
6. Technician A says that low and steady vacuum readings are commonly caused by a blown head gasket. Technician B says that using an incorrect PCV valve can cause low vacuum. Who is correct?
   A. Technician A only
   B. Technician B only
   C. Both Technician A and Technician B
   D. Neither Technician A nor Technician B
7. An engine has 0 in. Hg vacuum while cranking, and the engine spins over very quickly. Technician A says a likely cause is a broken timing chain. Technician B says the starter could be bad. Who is correct?
   A. Technician A only
   B. Technician B only
   C. Both Technician A and Technician B
   D. Neither Technician A nor Technician B
8. Each of the following is a likely cause of low but steady vacuum readings *except:*
   A. A cracked intake manifold
   B. Low engine compression
   C. Incorrect camshaft timing
   D. Faulty ignition components
9. Technician A says that a sticking valve can cause occasional drops in the vacuum reading at any engine rpm. Technician B says that valves tend to stick when they are cold. Who is correct?
   A. Technician A only
   B. Technician B only
   C. Both Technician A and Technician B
   D. Neither Technician A nor Technician B
10. Technician A says that an engine with a plugged exhaust will not idle, but it will run fine at higher rpm. Technician B says that an engine with a restricted exhaust will develop excessive vacuum. Who is correct?
    A. Technician A only
    B. Technician B only
    C. Both Technician A and Technician B
    D. Neither Technician A nor Technician B

Chapter 15 Spark Plug Evaluation and Power Balance Testing

Introduction

Spark plug reading and power balance testing are two more methods you will use to gather information about how an engine is operating. You may use these tests during a routine tune-up, when there is a drivability concern, or as part of gathering information about an engine mechanical failure. Spark plugs can give you tremendous insight into what has been occurring within the combustion chambers. When you have a drivability concern one of the first areas to begin investigating is the spark plugs. This can help identify whether one or more cylinders are acting up or whether all cylinders are affected. It can also help guide your diagnosis toward a fuel, ignition, or mechanical issue, for example. Similarly, a power balance test can determine how much each cylinder is contributing and pick out the malfunctioning cylinder(s).

> **Interesting Fact**
>
> *Some spark plugs are advertised to last for over 100,000 miles. This does not guarantee they will last that long; these platinum plugs have a high failure rate over 50,000 miles and often become welded into the cylinder head by 75,000 miles. A 60,000 mile service interval is a reasonable compromise that is likely to serve the customer better.*

SPARK PLUGS

Spark plugs are the workhorses of the ignition system. All the buildup of electrical energy in the coil(s) comes to a head at the spark plug gap **(Figure 1)**. The electrodes of the spark plug must provide an adequate electrical path. If the spark is well timed and strong enough it jumps the gap with a powerful burst. Then it is the conditions within the combustion chamber that determine whether the spark will ignite the air-fuel mixture well enough to foster good combustion.

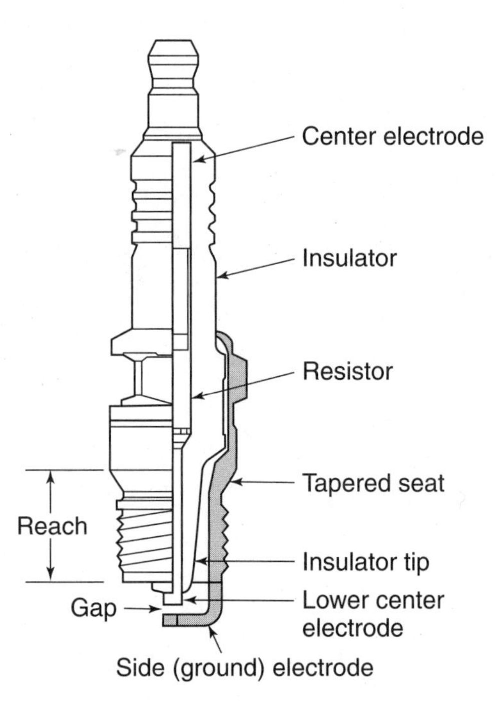

Figure 1. Spark plug construction.

The spark plug must be in good condition to allow the spark to jump the gap. If the electrodes are rounded from wear, if they are covered with oil or gas, or if the gap is too large the spark will not be strong enough to ignite a flame front if it sparks at all. Spark plugs are generally one of two basic types: the older-type plug with nickel alloy steel electrodes, and the newer, more common platinum-tip electrode plugs. Most new vehicles use the platinum plugs as original equipment. The platinum-tip plugs have a service interval of 60,000 miles or longer. The older-style plugs must generally be replaced at least every 30,000 miles. The term "tune-up" is out of date for modern vehicles. There is nothing to tune; for a maintenance service you will replace the spark plugs and the air and fuel filters, clean the throttle bore, and inspect the engine components and systems for proper condition and operation. Follow the specific service information guidelines to perform thorough maintenance services for your customers. Today's vehicles will often run for over 200,000 miles with proper maintenance and prompt repairs.

Many times customers do not bring their cars in for service until there is a drivability concern. Spark plugs have a high failure rate; check them early on in your diagnosis. As a spark plug wears, the gap between the electrodes becomes so wide it may not allow the spark to jump the gap. This can cause partial or total misfiring on that cylinder. Ignition misfire is usually noticeable at low rpms under heavy acceleration. The engine may buck and shudder when a cylinder misfires. It is often difficult to see wear on a platinum plug; the electrodes are small and do not show much visible degradation. Many technicians will replace rather than inspect and reinstall platinum plugs with over 30,000 miles on them. Given today's labor rates this is often more cost-effective for the consumer. If the engine is running poorly and you have the spark plugs out, it may be wise to perform a compression test. We will discuss compression testing in the next chapter. These tests can confirm the sealing of the combustion chamber and proper valvetrain action.

Spark Plug Removal and Installation

While spark plug removal is a very common task in the automotive repair shop it is not always a simple one. Many times you will have to disassemble parts of the engine to access the spark plugs. There are vehicles that require you to remove an engine mount or remove the turbocharger just to access the plugs. Every time you replace a set of spark plugs you want to be sure you do not damage the plugs or the threads while installing them. Always be sure to replace the whole set of plugs. If one has failed the others are likely not far behind.

If access to the plugs is not simple, read through the service information to pick up any tips that will make the job easier. This task often requires a universal joint and an assortment of extensions **(Figure 2)**. *Never* use air tools to remove or install spark plugs. It takes a lot longer to repair a spark plug hole than it does to loosen plugs with a ratchet! Blow away any sand and grit from around the spark plug holes before removing the plugs to prevent it from falling into the cylinder head. Remove the plugs, and keep them in order so you can evaluate the condition of each plug and correlate it with the correct cylinder. Keep the spark plug wires (if applicable) organized so you return them to their correct positions. Carefully inspect the ends of the plug wires. They can easily be damaged during removal and may require replacement. Also look closely at the plug wires for signs of electrical leakage or arcing to a ground path. If the wire's insulation is damaged you can often see a light gray or white residue on the wire where the high voltage arcs to a ground, on the valve cover, for example, rather than across the spark plug gap.

Figure 2. This spark plug is readily accessible; not every engine makes it this easy.

Many technicians will use nothing but original equipment manufacturer (OEM) spark plugs on late model vehicles. Today's PCMs and their monitoring systems are very sensitive; using aftermarket spark plugs is a very common cause of drivability problems. Manufacturers spend a lot of time and money designing the spark plug that will deliver the perfect spark for the individual combustion chamber. Many aftermarket companies produce excellent spark plugs. Unfortunately, they are usually designed to fit several applications. This does not guarantee that they will operate exactly like original equipment in each vehicle. If you do find quality aftermarket spark plugs that perform properly in an engine it is appropriate to use them. Many technicians have a spark plug brand they have come to trust and use it successfully on many if not all applications.

Interesting Fact

In one semester three students came to the instructor with a misfiring problem. In each case they had recently replaced spark plugs using a popular, low-cost brand of spark plug. Each of the three late model vehicles was repaired by installing higher-quality or OEM spark plugs.

Check the spark plug gap before installing the new set of plugs. Sometimes the gap gets closed if the plugs have been dropped. Note that some platinum spark plugs do not have adjustable gaps. Clean the area around the spark plug hole. If you get grease on the electrodes during installation you will likely wind up with a misfiring spark plug. Install the new plugs by hand; *do not* use air tools. If you cannot reach the plug itself, turn it in a few turns using only the extension. When the plugs are very difficult to reach it is helpful to put a piece of vacuum line on the end of the spark plug. This flexible connection allows you to reach the spark plug hole. Turn the plug in a few turns using the end of the vacuum line. If you start to cross thread the plug the vacuum line will spin on the spark plug and prevent thread damage. Finish tightening the plugs to the correct torque specification using a torque wrench. After your work is complete always give the vehicle a thorough road test to be sure the engine runs perfectly.

> You Should Know *Spark plug torque specifications are very light on many new engines, and a stiff elbow can easily strip the spark plug threads on an aluminum cylinder head.*

Spark Plug Reading

Each spark plug can tell you a story about what is occurring in the cylinder. You can use this information to help determine whether the engine is mechanically sound during a routine service. If one spark plug comes out caked with oil deposits from oil leaking past the rings or valve seals you will be able to let the customer know that the engine is in need of more serious work. When performing a drivability diagnosis the spark plug can help guide your diagnosis. If one spark plug is wet with fuel you should first confirm that the ignition system is delivering adequate spark to the plug. And whenever you are trying to narrow down the cause of an engine failure analyze the spark plugs as part of your preteardown investigation. The more information you have going into the job, the more confident you can be that you will find the real cause of the problem.

Spark plugs that are wearing normally should show a light tan to almost white color on the electrodes with no deposits **(Figure 3)**. The electrodes should be square, and the gap should be at the specified measurement. You will see plugs that have a red or orange tint on them. This is the result of fuel additives found in certain fuels or in fuel system cleaning additives.

Figure 3. This spark plug is wearing nicely; it has no deposits and is a light tan color.

Figure 4. Notice the wide gap and rounded electrodes on these worn spark plugs.

A spark plug that has many miles on it but is wearing normally will still be light tan in color. There should only be very light deposits, if any, on the plug **(Figure 4)**. On nonplatinum plugs you will be able to see that the corners of the electrodes are rounded. The gap will usually be wider than specified. On platinum plugs the tip may look brand new even when the plug is not firing well. Often you will have to judge wear by how many miles the vehicle has been driven since the last change. If the plug is not firing at all it will be gas soaked and smell like fuel.

> You Should Know *The classic symptoms of an ignition misfire are bucking and hesitation on hard acceleration. A fuel-related problem, such as a restricted fuel filter, is more likely to cause a steady lack of power as engine speed increases.*

Fuel-fouled plugs will be wet with gas when you remove them from the engine. They will smell like fuel and may have a varnish on the ceramic insulator from heated fuel. The most common cause of this is inadequate spark, though a leaking fuel injector will also flood the plug. If the valves are burned or the compression rings are badly worn the engine may lack adequate compression to make the air-fuel mixture combustible. A compression check will either verify or rule out that problem.

When a vehicle is burning oil the spark plugs will develop tan to dark tan deposits on and around the electrode. As the oil is heated during combustion it bakes onto the plug, leaving a residue. You will have to rule out any other possible causes of oil consumption, such as a plugged PCV system, badly worn valve seals, or a faulty turbocharger, but this is usually a telltale sign that the engine is mechanically worn. If you find this problem during a routine service it is important to warn the customer about the failing condition of his engine. If the oil is wet and thick on the plug look for a cause of heavy oil leakage into the cylinder. From time to time a head gasket will crack between an oil passage and the cylinder, allowing oil into just one cylinder. This can cause an oil-soaked plug. Similarly, a hole blown in a piston can cover a plug with oil, but you will have other indicators of a serious mechanical failure.

Black, soot-covered spark plugs are found when the air-fuel mixture is too rich. This means there is too much fuel and not enough air. This can be caused by something as simple as a plugged air filter or as involved in the powertrain control system as an inaccurate ECT sensor **(Figure 5)**. An engine that does not have good compression can also create carbon because combustion will not be complete. The key is to notice the condition. Locate the cause before considering a maintenance service complete. In addition, write down your observations if you are evaluating an engine for serious mechanical repairs; you will need to refer to all your observations and measurements when making repair decisions.

Blistered or overheated spark plugs are a sign of potentially serious trouble in the combustion chamber. The porcelain insulator around the center electrode will actually have blisters in it, or there may be pieces of metal welded onto the electrode. This can be a sign that the wrong spark plugs were installed in the engine. Otherwise, it is likely that pinging or detonation is occurring. If there is no engine damage yet, it is critical to find the cause of the problem before returning the vehicle to the customer. If you find this as you are diagnosing an engine mechanical failure it is essential to record this information and find the source of the problem during your repair work. A cooling system passage with deposits in it could be causing one corner of a combustion chamber to be getting so hot that detonation occurs. If you were to replace the damaged piston and rings but not repair the cause of the problem the customer would be back soon with a repeat failure. Your thorough evaluation of the engine and analysis of the causes of the symptoms will ensure customer satisfaction.

Figure 5. A faulty engine coolant temperature sensor can cause the engine to run rich.

POWER BALANCE TESTING

On a rough-running or misfiring engine you can perform a power balance test to identify a cylinder that is not equally contributing power to the engine. To perform the test you disable one cylinder at a time, while noting the change in engine rpm and idling condition. The more the rpm drops or the rougher the engine runs, the more the cylinder is contributing. If a cylinder is producing little or no power, the change in rpm and engine running will barely be noticeable. You must note the rpm drop quickly as the cylinder is initially disabled; today's PCMs will boost the idle almost immediately to compensate for the drop. You will also be able to feel and see the roughness of the engine as a functional cylinder is disabled. We will discuss a few different methods of performing a power balance test.

Power Balance Testing Using an Engine Analyzer or Scan Tool

Many engine analyzers can perform a manual or automated power balance or cylinder efficiency test. With the analyzer leads connected to the ignition system as instructed in the user's manual, select the power balance test from the menu **(Figure 6)**. The tester will automatically disable the ignition to one cylinder at a time for just a few seconds and display the rpm drop. The analyzer disables spark in the firing order; an engine with the firing order 1-3-4-2 would be tested on cylinder #1 first, then #3, #4, and #2. On some analyzers the power balance results will be displayed in bar graph format for easy comparison of cylinder contribution.

On some vehicles the PCM is programmed to be able to perform a power balance test. The manufacturers' and some aftermarket scan tools will be able to access this test.

Figure 6. This engine analyzer shorts one cylinder at a time and displays the engine rpm drop.

The only hookup required is to the diagnostic link connector (DLC). After connecting to the DLC select power balance test from the menu system. At the prompt the scan tool will communicate with the PCM to initiate its disabling of cylinders one by one. The rpm drop for each cylinder will be displayed after the test.

Manual Power Balance Testing

Even without an engine analyzer or a capable scan tool you can perform a power balance test. The procedure will require that you manually disable either fuel or spark to each cylinder. You must be careful not to disable spark for long periods of time, and you must allow the engine to run on all cylinders for 30 seconds between tests. Failure to follow these guidelines can overload the catalytic converter with raw fuel and damage it from overheating. It is also important that you check the vehicle for **diagnostic trouble codes (DTCs)** after performing this test and clear the codes if any are present. A DTC is a code that the PCM sets when it identifies a problem with a system or component. You can access DTCs using a scan tool; the code can help you locate the source of the problem. If the PCM detects a misfire it is likely to set a DTC. When the malfunction indicator light comes on a DTC has been set. Using the menu on a scan tool you can erase any DTCs when your testing is complete.

On vehicles that use spark plug wires to deliver the spark to the plug you can short the spark to ground on each cylinder in turn to test the power contribution of cylinders. One way to do this is to use an insulated pair of pliers **(Figure 7)**. Carefully pull the wire off the plug, and hold the wire to the block or head to allow it to find a ground path. It is important that the spark be able to find an easy path to ground; do not hold the wire more than half an inch away from a good metal conductor. The coil can be damaged if it uses all its possible power to find a path to ground.

Figure 7. Use insulated pliers to remove a live plug wire. Ground it to a metal component immediately to prevent harming the coil.

Sometimes the spark will track through the insulation of the coil and destroy it. As you remove the spark plug wire you should notice a clear change in rpm and idle quality if that cylinder is contributing adequate power.

Many newer engines use coil on plug ignition systems. Each spark plug has its own coil, and you cannot safely remove the coil and ground it. Instead, locate the connector going into the coil **(Figure 8)**. While the engine is running you can briefly remove the connector to prevent the coil from firing. As you unplug the coil monitor the rpm drop on a tachometer. Repeat this for each cylinder, and compare your results. This testing is very likely to trigger the MIL and set a diagnostic trouble code, so be sure to clear the DTCs after your testing.

Figure 8. Disconnect the low voltage primary wiring to the individual coils and record rpm drop. Be sure to clear any DTCs after your testing.

Another way to perform a manual power balance test is to unplug the connectors to the fuel injectors one by one while watching the engine rpm. By preventing fuel from entering the cylinder you can momentarily disable the cylinder. This allows you to note how much the cylinder is contributing to the overall power of the engine.

Analyzing the Power Balance Test Results

When an engine is running properly each cylinder should cause very close to the same rpm drop when disabled. Cylinders that are lower by more than 50 rpm should be analyzed. Look at the results below:

Cyl. #1	Cyl. #3	Cyl. #4	Cyl. #2
125 rpm	150 rpm	50 rpm	125 rpm

Cylinder number four is definitely not contributing equally to the power of the engine. The rpm changed very little when it was disabled, meaning that the engine speed is barely affected by that cylinder. There could be many causes of the problem, but you now know which cylinder to investigate. The problems could relate to fuel delivery; perhaps the fuel injector was not delivering adequate fuel to support combustion. An ignition system problem, something as simple as a spark plug with a very wide gap from worn electrodes, could also cause this. A cylinder with low compression will also produce less power and should be suspect during a power balance test. The next two chapters will describe compression testing and cylinder leakage testing. These tests will help you determine whether the weak cylinder has an engine mechanical problem. They can also provide information about what the specific cause may be.

Summary

- The spark plug provides the path for electrical energy from the coil(s). When a proper spark jumps the gap between the plug electrodes it ignites the air-fuel mixture to begin combustion.
- The older-type spark plugs, with nickel alloy steel electrodes, often require replacement every 30,000 miles.
- The newer platinum spark plugs have a longer service interval, typically 60,000 to 100,000 miles.
- Check spark plugs when diagnosing a drivability concern; they are a wear item and have a high failure rate.
- Replace spark plugs as a set; if one fails the others will likely follow. Remember to check the plug wires too; they will often require replacement at the same time.
- Use high-quality spark plugs designed for the engine.
- Keep the spark plugs in order when removing them so you can trace any problems with the plug to the correct cylinder.
- A spark plug that is operating properly in a healthy environment should be light tan, with few if any deposits.
- Check the ignition system and the fuel injector if a spark plug comes out wet with fuel.
- Light oil consumption can be detected by tan deposits building up on the spark plug. Heavier oil consumption will leave black wet deposits on the electrodes.
- A blistered spark plug is an indication of an overheated plug. Make sure the correct spark plug is installed. Detonation will blister, crack, or break spark plugs; the problem should be repaired immediately to prevent serious engine damage.
- Use a power balance test to find a cylinder that is not contributing equally to the power of the engine; the weak cylinder will not cause as significant an rpm drop as the healthy ones when disabled.
- Scan tools and engine analyzers may have a menu-driven, automated power balance test procedure.
- Manual power balance testing can be performed by carefully disabling either fuel or spark to each cylinder in turn.
- Investigate any cylinder that produces an rpm drop 50 rpm lower than the others.

Review Questions

1. Describe some problems with the spark plug that could prevent a good spark from occurring.
2. What are the signs of wear on a nickel alloy spark plug?
3. Describe what you will see on the spark plug electrodes if an engine is burning oil.
4. Describe the procedure to perform a manual power balance test by disabling the ignition.
5. Technician A says that nickel alloy steel plugs last longer than platinum plugs. Technician B says that spark plugs should be inspected at every oil change. Who is correct?
   A. Technician A only
   B. Technician B only
   C. Both Technician A and Technician B
   D. Neither Technician A nor Technician B

6. Technician A says to check spark plugs as one of the first steps in diagnosing a drivability concern. Technician B says that a spark plug gap that is too wide can cause engine misfiring. Who is correct?
   A. Technician A only
   B. Technician B only
   C. Both Technician A and Technician B
   D. Neither Technician A nor Technician B
7. Technician A says that black flaky carbon on a spark plug is caused by excessive oil consumption. Technician B says that a spark plug wearing normally will be light tan to white. Who is correct?
   A. Technician A only
   B. Technician B only
   C. Both Technician A and Technician B
   D. Neither Technician A nor Technician B
8. Technician A says to replace the thermostat if the spark plugs are blistered. Technician B says that detonation can cause plug blistering. Who is correct?
   A. Technician A only
   B. Technician B only
   C. Both Technician A and Technician B
   D. Neither Technician A nor Technician B
9. Technician A says that you must have a scan tool or engine analyzer to perform a power balance test. Technician B says the cylinder that produces the highest rpm drop is the weakest one. Who is correct?
   A. Technician A only
   B. Technician B only
   C. Both Technician A and Technician B
   D. Neither Technician A nor Technician B
10. A weak cylinder is found during a power balance test. Technician A says the fuel injector could be faulty. Technician B says it could be an engine mechanical problem. Who is correct?
    A. Technician A only
    B. Technician B only
    C. Both Technician A and Technician B
    D. Neither Technician A nor Technician B

Compression Testing

Introduction

Compression testing is an excellent way to determine whether an engine has serious mechanical problems. When an engine is running poorly it is important to establish that the mechanical function of the engine is sound before delving into testing the myriad of systems and components that could also cause drivability issues. Many technicians will perform a compression test on an engine that is running quite poorly before performing a major maintenance service. As we have discussed customers may bring their vehicle in for service only when it acts up. They may believe that a "tune-up" will fix the problem, but no amount of new parts will make an engine run well if it has a burned valve. It is important to perform a compression test before undertaking any major engine work. You should know whether you will be performing work on the cylinder head and valves only or if the entire engine needs to be overhauled or replaced. A valve job, replacing or machining the valves and seats to reestablish proper sealing, may cost upwards of a thousand dollars on many engines. A complete engine overhaul of the head and block or installing a crate engine is likely to set the customer back at least twice that amount. Compression testing can help you provide you and the customer with information to make an appropriate repair choice.

Interesting Fact

A vehicle that was running very poorly had been to two shops that tried unsuccessfully to locate the cause of the problem. A full maintenance service had been performed, and a new PCM, ECT, MAF, and turbocharger were installed to no avail. A running compression test showed very low readings on two cylinders. Careful inspection of the overhead camshaft showed well-worn lobes preventing adequate airflow into the engine.

CRANKING COMPRESSION TEST

A cranking compression test is the most commonly used type of compression testing. It can clearly determine if the valves, rings, and head gasket(s) are sealing the combustion chamber properly **(Figure 1)**. When any of those components are faulty the engine may not be able to develop adequate compression. The manufacturer will provide a specification for the appropriate cranking compression pressure. Low compression will cause less productive combustion. You and the customer will notice that the engine lacks power or misfires on one or more cylinders.

To perform a cranking compression test:

1. Let the engine warm up to normal operating temperature. (This ensures that the pistons have expanded to properly fit and seal the cylinder.)
2. Let the engine cool for 10 minutes so you do not strip any threads; then remove the spark plugs. Remember to keep them in order so you can read them.
3. Block the throttle open at least partway so the engine can breathe.
4. Disable the fuel system by removing the fuel pump fuse or relay to prevent contamination of the engine oil.

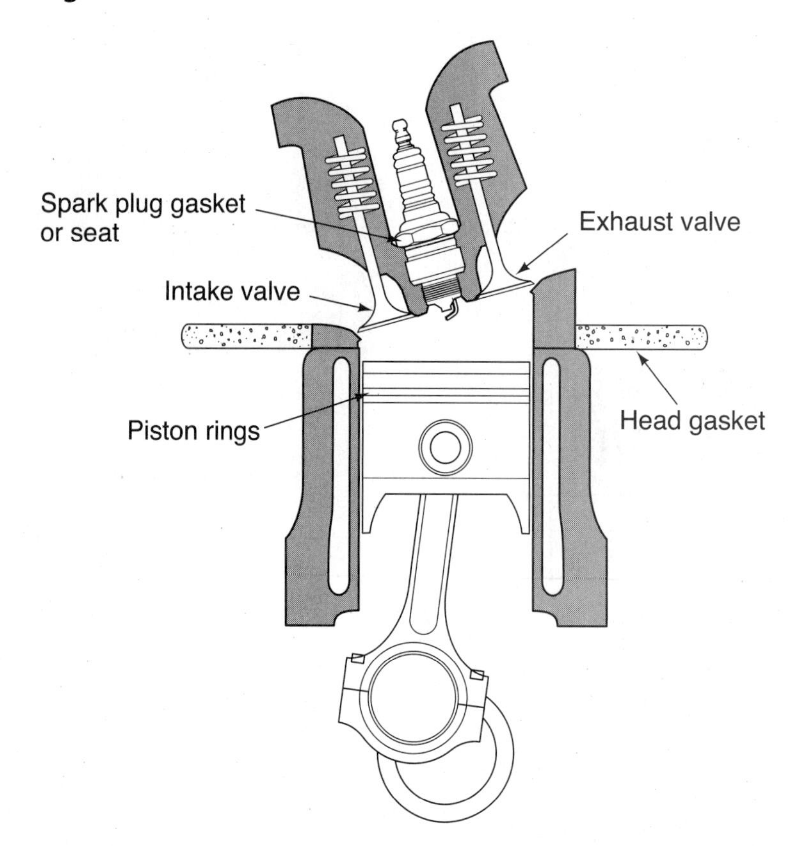

Figure 1. A cranking compression test will detect leaks at the valves, rings, pistons, and head gasket.

5. Disable the ignition system so you do not damage the coils. You can either remove the ignition fuse or remove the low voltage connector to the coil(s). Do not simply remove the plug or coil wires and let them dangle. The coil can be damaged as it puts out maximum voltage trying to find a ground path for the wires.
6. Install a remote starter, or have an assistant help you.
7. Carefully thread the compression tester hose into the spark plug hole, and attach the pressure gauge to the adapter hose **(Figure 2)**.
8. Crank the engine over through five full compression cycles. Each cycle will produce a puffing sound as engine compression escapes through the spark plug ports.
9. Record the final pressure.
10. Repeat the test on each of the other cylinders.

Analyzing the Results

With all the pressure readings in front of you compare them to the manufacturer's specifications. Typical compression pressures on a gas engine range from 125 to 200 psi. The readings should be close to the specification and within 20 percent of each other. Some manufacturers specify only that the compression be over 100 psi and that each reading be within 20 percent of the others. To calculate 20 percent easily, take the normal reading (let us say 160 psi)

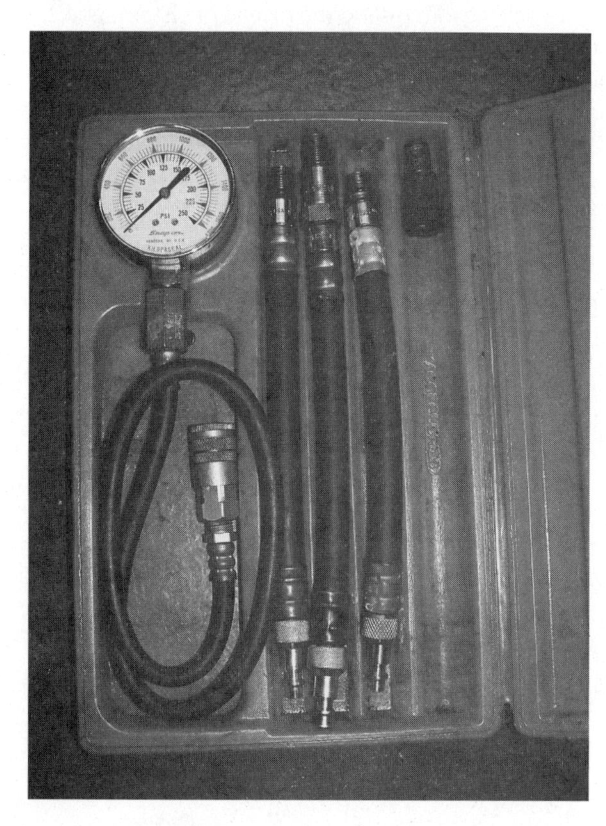

Figure 2. This style compression tester has adaptors that thread into the spark plug hole.

and drop the last digit (16 psi); then multiply by 2 (32 psi), and that is 20 percent. If the readings are 160, 150, 100, and 155 the third cylinder is more than 32 psi different from the other three. That indicates a significant problem with cylinder #3 that must be identified and corrected.

If one or more cylinders are below specification it is likely caused by worn rings, burned valves or valves sticking open from carbon on the seats, a faulty head gasket, or a worn or broken piston. Perform a wet compression test to help find the probable cause. When all the cylinders are slightly low it usually indicates a high-mileage engine with worn rings. Two adjacent cylinders with low compression readings may indicate a blown head gasket leaking compression between two cylinders. If all the cylinders are near zero or very low suspect improper valve timing. A timing belt, chain, or gears that have "jumped" time by slipping a tooth will allow the valves to be open at the wrong time, causing low compression readings.

You Should Know *A "blown" head gasket that is allowing coolant into the combustion chamber may be doing an adequate job of sealing compression. If an engine passes a compression test it does not prove that the head gasket is sealing properly.*

WET COMPRESSION TEST

To help determine the cause of a weak cylinder perform a wet compression test **(Figure 3)**. Use an oil can to put two squirts of oil into the weak cylinder. Reinstall the compression gauge, crank the engine over five times, and record the reading. If the low-reading cylinder increases to almost normal compression pressure it is likely that worn rings are causing the cylinder leakage. The extra oil in the cylinder helps the rings to seal better during the test. If a valve is burned or being held open by carbon, or if there is a hole in the piston, a film of oil will not dramatically improve the compression pressure. Note that a wet compression test may not be as effective on horizontally opposed engines because the oil will seal only the lower half of the cylinder wall.

Look at the following examples:

Compression test results:

Cyl. #1	Cyl. #2	Cyl. #3	Cyl. #4	Cyl. #5	Cyl. #6
175 psi	165 psi	170 psi	80 psi	170 psi	170 psi

Wet compression test:

Cyl. #4
160 psi

This is a clear indication that the rings are severely worn on cylinder #4. If the wet compression test had shown a rise to only 105 psi you would suspect a burned or leaking valve. To definitively determine the cause of low cranking compression perform a cylinder leakage test as described in the next chapter **(Figure 4)**.

RUNNING COMPRESSION TEST

It is possible for the cranking compression to show good results even if the engine is not mechanically sound.

Figure 3. Hold this compression gauge firmly against the spark plug hole to measure compression.

Figure 4. This cylinder showed low compression of only 100 psi while cranking dry and wet. A leakdown test will verify a leaking valve.

A cranking compression test does not do a good job of testing the valvetrain action. There is a lot of time for the cylinder to fill with air when the engine is spinning at cranking speeds of only about 150–250 rpm. A more accurate way to test the loaded operation of the valvetrain is to perform a running compression test. Running compression readings will be quite low compared to the cranking compression pressures because there is so much less time to fill the cylinders with air.

If you have identified a weak cylinder through power balance testing check that cylinder and a few others to gain comparative information. In most cases it is worth the time to check all the cylinders. To perform the running compression test:

1. Remove the spark plug from only the test cylinder.
2. Disable the spark to that cylinder by grounding the plug wire or removing the primary wiring connector to the individual coil.
3. Disable the fuel to the test cylinder by disconnecting the fuel injector connector.
4. Install the compression gauge into the spark plug port.
5. Start the engine and let it idle.
6. Release the pressure from the gauge; that reading reflects cranking compression. Record the new pressure that develops at idle.
7. Rev the engine to 2500 rpm. Release the idle pressure, and record the new pressure that develops at 2500 rpm.
8. Remove the gauge, and reinstall the spark plug. Remember to reattach the plug wire or coil and the fuel injector connector.
9. Repeat for several or all cylinders.

Analyzing the Results

Generally, the running compression will be between 60 and 90 psi at idle and 30 to 60 psi at 2500 rpm **(Figure 5)**. The most important indication of problems, however, is how the readings compare to each other.

Look at the following readings:

	Cyl. #1	Cyl. #2	Cyl. #3	Cyl. #4	Cyl. #5
Idle	70 psi	75 psi	45 psi	70 psi	75 psi
2500	40 psi	40 psi	20 psi	45 psi	40 psi

Cylinder #3 is barely below the general specifications but it is clearly weaker than the other cylinders. These results warrant investigation of the problem in cylinder #3. The most likely causes of low running compression readings are:

- Worn camshaft lobes
- Faulty lifters

Figure 5. The same cylinder showed low running compression at idle.

- Excessive carbon buildup on the back of intake valves (restricting airflow) **(Figure 6)**
- Broken valve springs
- Worn valve guides
- Bent pushrods

Use a stethoscope, vacuum testing, and visual inspection under the valve covers to help locate the reason for the low reading.

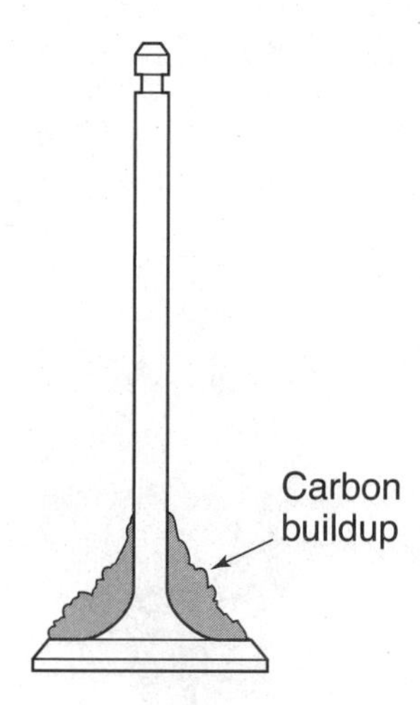

Figure 6. Excessive carbon buildup on the back of intake valves caused by a bad valve seal or guide leakage prevents the engine from pulling in enough air to make adequate power. The added weight can also cause the valve to float at higher rpms.

Summary

- Compression testing can help you detect mechanical faults in the engine.
- Use a compression gauge in the spark plug port to check compression.
- The cranking compression test is used regularly to assess the condition of the rings, pistons, valves, and head gasket(s).
- Cranking compression pressures should meet manufacturers' specifications and be within 20 percent of each other.
- Burned valves or pistons, worn rings, a blown head gasket, or improper valve timing can all cause low compression pressures.
- Use a wet compression test to help determine whether low readings are caused by worn rings or by a burned valve or piston. If the low reading rises substantially with oil in the cylinder, suspect worn rings.
- A running compression test helps you evaluate the condition of the valvetrain.
- Running compression pressures are usually between 60 and 90 psi at idle and 30 and 60 psi at 2500 rpm. All readings should be nearly equal.
- Worn camshaft lobes, faulty lifters, carbon buildup, broken valve springs, worn valve guides, and bent pushrods are typical problems causing low running compression pressure.

Review Questions

1. Outline the procedure to perform a cranking compression test.
2. Running compression pressures are generally between __________ and __________ psi at idle and between __________ and __________ psi at 2500 rpm.
3. List three likely causes of low running compression.
4. Technician A says to check compression before disassembling an engine. Technician B recommends performing a compression test before completing a major maintenance service if the engine is running very badly. Who is correct?
   A. Technician A only
   B. Technician B only
   C. Both Technician A and Technician B
   D. Neither Technician A nor Technician B
5. Technician A says that cranking compression pressures are usually between 200 and 250 psi. Technician B says running compression pressures are lower than cranking compression pressures. Who is correct?
   A. Technician A only
   B. Technician B only
   C. Both Technician A and Technician B
   D. Neither Technician A nor Technician B
6. A compression test is performed on an engine that is misfiring. All the pressures are between 175 and 185 psi, except cylinder #5: it is 80 psi. Technician A says a valve seal could be bad. Technician B says the rings could be worn or broken. Who is correct?
   A. Technician A only
   B. Technician B only
   C. Both Technician A and Technician B
   D. Neither Technician A nor Technician B
7. An engine produced the following compression pressures: Cyl. #1: 150 psi, Cyl. #2: 75 psi, Cyl. #3: 80 psi, Cyl. #4: 145 psi. Which is the most likely cause?
   A. A hole in the pistons
   B. A worn camshaft
   C. Bad rings
   D. A blown head gasket
8. An engine produces the following compression pressures: Cyl. #1: 95 psi, Cyl. #2: 140 psi, Cyl. #3: 145 psi, Cyl. #4: 135 psi. When a wet test is performed on cylinder #1 the pressure rises to 160 psi. The most likely cause of the problem with cylinder #1 is:
   A. A blown head gasket
   B. Worn rings
   C. A bad oil pump
   D. A burned valve
9. Technician A says to put 1/2 quart of oil in a cylinder to perform a wet compression test. Technician B says to run the vehicle at idle to perform a wet compression test. Who is correct?
   A. Technician A only
   B. Technician B only
   C. Both Technician A and Technician B
   D. Neither Technician A nor Technician B
10. An engine has low running compression on two out of four cylinders; the cranking compression is normal. Technician A says the camshaft lobes may be worn. Technician B says the head gasket may be blown. Who is correct?
    A. Technician A only
    B. Technician B only
    C. Both Technician A and Technician B
    D. Neither Technician A nor Technician B

Chapter 17 Cylinder Leakage Testing

Introduction

Once you have identified one or more cylinders with low cranking compression you can use a cylinder leakage test to pinpoint the cause of the problem. To perform a cylinder leakage test you put pressurized air in the weak cylinder through the spark plug hole when the valves are closed. You can then listen for leakage of air from the combustion chamber at various engine ports to determine the source of the leak. This is an excellent way to gather more information before disassembling the engine or when deciding whether a replacement engine should be installed. A cylinder leakage test is also a very thorough way of evaluating an engine when someone is considering purchasing a vehicle. You can even perform a cylinder leakage test on a junkyard engine to assess its value before purchasing it.

> **Interesting Fact**
>
> *A student, anxious to buy a good-looking, performance-modified Honda, was convinced to perform a cylinder leakage test after noticing a bit of blue smoke coming from the tailpipe. A leakdown test showed very serious ring wear on two cylinders. While he still bought the car he paid $1500 less for it than he had planned.*

CYLINDER LEAKAGE TEST

The cylinder leakage test can help you evaluate the severity of an engine mechanical problem and pinpoint the cause. You can test one low cylinder to find a problem, or you can test each cylinder to gauge the condition of the engine. Like a compression test, the cylinder leakage test detects leaks in the combustion chamber from the valves, the rings or pistons, and the head gasket. Using this test, however, you can clearly identify which is the cause of the leak **(Figure 1)**.

Figure 1. A cylinder leakage tester applies regulated air to the cylinder and measures the percentage of leakage.

To perform a cylinder leakage test:

1. Let the engine warm up to normal operating temperature.
2. Let the engine cool for ten minutes, and then remove the spark plug from the cylinder to be tested.
3. Remove the radiator cap and the PCV valve. This prevents damage to the radiator or engine seals if excessive pressure leaks into the radiator or the engine's crankcase.
4. Turn the engine over by hand until the cylinder is at TDC on the compression stroke. This ensures that the valves will be closed and the rings will be at their highest point of travel. The cylinders wear the most at the top because of the heat and pressure of combustion, so if the rings are leaking they will show the greatest evidence at the top. There are a few different methods of getting a cylinder to TDC compression, depending on the engine.
   A. Have an assistant turn the engine over slowly while watching the piston come up to top dead center. You should be able to feel and hear air blowing out of the spark plug hole if it is coming up on the compression stroke. If the test shows 100 percent leakage at the beginning you are on TDC exhaust. Rotate the engine 360° and retest.
   B. Remove the valve cover, and watch the rockers or camshaft as you turn the engine over. On the intake stroke the intake valve will open and then close. Then both the valves will remain closed while the piston comes up to TDC. If the engine has rocker arms move them up and down to be sure they both have lash (clearance) between the arm and the pushrod. If the engine has one or more overhead camshafts the followers will be on the base circle, not the lobes of the camshaft.
   C. If the vehicle has a distributor, remove the cap, and rotate the engine until the rotor points at the test cylinder's firing point.
5. With the cylinder at TDC compression thread the cylinder leakage tester adapter hose into the spark plug hole **(Figure 2)**.
6. Apply air pressure to the leakage tester, and calibrate it according to the equipment instructions. On the tester shown you turn the adjusting knob until the gauge reads 0 percent leakage when it is not attached to the cylinder.
7. Connect the leakage tester to the adapter hose in the cylinder while watching the crankshaft pulley. If the engine rotates at all reset it to TDC. When the engine is right at TDC it will not rotate when air pressure is applied.
8. Read the percentage of leakage on the tester and gauge as follows:
   Up to 10 percent leakage is excellent; the engine is in fine condition.

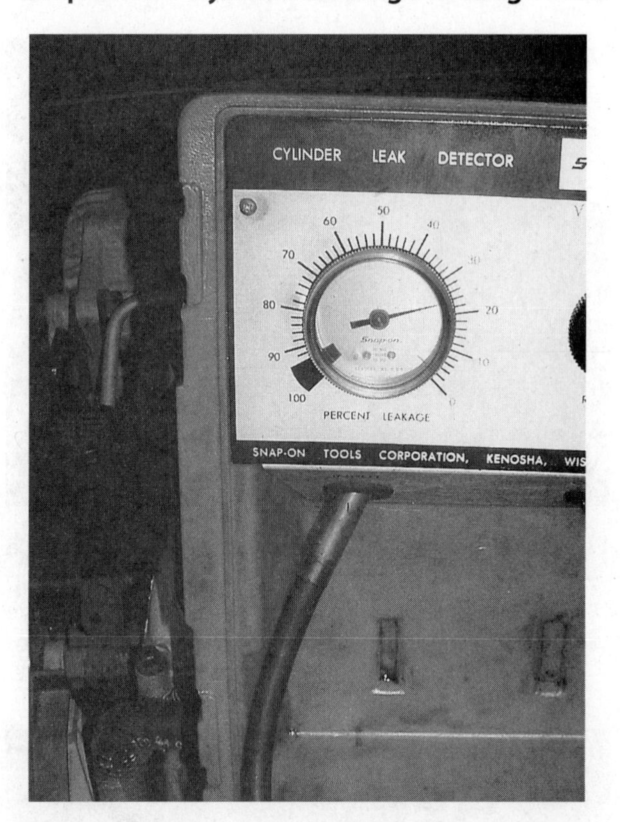

Figure 2. This cylinder shows 25% leakage, which is just within the acceptable range.

   Up to 20 percent leakage is acceptable; the engine is showing some wear but should still provide reliable service.
   Up to 30 percent leakage is borderline; the engine has distinct wear but may perform reasonably well.
   Over 30 percent leakage indicates a significant concern that warrants repair.
9. With the air pressure still applied use a stethoscope to listen at the following ports:
   At the tailpipe: Leakage here indicates a burned or leaking exhaust valve **(Figure 3)**.
   At the throttle or a vacuum port on the intake manifold: Leakage here indicates a burned or leaking intake valve **(Figure 4)**.
   At the radiator or reservoir cap: Leakage here proves that the head gasket is leaking.
   At the oil fill cap or dipstick tube: Leakage here indicates worn rings **(Figure 5)**.
   If the leakage at the oil fill is near 100 percent suspect damage to a piston **(Figure 6)**.
10. Release the pressure, remove the adapter, and rotate the engine over to the next test cylinder. Repeat as indicated.

While a cylinder leakage test can identify the primary cause of low compression readings remember to keep your mind and eyes open to other problems when repairing the engine. Do not automatically inform a customer

Figure 3. Listen for air at the tailpipe indicating a burned exhaust valve. Be sure the shop exhaust equipment is not attached to the exhaust; it will produce the same sound of air rushing past the stethoscope.

Figure 4. Pull any line off the intake manifold to detect leakage from an intake valve.

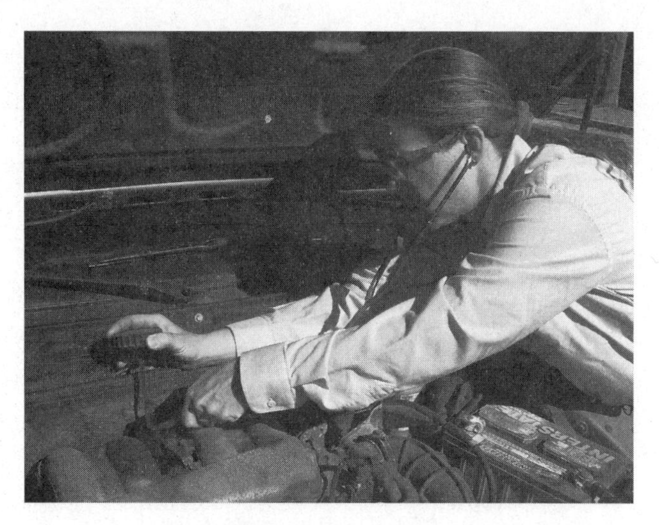

Figure 5. Worn rings will produce airflow past the rings into the crankcase. Check for air leaking from the oil fill or dipstick.

Figure 6. This destroyed piston showed 100% leakage.

> **You Should Know** *There will always be some leakage past the rings; a light noise is to be expected. Excessive ring wear will cause distinct blowing out of the cap or dipstick hole. You will be able to hear the noise and feel the pressure.*

that an exhaust valve is leaking and a valve job will cure the problem. Consider the engine mileage; oil condition, pressure, and usage; and degree of leakage past the rings. If the engine has 150,000 miles on it, for example, it is very likely that the rings are significantly worn. By replacing the faulty valve(s) and restoring the others to like-new condition you will increase the compression and combustion pressures. The change may be significant enough that the old rings will not seal as well as they did before the valve job. In this case the customer would be back complaining about blue smoke from oil consumption or about the engine still lacking adequate power. This may very well be an example of when a replacement engine would be the most cost-effective repair for the customer. Evaluate the whole situation, and make your recommendation to the customer. If you offer him repair options be very clear about explaining the possible consequences.

Summary

- A cylinder leakage test can pinpoint the cause of improper combustion chamber sealing.
- A cylinder leakage test can detect faults with the valves, rings, pistons, and head gasket.
- Make sure the engine is warmed up when performing a leakage test to ensure that the rings can seal properly.
- The engine must be on TDC of the compression stroke so the valves are closed on the test cylinder.
- Over 30 percent leakage indicates a significant problem that should be repaired.
- Air heard at the tailpipe indicates a burned exhaust valve.
- Air escaping from the intake manifold indicates a burned intake valve.
- Air heard coming from the oil fill cap or dipstick tube points to worn rings.
- A blown head gasket will allow air pressure to leak into the cooling system.
- Evaluate the results of all your engine tests to form your recommendations to the customer.

Review Questions

1. Air heard escaping from the tailpipe during a cylinder leakage test indicates ___________.
2. Air heard coming from the oil fill during a leakage test indicates ___________.
3. What leakage test results prove that a head gasket is leaking?
4. Cylinder leakage over ___________ percent will usually cause noticeable drivability concerns.
5. Technician A says that a cylinder leakage test can find a worn ring. Technician B says a cylinder leakage test can detect a worn main bearing. Who is correct?
   A. Technician A only
   B. Technician B only
   C. Both Technician A and Technician B
   D. Neither Technician A nor Technician B
6. Technician A says the test cylinder must be on TDC exhaust to test the exhaust valve(s). Technician B says the test cylinder must be on BDC intake to test the intake valves. Who is correct?
   A. Technician A only
   B. Technician B only
   C. Both Technician A and Technician B
   D. Neither Technician A nor Technician B
7. Technician A says that the engine should be fully warmed up to perform a leakage test. Technician B says that a cylinder wears the most at the top. Who is correct?
   A. Technician A only
   B. Technician B only
   C. Both Technician A and Technician B
   D. Neither Technician A nor Technician B
8. Technician A says that you can find TDC compression by watching the ignition rotor. Technician B says that when a cylinder is at TDC compression the camshaft lobes open both the valves. Who is correct?
   A. Technician A only
   B. Technician B only
   C. Both Technician A and Technician B
   D. Neither Technician A nor Technician B
9. An engine has 45 percent leakage on cylinder #3, and bubbles are seen in the radiator. Technician A says the head gasket should be replaced. Technician B says that a faulty water pump could be the cause. Who is correct?
   A. Technician A only
   B. Technician B only
   C. Both Technician A and Technician B
   D. Neither Technician A nor Technician B
10. An engine with 135,000 miles is running poorly. The cause of low compression on the third cylinder is being diagnosed using a cylinder leakage test. The results show 55 percent leakage. Most of the leakage is past an intake valve, but there is significant air coming out the oil fill port. Technician A says that he would perform a valve job. Technician B says that he would recommend an engine replacement. Who is correct?
    A. Technician A only
    B. Technician B only
    C. Both Technician A and Technician B
    D. Neither Technician A nor Technician B

Chapter 18 Engine Smoke and Noise Diagnosis

Introduction

Smoke coming from the tailpipe is never a good sign. Some causes may be relatively simple and inexpensive to rectify, while others may require a complete engine overhaul or replacement. You will need to correctly diagnose the cause of smoke to avoid costly and unnecessary repairs. Look carefully for smoke before disassembling an engine for overhaul; during your repairs you will need to be sure to correct the problem causing the smoke. The color, odor, and sound of the exhaust should all be evaluated.

As engine components wear, clearances become greater, parts break, or noises develop in the engine. These can be warning signs that a component is wearing, or they can predict imminent engine failure. Training, experience, and careful listening can help you narrow down the possible causes of unusual noises and recommend effective repairs.

> **Interesting Fact**
>
> *A customer came in distraught convinced that her engine was "blown" because it started blowing blue smoke out of the tailpipe. Careful checking of the specific times when the smoking occurred pointed to worn valve seals. Less than $200 later the customer was relieved to pick up her vehicle, which was running fine* **(Figure 1)**.

SMOKE DIAGNOSIS

A properly running, mechanically sound engine should emit very little if any smoke from the tailpipe. When you are evaluating the condition of an engine, check carefully for excessive smoke at startup, when idling, under load, and after a period of idling or deceleration. The color and odor of the smoke and the times when it occurs can lead you toward the correct diagnosis.

BLUE SMOKE

Blue smoke coming from the tailpipe occurs when the engine is burning oil in the combustion chamber. This is called oil consumption; the engine is burning oil, not leaking it externally. The customer may notice the smoke as well as the need to add oil in between oil changes. There are several possible causes of blue smoke:

- A plugged PCV system
- Worn valve seals or guides

Figure 1. The brittle intake valve seal on the left crumbled while it was being removed.

- A blown turbocharger
- A leaking head gasket
- Worn rings or cylinders

Obviously, the difference in parts and labor to repair a PCV system as opposed to replacing faulty rings is dramatic. Be very careful to check each of these possible causes before disassembling or replacing an engine.

A clogged or frozen PCV valve or hose can cause a tremendous amount of blue smoke to pour out of the tailpipe at all times when the engine is running. The more you accelerate, the greater the smoking. Shake the PCV valve; it should rattle. Replace it if there is any possibility that it could be faulty. Blow through the larger-diameter hose attached to the PCV valve that connects to the air filter housing. If it is clogged clean it thoroughly, and road test the vehicle again to confirm your repair **(Figure 2)**.

Worn valve seals will cause smoking at predictable times. At startup after the engine has been sitting for a period you will notice a cloud of blue smoke. This occurs because when the engine is shut off the oil sitting under the valve cover can leak past the worn seals into the combustion chamber. When the engine starts it burns this oil, which produces blue smoke. It will clear up significantly after running for a few minutes. Once the engine is running the oil is constantly circulating through the engine and not pooling up above the seals, so the smoke will dissipate.

Figure 2. This PCV line can easily get plugged or frozen with sludge from the blowby gasses.

Figure 3. With air pressure applied to the cylinder you can remove the spring to access the valve seal for replacement.

Also, when the engine is at idle or decelerating for a period of time the engine vacuum is high. This can draw oil into the combustion chamber past the weak seals. When you accelerate after idling or decelerating you may notice a puff or cloud of blue smoke. Valve seals can be replaced relatively inexpensively; the cylinder head does not need to be removed **(Figure 3)**.

A turbocharger with blown seals can produce huge clouds of smoke on acceleration. This could easily be confused with worn rings, but an engine overhaul would not cure the problem. Whenever a turbocharged engine is smoking check the turbo seals first. Remove the boot on the intake side of the compressor, and look for oil. If plenty of oil is sitting in the turbo housing and the boot, replace the faulty turbocharger. Shake the impeller up and down to help confirm this diagnosis. Usually it is wear in the bearings that allows the shaft to rock and destroy the turbo seals.

While this is not a common problem it is possible for a head gasket to split by an oil passage and allow oil into the combustion chamber. Check the spark plugs; if one is drenched in oil, literally dripping, suspect a blown head gasket. Perform a compression test. Normal or just slightly low readings point to a head gasket. If the rings are badly worn you will usually get low compression results. Look very closely at the head gasket once the head is off to be sure that the head gasket is the cause of the blue smoke.

You should be able to see a crack or path in the gasket from an oil passage into the combustion chamber.

Worn rings are a common cause of oil consumption. As the oil rings wear they cannot scrape all the oil off the cylinder walls. This leaves too much in the combustion chamber, and it is burned during combustion. The engine may smoke all the time, but it will usually emit more blue smoke while accelerating. Perform a cylinder leakage test to confirm your suspicions. When the rings are badly worn you will need to rebuild or replace the engine. Fully evaluate the condition of the engine, including valve seating, oil pressure, noises, cylinder leakage, and oil consumption. This will help you determine whether it would be more cost effective to rebuild an engine or replace it with a rebuilt unit. If the engine has valve problems, knocking noises, and excessive cylinder leakage, for example, it will likely be more efficient and inexpensive to replace the engine. In other cases you may conclude that a set of rings and engine bearings will cure the problems, and repairing the engine will be the best option.

WHITE SMOKE

Excessive white smoke (steam) coming from the exhaust is an indication that coolant is getting into the combustion chamber. Some light smoking, particularly at engine startup, is normal as the engine burns off the condensation in the exhaust system. If there is coolant in the exhaust the smoke will have a distinctive sweet smell to it, similar to when an engine overheats and the coolant boils over. The leak is most commonly past the head gasket, but a crack in the cylinder head or block will also cause coolant consumption. A distinct puff or small cloud of white smoke at startup coupled with rough running may be the result of a small crack in the head gasket or the cylinder head. Pressure test the cooling system overnight as described in Chapter 11 to verify a problem. When clouds of white smoke are pouring out of the tailpipe it is usually that the head gasket is blown; otherwise there is a significant crack in the block or head.

On an older vehicle with an automatic transmission it is possible that a leaky vacuum modulator could cause these symptoms. On transmissions with a vacuum modulator it is possible for the diaphragm inside it to rip. This allows the vacuum line attached from the intake manifold to the modulator to pull in high quantities of automatic transmission fluid (ATF). The ATF introduced into the intake manifold is then burned in the cylinders and can also cause excessive white smoke from the tailpipe. A simple test to eliminate the modulator as the problem is to remove the vacuum line from the modulator and check to be sure there is no ATF present in the line. If the modulator is faulty the smoking will stop when the vacuum line is disconnected.

Remove the cylinder head, and closely inspect the head gasket, head, and block for signs of failure and cracks. The piston top of the cylinder that coolant was leaking into will be perfectly clean. The other pistons will have varnish and carbon on them. If no problems can be detected visually, or if the engine has overheated, send the head to a machine shop to check it for cracks.

When coolant has gotten into the combustion chamber you must also check the condition of the oil. When coolant mixes with the oil it produces a liquid very similar to a coffee milkshake. The very thick mixture of oil and coolant will block oil passages and quickly destroy the engine bearings. If the oil has been contaminated with coolant warn the customer that major bottom end damage may have occurred. If the engine has been run for any length of time with this "lubrication" you will likely hear knocking from the bottom end of the engine. When the engine has high mileage it may be advisable to check the price and availability of a rebuilt engine to repair the extensive engine damage this can cause.

If no knocking is heard and you replace the head gasket, flush the lubrication system thoroughly as part of your repair. Run the engine up to normal operating temperature, and change the oil. Road test the vehicle, and check for oil contamination and unusual noise before returning the vehicle to the customer.

Check that the malfunction indicator light is not on and that there are no diagnostic trouble codes indicating problems with the oxygen sensors. Oxygen sensors fail rapidly when the engine burns coolant. Many technicians will automatically replace the oxygen sensors after a repair is made for coolant consumption. This action can prevent a customer from returning shortly after one major repair only to incur another significant expense.

BLACK SMOKE

Black smoke is caused by a rich air-fuel mixture or incomplete combustion. Eliminate engine breathing problems first. Check the air filter and the exhaust system for restrictions. You may also need to check the fuel delivery system for faults. Possible problems may be a faulty mass airflow sensor **(Figure 4)** or **fuel pressure regulator**, a leaking injector or injectors, an inaccurate engine coolant temperature sensor or oxygen sensor, or faulty engine control wiring. You can use a scan tool to help verify proper operation of the sensors. Follow the manufacturer's diagnostic troubleshooting chart to locate the cause of the problem. One or more cylinders with low compression can cause the air-fuel mixture to be less combustible. This can result in imperfect combustion and the emission of black smoke from the tailpipe. Check compression to verify that the engine is mechanically sound. If an engine has been running rich for an extended period of time the rings may have been damaged. The catalytic converter also suffers from rich running and could become partially restricted or plugged. A faulty converter could also trigger the MIL and set a DTC for low catalyst efficiency on OBD II vehicles.

Figure 4. The MAF sensor is in the air intake duct work between the air filter and the throttle housing. The MAF is the primary sensor for fuel delivery once the engine is running.

Figure 5. These rod journals were destroyed when a bearing spun. The knocking sounded as serious as the fault.

EXHAUST NOISES

Exhaust should be emitted in steady pulses from the tailpipe. When a cylinder is misfiring you will be able to feel the irregular pulses in the exhaust stream. Remove a spark plug wire from a plug, and ground it. Start the engine, and feel the exhaust so you can experience what this feels like. This can be a useful part of your engine drivability diagnosis. If you know that one cylinder is not firing you can use a power balance test to locate the faulty cylinder.

Backfiring from the exhaust usually indicates improper camshaft or ignition timing. If an engine has a timing belt that has jumped a tooth the engine will run poorly and often backfire on acceleration. The engine also uses a camshaft position (CMP) sensor to determine which cylinder should fire next. Occasionally this sensor can fail and provide improper information to the PCM. The PCM may then fire the spark plug at the incorrect time, causing the engine to run poorly and backfire through the exhaust. When an engine has spark plug wires they can be misrouted, and this will cause backfiring as well. Check for proper spark plug wire routing first, and then inspect the camshaft timing.

ENGINE NOISE DIAGNOSIS

Irregular engine noises may be caused by major or minor engine problems. Your job is to diagnose the noise as accurately as possible so you can repair the engine in the most effective and efficient way. Low oil pressure can cause or increase the level of engine noises, so check that the engine has adequate oil and pressure before drawing conclusions. Some bottom end noises clearly indicate an engine that requires a thorough overhaul or replacement. Other noises may be resolved by much less drastic measures. While it is difficult to describe noises on paper a few guidelines should help you narrow down the possible problem(s). A deep knock within the engine will require engine disassembly or replacement. If you disassemble the engine be sure to look for all potential problems. Noise diagnosis should help you identify problems but not limit your investigation once the engine is apart.

BOTTOM END KNOCK

A loud knocking noise from the bottom end of the engine portends imminent failure of the engine. This is a deep knocking sound caused by excessive clearance between the main or rod bearings and the crankshaft **(Figure 5)**. Place a stethoscope on the block just above the oil pan, and listen while the engine idles and when you open the throttle. A loud rapping noise indicates excessive bearing wear. Check the oil pressure at idle; low readings confirm your diagnosis. Excess bearing clearances allow oil to leak through the bearings, lowering pressures and allowing the crankshaft journals to crash against the bearing face without a soft cushion of oil. Engine knocking that is repaired promptly may require only engine bearing and ring replacement. If the engine is allowed to run for extended periods of time with worn bearings the crankshaft and connecting rods will likely be damaged. This will often warrant a replacement engine rather than an overhaul. Be certain that the **flexplate** or **flywheel** is not cracked or loose and that the catalytic converter is not rattling; these problems can cause knocking that sounds very much like a bearing knock. Use your stethoscope on the bell housing and on the catalytic converter to clearly identify the source of the noise.

CRANKSHAFT ENDPLAY

When an engine knocks or clunks loudly on acceleration suspect excessive **crankshaft endplay**. The repair

Figure 6. This thrust bearing was worn down to the copper, causing a clunking on acceleration. It did its job of protecting the crank and block; a new set of bearings cured the problem.

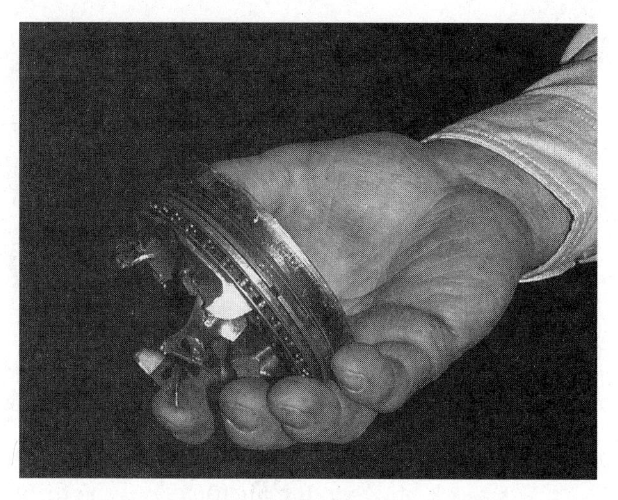

Figure 7. The piston let loose due to excessive pin wear. It caused extensive cylinder and head damage.

requires replacement of the engine main bearings, but while you are disassembling the engine you must pay particular attention to the crankshaft and block. If the crankshaft has been slamming against the block without the benefit of a cushion from the **thrust bearing** significant damage may have occurred. The thrust bearing is designed to minimize forward and backward movement of the crankshaft to protect the crankshaft and block from wear; look for scoring along the thrust surfaces of the block and crankshaft **(Figure 6)**. Minor damage can be corrected by a different thrust bearing, but serious damage requires component replacement. If left unrepaired excessive crankshaft endplay can result in the crankshaft breaking through the block.

Many other drivetrain problems can cause clunking on acceleration. Make sure that the noise is coming from the engine rather than the differential or a U-joint, for example. You can often feel excessive crankshaft endplay with the engine installed. Put the engine on a lift and pry, and push th crankshaft pulley forward and backward in the engine block. Movement and noise should be minimal. If the crankshaft moves significantly—that is, more than a few thousandths of an inch—and you can hear a knocking noise the enginewillrequire repairs.

PISTON PIN KNOCK

When a piston pin develops clearance within the piston pin bore you will hear a higher-pitch double-time knock at the top of the block. The noise is deeper than **valvetrain clatter** but not as deep as bottom end knocking. Place a stethoscope near the top of the block, and listen for two knocks close together. The pin will knock as the rod pushes the piston up and then again as combustion forces the piston down. It is important to repair this problem immediately. Excessive pin clearance will often break the piston off the rod **(Figure 7)**. This can cause extensive damage to the block and cylinder head.

PISTON SLAP

Piston slap occurs when there is too much clearance between the piston and the cylinder. It usually begins as a knocking noise when the engine is cold that goes away as the engine fully warms up. Prompt repair can prevent piston breakage and more serious engine damage. Once the piston is slapping continuously even when the engine is warm, the engine is at risk of a catastrophic failure. If you rebuild the engine carefully measure the cylinder bore diameter, out of round, and taper. Often the repair for piston slap includes boring the cylinders out oversize and installing new pistons. In many cases these extensive repairs will cost more than a crate engine, so a rebuilt engine assembly will be installed. Some manufacturers have been having problems with excessive piston slap on very low mileage engines. These engines are being replaced under warranty.

You Should Know

Generally, the deeper the engine noise or knocking is, the more extensive and expensive the repairs will be. Light ticking from the valvetrain may be relatively simple and inexpensive to repair. A bottom end knock usually requires serious engine repairs or replacement.

TIMING CHAIN NOISE

A loose timing chain or worn gears will make a clacking or rattling noise, particularly when you allow the engine to decelerate after giving it some throttle. A chain will often slap on the timing chain cover when it wears. You will learn to identify this noise, but you can also verify it with a stethoscope on the timing cover. If the engine has gears the noise will be less of a rattle and more of a knock. It will still be worst on deceleration.

VALVETRAIN CLATTER

When there is excessive clearance between components in the valvetrain you will hear a tinny clattering sound from the top of the engine. A common cause of valvetrain clatter is a collapsed hydraulic lifter. You may hear this in the morning, after a car has been sitting for an extended period, or after an oil change. The noise should go away within a few seconds. If the noise persists the lifters are probably worn. Be sure that the engine is properly filled with clean oil before dismantling the engine to replace the lifters. Metallic clicking noises can also be caused by valves that are too loose, a worn camshaft, worn lifters, worn or broken rocker arms, or worn rocker arm shafts. Remove the valve cover(s) to inspect the valvetrain components, and look closely for signs of wear **(Figure 8)**. Again, a stethoscope can help point you to the right area of the cylinder head or even to the affected cylinder.

Figure 8. This badly worn lifter was causing serious valve train clatter. The other lifters were not much better.

Summary

- Smoke from the tailpipe indicates an engine drivability or mechanical problem.
- Blue smoke is created when oil is burned in the combustion chamber.
- Oil consumption can be caused by a plugged PCV system, worn valve seals, worn turbocharger seals, a leaking head gasket, or worn rings.
- An engine will emit white, sweet-smelling smoke (steam) when it is burning coolant.
- Coolant consumption is most commonly caused by a blown head gasket, though a cracked cylinder head or block can also allow coolant into the combustion chamber.
- Black smoke is caused by too rich an air-fuel mixture or incomplete combustion.
- Low compression is a possible cause of light black smoke, but more common causes are airflow restrictions and fuel injection and engine sensor failures.
- Listening closely to engine noises can help you diagnose engine problems correctly.
- Deep bottom end knocking is caused by excessively worn engine bearings; this will usually be accompanied by low oil pressure.
- Excessive crankshaft endplay will typically cause knocking on acceleration. If ignored the crankshaft can punch a hole through the end of the block.
- Piston pin knock occurs when there is clearance between the piston and the pin; it creates a double-time knock from the top of the block.
- Piston slap is worst in the morning and may go away when the engine is warmed up. It is a warning sign that the engine is significantly worn.
- A worn timing chain will cause a rattling or clacking noise as it slaps the timing cover on deceleration.
- Valvetrain clatter is heard when there is looseness in the valvetrain. Possible causes include collapsed hydraulic lifters, valves out of adjustment, and worn rockers or a worn camshaft.

Review Questions

1. List three causes of excessive blue smoke being emitted from the tailpipe.
2. What color smoke will a blown head gasket cause?
3. What are three possible causes of excessive black smoke being emitted?
4. A deep bottom end knock may be caused by __________.
5. A customer started up her car one extremely cold morning, and within a few minutes a cloud of blue smoke was coming from the tailpipe. She stated that the engine had been running perfectly with no smoking at all the day before. Which is the most likely cause?
   A. A blown head gasket
   B. Worn rings
   C. Bad main bearings
   D. A plugged PCV hose
6. Technician A says to check the rings if oil is leaking from the engine. Technician B says that a blown head gasket will usually cause blue smoke. Who is correct?
   A. Technician A only
   B. Technician B only
   C. Both Technician A and Technician B
   D. Neither Technician A nor Technician B
7. An engine has a plugged air filter. This may cause:
   A. Black smoke
   B. White smoke
   C. Oil leaks
   D. Blue smoke
8. A loud clattering is heard from the top of the engine. Technician A says it may be caused by collapsed lifters. Technician B says the rocker arms could be worn. Who is correct?
   A. Technician A only
   B. Technician B only
   C. Both Technician A and Technician B
   D. Neither Technician A nor Technician B
9. Technician A says that a loud clunk on acceleration may be caused by a loose timing chain. Technician B says that a metallic clacking noise on deceleration is usually caused by worn camshaft bearings. Who is correct?
   A. Technician A only
   B. Technician B only
   C. Both Technician A and Technician B
   D. Neither Technician A nor Technician B
10. An engine has a knocking noise when cold that goes away when the engine warms up. The oil pressure is good. Which is the most likely cause?
    A. Worn main bearings
    B. Excessive piston-to-cylinder clearance
    C. A worn timing chain
    D. A cracked flexplate

Section 4

Engine Block Construction, Inspection, and Repair

Some engine analysis may require you to measure as precisely as .0005 in. or five ten-thousandths of an inch. This is roughly one-quarter of the thickness of a strand of hair.

SECTION OBJECTIVES

After you have read, studied, and practiced the contents of this section you should be able to:

- Remove the engine from the vehicle to perform a thorough engine overhaul or engine replacement.
- Understand that labeling connectors, hoses, and brackets during removal will simplify installation.
- Understand the importance of careful inspection during disassembly.
- Remove the ring ridge before removing pistons.
- Mark the rod and main caps before disassembly.
- Realize that precise and thorough measurements and inspections are critical to a quality engine evaluation or overhaul.
- Inspect the block for damage, warpage, cracks, and main bore alignment.
- Measure and analyze the condition of the cylinder bores.
- Examine the main, rod, and cam bearings for irregular wear.
- Check the piston for excessive or irregular wear, the ring lands and grooves for damage, and the pin boss for cracks.
- Measure the piston diameter and calculate piston clearance.
- Inspect the crankshaft for cracks and straightness and its journals for scoring and proper size.
- Determine the repairs needed to correct worn or damaged components of the bottom end.
- Use an organized method of repair and reassembly to ensure thorough, quality work.
- Properly fit engine bearings and measure bearing clearances.
- Torque the crankshaft into place, and measure crankshaft end play.
- Check for correct ring end gap and side clearance; install rings with proper gap spacing.
- Fit the pistons in their proper bores in the correct direction.
- Reassemble the timing mechanism following the manufacturer's alignment requirements.
- Mount the oil pump and oil pan.

Chapter 19 Engine Removal

Introduction

When an engine needs to be overhauled or replaced you will need to remove it from the vehicle. Some technicians will replace rings and bearings with the head and oil pan off but the block still in the vehicle. It is almost always worth the time to remove the engine to ensure that you can complete the job thoroughly. It can be very difficult to replace core plugs or clean oil galleries with the engine still in the car, for example. When a complete overhaul is indicated there is no question; you will need to remove the engine. Engine removal is not particularly complicated, but it does take some time to remove the wiring, hoses, lines, mounts, and hardware in preparation.

I know one technician who can remove an engine in just over an hour even though the labor guide allows 3.2 hours. This took a lot of practice. He does not have to rush; he just knows exactly which tool to grab for every step **(Figure 1)**.

PREPARATIONS FOR ENGINE REMOVAL

The best procedure to follow when removing an engine is the manufacturer's. There may be a special tool required to decouple lines, or there may be a hidden bolt or a shortcut that you could miss if you do not read through the specific instructions for the vehicle. If you work on the same types of vehicles regularly you may become familiar enough with the procedure that you no longer need to read along step by step. Until that time, read through the service information so you do not damage components or perform any unnecessary tasks. The steps provided here are general in nature; they are no substitute for the specific procedure. They can familiarize you with the process so that

Figure 1. An angled open-end wrench can sometimes fit where nothing else will.

Figure 2. Do not let a careless mistake ruin your excellent engine work.

when you remove an engine the procedure will make sense. Transversely mounted engines will often be removed from the bottom of the vehicle after some suspension components are disconnected. Vehicles with a longitudinally mounted engine usually call for the engine to come out through the top. Be sure you know which method will be used; there are special steps for either removal procedure.

Before removing the engine from the vehicle carefully inspect the powertrain for fuel, oil, coolant, and other leaks. You will need to identify the source of these leaks, determine the action needed to repair them, and correct the problems while the engine is out.

Install fender covers so they completely cover the paint on the front and sides of the engine compartment **(Figure 2)**. A big bill and a small scratch will not please the customer! Some technicians like to clean the exterior of the engine and the inside of the engine compartment before beginning the job. If you choose to wash the engine be sure to cover the starter, generator, and ignition wiring to prevent water damage to these electrical components.

If you will remove the engine through the top of the engine compartment, mark the hood brackets with a grease pencil to make installation and adjustment easier. Carefully remove the hood, and store it in a safe spot in the shop or in the vehicle.

REMOVING A REAR-WHEEL DRIVE ENGINE

Most rear-wheel drive (RWD) engines are removed through the top of the engine compartment; check the manufacturer's service information to be sure. You may need to remove the transmission with the engine, though many times you will decouple the two components and leave the transmission in the vehicle. The steps for removing an RWD, longitudinally mounted engine typically include:

1. Remove the battery or its connections, starting with the ground side clamp, to protect against electrical arcing.
2. Disconnect the engine wiring. Sometimes this can be accomplished by disconnecting one large connector that feeds all the engine electrical components **(Figure 3)**. Many times each connector must be disconnected. Carefully label the connectors so you know where to plug them back in. Use masking tape or marking tags. Do not rely on your memory; it can take a lot longer to figure out where wires go than it does to mark them during removal **(Figure 4)**.
3. Disconnect the engine ground strap(s).

Figure 3. Look for a main engine connector to save you the time of disconnecting each smaller connector.

Figure 4. Labeling connectors as you disconnect them will save you time during reassembly.

Figure 5. These fuel line fittings come apart easily with the proper tool.

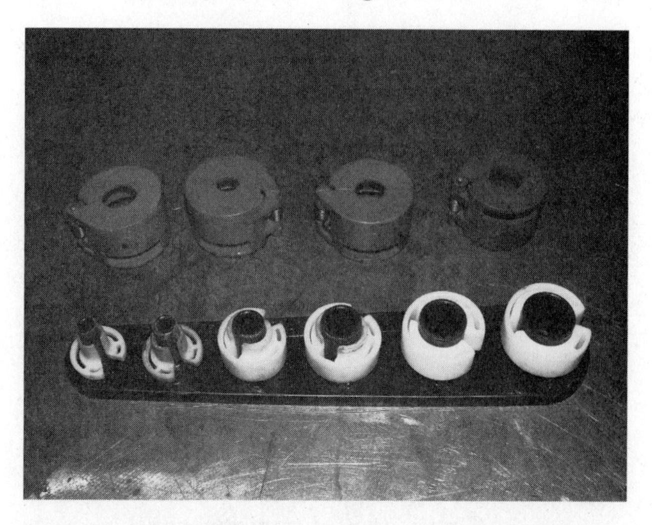

Figure 6. Fuel line disconnect tools quickly unlock the fuel line clips for easy removal.

4. Unplug the vacuum lines, and label them clearly for easy installation. If there are multiple lines it may be easiest to number both sides of each connection.
5. Remove the intake ducting from the throttle housing, and be sure the duct work is out of the way.
6. Remove the throttle cable.
7. Remove the clutch cable or any hydraulic lines. Plug the opening so the brake fluid does not drain out over the engine. Brake fluid will destroy paint and cause corrosion. Many times the clutch reservoir is part of the brake master cylinder; losing all the fluid will necessitate brake bleeding.
8. Relieve the fuel pressure, and disconnect the fuel feed line and return line if present. Plug the lines on both ends to prevent leakage or contamination. Newer vehicles often have fuel lines fitted with quick disconnect fittings. Use disconnect tools to depress the locking tabs and pull the ends apart easily **(Figure 5** and **Figure 6)**.

> **You Should Know** *Always wear safety glasses and put a shop rag around a fuel fitting when loosening it; fuel is likely to spray out.*

9. Drain the coolant from the radiator and the block. Be sure to properly recover the used fluid, and deal with it appropriately. If the vehicle has an automatic transmission remove the transmission oil cooler lines.
10. Drain the engine oil. Reinstall the drain plug.
11. Disconnect the exhaust front pipe from the exhaust manifold. The bolts will often be rusty and difficult to remove. Sometimes you will need to heat the nuts or the flange with a torch to prevent breaking studs or bolts in the manifold. Support the exhaust system with mechanic's wire to prevent damaging it.

> **You Should Know** *When bolts or nuts will not loosen easily, try to tighten them slightly, and then loosen them. Alternatively, give a solid blow to the bolt or nut with a hammer. Be careful not to damage the hardware; use a wide punch if necessary. Either method is likely to snap the rust bond between the threads and facilitate removal. If these tricks do not help use a torch to heat the joint.*

12. Remove the upper and lower radiator hoses and the heater hoses.
13. It is usually necessary to remove the cooling fan and radiator. If the fan is pump driven on the front of the engine, remove it.
14. Disconnect the power steering pump or lines, and tie them out of the way with mechanic's wire.
15. Remove the air-conditioning compressor, and tie it out of the way, being careful not to damage the lines. If it is necessary to disconnect the lines from the **AC compressor** evacuate the system of its refrigerant first.

> **You Should Know** *It is illegal and cost prohibitive to vent refrigerants to the atmosphere. Use appropriate air-conditioning reclaiming equipment to recover the AC system refrigerant. You will need to be certified for AC work before you can legally perform this procedure* **(Figure 7)**.

Figure 7. Shops must have equipment to recover AC refrigerant. Technicians must be certified to work on the AC system.

16. Remove any brackets that will be in the way. Label them carefully, and keep the bolts and nuts taped to them to make installation easier. There are no instructions given for how or where to reinstall brackets.
17. Remove the starter wiring and starter.
18. Thoroughly examine the top of the engine for any remaining connections or lines. Be sure the engine accessories are out of the way.
19. Remove the bolts holding the transmission bell housing onto the engine if the engine is coming out without the transmission.
20. If the transmission will come out with the engine disconnect the driveshaft and shift and clutch linkage. You will also need to remove any electrical connections on the transmission. Older vehicles may have a speedometer cable that should be removed.

REMOVING A FRONT-WHEEL DRIVE ENGINE

The initial steps for removing a front-wheel drive (FWD) engine are very similar to those for removing an RWD engine. Disconnect the battery first to be sure no electrical sparking or damage occurs. Drain the fluids, and disconnect and label electrical connections, vacuum lines, hoses, and brackets. It is usually unnecessary to remove the fan or radiator, but check the manufacturer's information for its recommendation. Remove the intake ducting and the front exhaust pipe. Disconnect the throttle cable. Remove any accessories that will get in the way of removal, including the AC compressor if applicable.

Determine whether the engine is coming out alone or if the transaxle should come out with the engine. Follow the manufacturer's recommendation on this; the manufacturer has tried it both ways already. Often you will need to remove the transaxle with the engine. In this case you will need to remove the drive axles, the shift and clutch linkage,

Figure 8. Attach the chain to the engine lifting points to lift the engine.

and any electrical connectors from the transaxle. To remove the axle shafts you will have to separate the **ball joints** and sometimes the steering linkage from the **hubs** so you can swing the hubs off the shaft. Remove the axle nut from the hub, and remove the axle from the vehicle.

If the engine is coming out the top, bolt a chain to the lifting hooks on the engine or to two good-sized mounting bolts with washers **(Figure 8)**. Support the engine lightly with an engine crane. Remove the engine mounts and transaxle mounts if applicable **(Figure 9)**. Make a final check that the engine is loose and no other connections, hoses, or linkage remain. Guide the engine out as you lift it with the crane.

If you are removing the engine through the bottom, you will have to disconnect part or all of the **engine cradle**

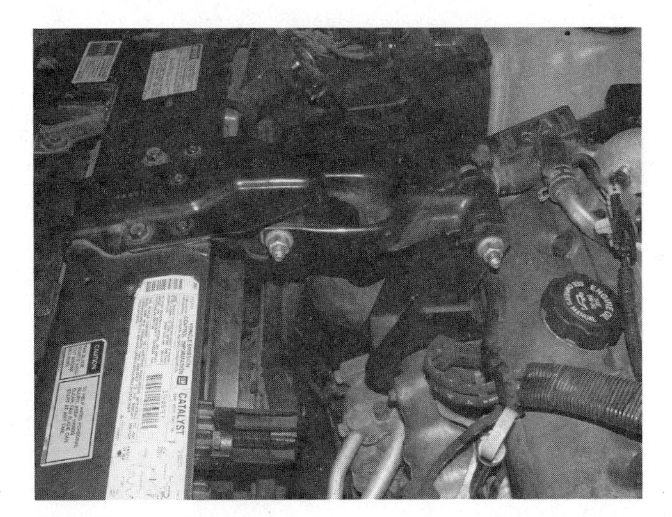

Figure 9. Remove the upper engine mounts in preparation for dropping the engine through the bottom.

Figure 10. The cradle mounts to the body and forms a frame for the suspension. The engine often mounts to the cradle.

or cross members from the body **(Figure 10)**. Sometimes the engine will come out on the engine cradle. Other times you will remove the engine cradle and drop the engine onto a special dolly. In either case make a final check that all the attaching bolts are free, and set the engine on the dolly or the cradle on jack stands.

Slowly raise the vehicle while you watch closely for any obstructions or remaining connections. If the transaxle is still attached to the engine remove the connecting bolts and nuts, and separate the two units. With the engine on the ground use an engine crane to raise the engine **(Figure 11)**.

MOUNTING THE ENGINE ON A STAND

To disassemble and repair the engine you will need to mount it on an engine stand. This will allow you access to all parts of the engine. You can rotate it to work on the top or the bottom of the engine. Remove the flywheel and clutch

Figure 11. Removal accomplished; next mount the engine on a stand.

or flexplate from the engine so you will have access to core plugs and the rear main seal. Mount the flange from the engine stand onto the back of the engine using the transmission/transaxle mounting holes. Engine stands come with an assortment of mounting bolts. Use bolts long enough to thread fully into the engine. With the mounting flange secure on the engine lower the crane until the snout of the flange is close to the receiving hole in the engine stand. Maneuver the snout into the stand; sometimes you may have to lift the stand slightly to match the angle of the engine. Tighten the bolt, or install the pin that secures the stand to the flange. Slowly lower the engine crane, and let the engine settle on the stand. Remove the crane and the chain from the engine.

Summary

- Remove the engine from the vehicle to complete an engine overhaul or to replace it.
- Read through the manufacturer's procedure for engine removal; it will outline the most efficient way to complete the task.
- Cover the front end of the vehicle with fender covers to prevent scratches and solvents from damaging the paint.
- Remove the battery first to prevent arcing or shorts as you disconnect wiring and remove large components.
- Clearly label electrical connectors, vacuum lines, hoses, and brackets so you will be able to reinstall the engine efficiently several days after removing it.
- Drain the fluids from the engine and accessories so you do not create a hazardous work area during the job.
- Remove all cables and linkage from the engine and from the transaxle if you are to remove it with the engine.
- Disconnect the intake and exhaust from the engine.
- In most rear-wheel drive vehicles with a longitudinally mounted engine you take the engine out of the top of the hood.

- Many transversely mounted engines require that the engine be dropped through the bottom, often with the transaxle attached. This requires that you remove some suspension and cradle components.
- Once the engine is removed mount it on an engine stand so that you can easily access all parts of the engine.

Review Questions

1. List three reasons why it is beneficial to refer to the manufacturer's service information for the engine removal procedure.
2. Is it true or false that you should remove the battery positive cable first to remove power from the battery?
3. List two ways to try to loosen a seized bolt.
4. When removing the hood what should you do to make reinstallation and adjustment easier?
5. Technician A says that each electrical connector can fit in only one spot. Technician B says to check the engine compartment for one main engine harness connector. Who is correct?
   A. Technician A only
   B. Technician B only
   C. Both Technician A and Technician B
   D. Neither Technician A nor Technician B
6. Technician A says to remove the fuel inlet and return lines. Technician B says to remove the electric fuel pump. Who is correct?
   A. Technician A only
   B. Technician B only
   C. Both Technician A and Technician B
   D. Neither Technician A nor Technician B
7. Technician A says to remove the hoses and leave the AC compressor on the engine. Technician B says you must be certified to remove hoses from the AC compressor. Who is correct?
   A. Technician A only
   B. Technician B only
   C. Both Technician A and Technician B
   D. Neither Technician A nor Technician B
8. Technician A says it is often necessary to remove the radiator when removing an RWD engine. Technician B says that if the engine has a belt-driven fan it should be removed. Who is correct?
   A. Technician A only
   B. Technician B only
   C. Both Technician A and Technician B
   D. Neither Technician A nor Technician B
9. Technician A says to remove the hood before pulling the engine out the top. Technician B says to ask at least two other technicians to help you lift the engine out of the top of the engine compartment. Who is correct?
   A. Technician A only
   B. Technician B only
   C. Both Technician A and Technician B
   D. Neither Technician A nor Technician B
10. Technician A says that many transversely mounted engines are removed through the bottom of the engine compartment. Technician B says that you will often need to remove the ball joints and axles. Who is correct?
   A. Technician A only
   B. Technician B only
   C. Both Technician A and Technician B
   D. Neither Technician A nor Technician B

Chapter 20 Engine Disassembly

Introduction

The engine disassembly procedure is an important part of your engine repairs. Work like a detective, carefully inspecting each component as it is removed. You need to pay attention to all the clues to ensure that you find every fault and each potential problem during your service. You should only have to remove an engine once to make all the necessary repairs. Speed is not the most important factor when disassembling the engine. It will come apart easily, but being organized is what will allow you to put it back together efficiently as well. Separate and label bolts, and mark brackets. Write down problems as you see them so you will remember to repair them and so you can refer to them as you evaluate the extent of the damage. Save the old parts and gaskets until the overhaul is complete. You may need to compare new parts with the old parts, and the customer may want to look at damaged components.

When you are replacing an engine it is easiest to have the replacement engine next to the old one. That will make it clear which components need to be transferred to the new engine and show you how they are mounted.

Interesting Fact

The best engine repair technician I have met took some ribbing for how much time he spent disassembling an engine. He carefully marked each set of bolts or threaded them lightly back into place; everything was very well organized. Not once in the years I worked with him did he have an engine come back for as much as an oil leak or a rattle.

ENGINE DISASSEMBLY

Before beginning the engine disassembly clean the outside of the engine well enough that you will not get dirt or grime into it. Make a clean work area for yourself with enough room for proper storage and to lay parts out for examination. Many parts, if reused, will have to be installed back into their original location. Have coffee cans, Tupperware, or plastic bags available for storing the many sets of bolts you will remove. Use masking tape and a permanent marker for labeling **(Figure 1)**. Do not mix bolts from the oil pan, valve cover, timing cover, rockers, and main and rod

Figure 1. Separate and label the bolts to help yourself with reassembly. Many parts such as lifters and rocker arms should be reinstalled in their original location if they are to be reused.

caps into one big bin. Digging through a pile of bolts trying to find the correct ones is not an efficient or reliable way to reassemble an engine. Spray the exhaust manifold studs, the seam from the crankshaft snout to **harmonic balancer**, and any other rusty bolts with penetrating oil, and allow them to soak while you proceed. Follow the manufacturer's procedure for complete disassembly. The instructions will provide torque sequences for removing major components such as the head, main caps, and intake manifold.

CYLINDER HEAD REMOVAL

You can remove the cylinder head with the engine either in or out of the vehicle. The procedure is similar either way. You may have to remove some of the hoses, connections, and cables that were discussed in the preceding chapter to gain access to the valve cover(s) and head(s). Be sure the engine is cool when loosening intake or head bolts to prevent warpage. The following is a general outline of the procedure to remove a cylinder head:

1. Remove the intake manifold using the reverse of the tightening sequence (unless a loosening sequence is provided). You can usually leave the injectors, fuel rail, and pressure regulator attached to the manifold. On some vee engines you will have to remove the pushrods before you can lift the manifold out; see step #5 below.
2. Remove the exhaust manifold(s). In some areas of North America, those that use salt to clear snowy roads, the exhaust studs and nuts may be quite rusty **(Figure 2)**. Even though you will usually replace the studs take care not to break them; they will be easier to remove. Many times it is safest to use some heat on the nuts before attempting to loosen them.
3. Remove the valve cover(s).
4. If the engine has an overhead cam you will have to remove the timing belt or chain. Rotate the engine to TDC #1, and locate the timing marks on the camshaft pulley(s). Make notes about the TDC markings to facilitate installation. Some manufacturers call for removal of the timing cover to access the tensioner. On other vehicles it is possible to remove the tensioner from the outside of the cover. Release the tension on the chain or belt. Remove the camshaft sprocket from the camshaft, and inspect it for wear.
5. On an engine with the camshaft in the block loosen the rocker arms, and remove the pushrods. It is a good idea to mark their position so they can be installed in the same location. Inspect the pushrods for bends or wear.
6. Remove the lifters from their bores. If you might reuse them mark their correct location with a marker or masking tape; they should be returned to their original bore and contact the same lobe on the camshaft. You may need to spray some penetrating oil onto the lifters to remove them; varnish can build up and make them difficult to remove **(Figure 3)**.
7. If you are working on a vee engine mark the cylinder heads left and right. Left and right are determined by looking at the engine from the flywheel end.

Figure 2. These manifold nuts will require heat to loosen them. Replace the studs while the cylinder head is off.

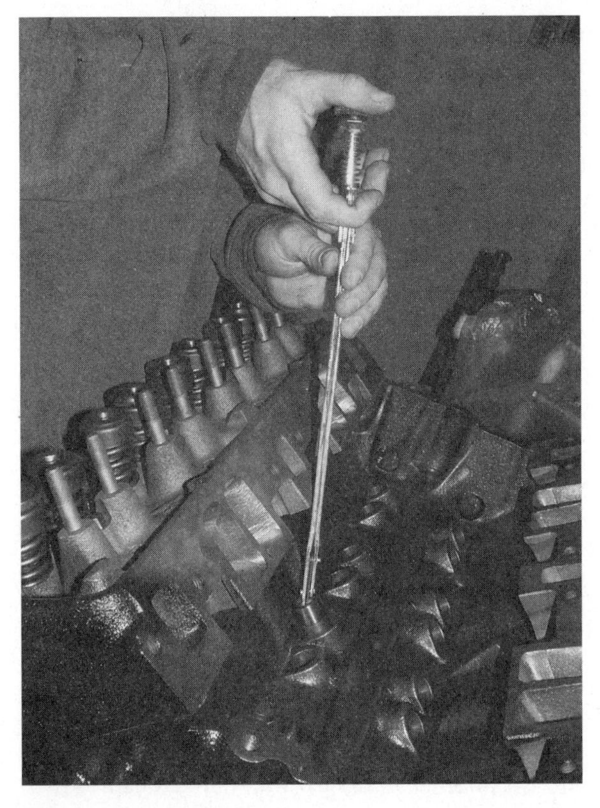

Figure 3. This lifter removing tool can help you remove the lifters without damaging them.

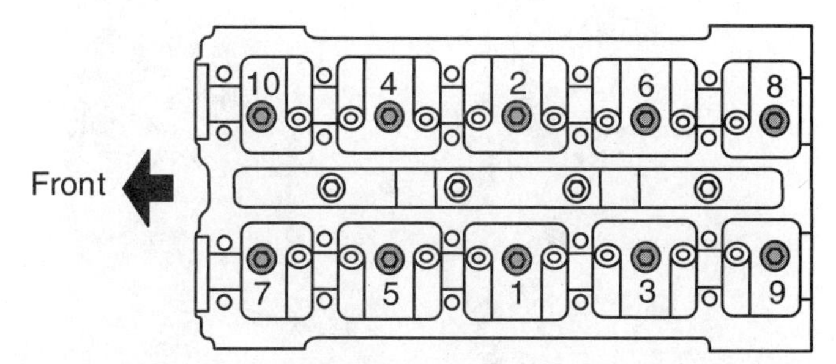

Figure 4. Reverse this tightening sequence to loosen the head bolts.

Interesting Fact

One time a new technician had a V6 engine almost fully put together after an overhaul. He started putting the brackets and accessories back on, but two of the brackets would not bolt up. On one head there were no mounting holes showing at all. After a period of frustration he realized the heads were installed on the wrong sides.

8. Loosen the head bolts using the loosening sequence. If one is not specified, use the reverse of the tightening sequence **(Figure 4)**. Check for any differences in the bolts, and if they differ mark their proper locations. If the bolts are to be reused clean them, and inspect the threads for damage. On many newer engines the manufacturers use torque-to-yield bolts and specify replacement after one use.
9. Knock the cylinder head or heads with a plastic dead blow hammer to try to loosen them. You may have to use a pry bar to release the head. Be careful not to damage the sealing surface of the head. If the head does not pry off easily check to be sure all the bolts are removed. One cylinder head has bolts coming up from the bottom that you cannot see from the top! Use the service manual to prevent damaging the head.
10. Remove the cylinder head and gasket. Carefully inspect the gasket for signs of leakage or detonation. Detonation will often distort the steel fire ring on the gasket that surrounds the cylinder bore. We will discuss thorough assessment of the cylinder head in Chapter 26, "Cylinder End Component Inspection."

TIMING MECHANISM DISASSEMBLY

To uncover the timing mechanism you will have to disassemble the front end of the engine, including the timing cover, any remaining accessories or brackets, and the harmonic balancer. Remember to mark the position of brackets and accessories, and mark their bolts. Use a puller to remove the harmonic balancer **(Figure 5)**. A few balancers slide off and do not require a puller; pry on opposite sides to remove it. Usually there are holes threaded in the end of the balancer, so you can mount the puller to the balancer and then push on the crankshaft. Sometimes you will have to use jaws on the outside of the balancer. In either case be sure to protect the internal threads of the crankshaft. Install a spacer button on the end of the crankshaft so the puller cannot thread into the crank.

Figure 5. The right puller makes short work of removing the harmonic balancer safely.

With the balancer off remove the timing cover. Be sure to mark where bolts go when they are different lengths; you may want to insert them into the proper hole of the cover and tape them into place. If the oil pump is mounted on the front of the crankshaft remove it, and set it aside for evaluation later **(Figure 6)**.

Figure 6. This crank-driven oil pump must be removed before the bottom end can be disassembled.

You Should Know

Many late-model overhead-camshaft engines have complex timing chain mechanisms and variable cam timing components that require special service tools and procedures. Refer to the manufacturer's specific procedures before disassembling the timing mechanism.

Before disassembling the timing mechanism rotate the engine to TDC #1 compression so that the timing marks line up. It is helpful to see how the marks should line up when the timing is correct. Check the chain or gears for excess slack or backlash. Remove the tensioner and guides if applicable. Remove the camshaft or crankshaft sprocket to release the chain or belt. If the engine has timing gears remove them now. Carefully inspect the timing mechanism and sprockets to determine whether they should be reused. When performing a thorough overhaul on a high-mileage engine you should generally replace the timing components. Record your recommendation.

ENGINE BLOCK DISASSEMBLY

Proceed methodically through the block disassembly. There are several special procedures and critical inspections to perform. Keep your eyes open and your mind on the job; pay attention to details. Be careful not to cause any damage as you disassemble the engine.

The engine should still be upright. Remove the oil pan using the reverse of the tightening sequence. If none is specified be sure to work back and forth across the ends of the oil pan while loosening. Removing the bolts one after the other in a circle around the pan can cause distortion. Many engine stands have drain pans that rest on their frame to contain the engine oil. If not place some rags underneath the engine to keep your work area safe.

Ring Ridge Removal

When the engine is running the top ring does not reach the top of the cylinder bore. It is constantly cleaning and scraping most of the cylinder wall, but at the very top it is pushing carbon up into a ridge. Remove this **ring ridge** before attempting to get the pistons out. If the pistons are driven up past the ring ridge they can easily break. Also, if new rings and bearings are installed without removing the ridge the top ring can hit the bottom of the ring ridge and snap the land or break the ring. Either occurrence would be disastrous after a fresh rebuild. If you can feel the ridge with your fingernail remove it before extracting the pistons.

Figure 7. Removing the ring ridge. Notice the farthest cylinder with a clean cut at the top of the cylinder.

Move the piston down toward bottom dead center, and place the ring ridge remover on top of the piston **(Figure 7)**. Tighten the top bolt down to center, and tighten the tool in the cylinder. Adjust the cutting bit so that it touches the cylinder wall just below the ridge. Then rotate the nut and cutting tip up and out through the top of the cylinder. Do not take a second pass; you could remove too much material from the cylinder wall. Repeat the procedures for each of the cylinders.

Piston Removal

Before removing the pistons check the connecting rod caps for markings. They may be numbered or have punch marks on both halves of the cap. If they are not marked use a paint stick or correction fluid to mark the rods on the parting faces of the cap halves. Number one is at the front (harmonic balancer) end of the engine **(Figure 8)**. If the rod caps are installed on the wrong connecting rods, or even backward on the same rod, the bore will not be perfectly round. This can cause crankshaft and bearing damage as the crank is forced to turn through a distorted bore.

Remove one connecting rod cap. You may need to tap the rod bolts and side of the cap lightly with a brass hammer to loosen the caps. Place a piece of vacuum line over the rod bolts so that they do not scrape the crankshaft or cylinder wall as you tap the piston out **(Figure 9)**. Use a plastic mallet to drive the piston out. You will be replacing the bearings, so it is appropriate to tap on the bearing half in the rod. Do not let the piston drop; pull it out the rest of

Figure 8. Mark the rod and the cap with a paint stick or correction fluid. The caps must be returned to the same rod and bolted in the same direction.

Figure 9. Protect the crankshaft and cylinder walls by using rubber hose on the ends of the connecting rod bolts.

the way once its head is out of the top of the block. Leave the bearing halves in so you keep them with the correct connecting rod for inspection. Lightly thread the cap back on the rod, and store it in a safe place where the pistons will not get scratched. Remove all the pistons.

Crankshaft Removal

Once the connecting rods and pistons have been removed the crankshaft can come out. The **main bore** in which the crankshaft lies was drilled out with one long boring bar. This is called line boring. As a result of this process the caps and block will form round bores only if the caps are installed in the same spot and in the same direction. If a main cap is put on backward it can cause the crank to seize **(Figure 10)**. Check the main caps for markings that indicate number and direction. If there are none use a paint stick or correction fluid to mark the block and caps.

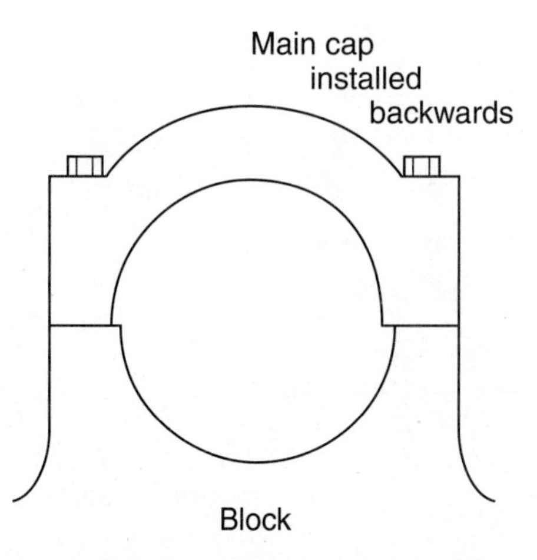

Figure 10. With the main cap on backwards the main bore is no longer round. This can seize the crankshaft.

Loosen the main caps in two stages in the reverse of the tightening sequence. In the first stage turn each main cap bolt about one turn to loosen it. Then follow the proper sequence again and fully loosen the bolts. Remove the main caps, leaving the bearings in for proper inspection. If the bearings are in their proper spot you will be able to see whether there is any pattern to the wear. You may be able to pick out something as serious as a twisted main bore or a bent crankshaft by looking at the bearings. Remove the crankshaft, and store it on end to prevent it from bending or sagging.

Camshaft Removal

If the camshaft is in the block remove it now. Loosen the screws or bolts attaching the cam thrust cap on the end of the block. You may need to use an **impact screwdriver** on countersunk Phillips head screws. Install the cam gear so you can hold the camshaft solidly to remove it without damaging the lobes. The lobes will chip if you knock them against the block; go slowly and carefully. If any lifters are still stuck in their bores be sure they are at the top of their travel so the camshaft does not hit them on its way out. Set the camshaft safely aside for later examination. Look at the cam bearings for unusual wear patterns; then remove them using a cam bearing tool.

Core and Oil Gallery Plugs

Many times you will send the engine block out to a machine shop to have it thoroughly cleaned and perhaps have the cylinders bored. The machine shop technicians will then remove the oil gallery plugs and clean the passages. They will also remove and replace the freeze plugs on any old or high-mileage engine. Your shop may also choose to perform these operations in-house if properly equipped.

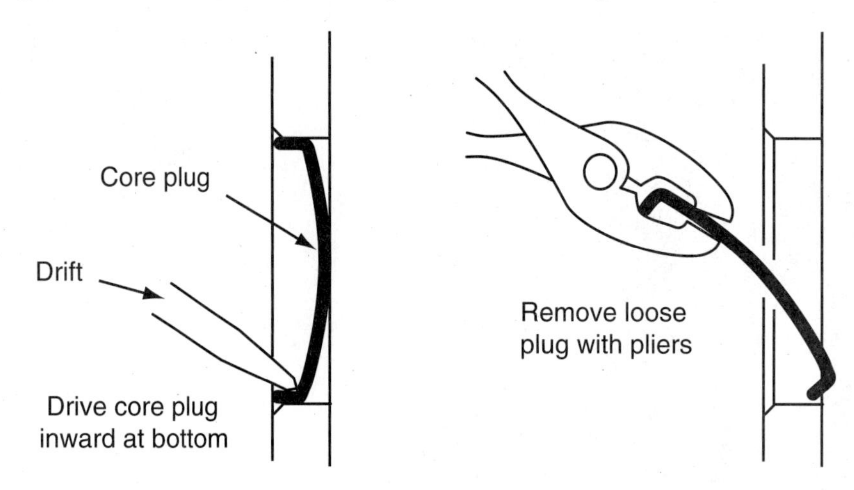

Figure 11. Remove the core plugs and oil gallery plugs so the block can be thoroughly cleaned.

To remove core plugs you can use a punch or chisel to hit one side of the plugs. Be careful not to drive the plug into the block. With the plug turned out of the block grab it with pliers or vise grips, and pull it out of the block **(Figure 11)**. Alternatively, thread a sheet metal screw into the center of the plug. Put vise grips on the screw, and tap or pull it out with the plug attached. Be sure to lightly sand the corrosion out of the sealing surface of the bore with emery cloth before installing new core plugs.

Oil gallery plugs have tapered threads that seal very tightly. Though you would think they would be constantly lubricated with oil the reality is that they can become quite rusty and difficult to remove. Use the proper tool to remove them; this may be a square plug, an Allen head, or an eight-point socket. If they are particularly difficult to remove heat them with a torch, drop a little wax on them, and try again. It may help to use an impact gun; just be sure to apply plenty of force against the plug to prevent stripping the head of the plug. Use stiff-bristle, long gallery brushes and hot soapy water to remove all sludge and debris from the oil galleries. Blow the passages out with pressurized air to get any residual particles out and to dry the galleries.

With the engine fully disassembled and the parts well organized you will be able to inspect each of the components thoroughly. With this information you will be able to make appropriate repair decisions. It is common to see significant damage to the cylinder bores and crankshaft, for example, and determine during disassembly that engine replacement would offer the most efficient and cost-effective resolution.

Summary

- Label and sort bolts, brackets, and components during disassembly to ensure efficient and correct reassembly.
- Prepare a clean workspace for yourself, and have small storage bins or cans, bags, masking tape, and a marker available.
- Remove the cylinder head when it is cool. Use the specified loosening sequence or reverse the tightening sequence to prevent warping the head.
- Use heat to prevent breaking the exhaust manifold studs off in the head.
- Spray penetrating oil on the lifters, and use a lifter removal tool to extract the lifters from the block. Mark their location if you might reuse them.
- When working on a vee engine mark the cylinder heads left and right to prevent incorrect installation.
- Rotate the engine to TDC #1 before removing the timing mechanism. Check the timing marks on the sprockets or gears, and note their proper position before dismantling.
- Use a puller with a spacer on the end of the crankshaft threads to remove the harmonic balancer.
- Use a ring ridge remover before removing pistons if you can feel the ridge with your fingernail.
- Mark the connecting rod and main caps before disassembly. They must go back in the same spot and in the same direction.
- Leave the bearings in their bores so you can analyze their wear patterns later.
- Remove the camshaft very carefully to prevent chipping a lobe on the way out.
- If you will clean the block in-house remove the oil gallery plugs, and clean the passages thoroughly.

Review Questions

1. How should you store engine hardware to make reassembly most efficient?
2. What step should you take to prevent damaging the pistons when removing them?
3. Why should you keep the engine bearings in their proper order during disassembly?
4. What problems can occur if the connecting rod or main caps are not installed in their original location?
5. Technician A says to reverse the tightening sequence when loosening the cylinder head. Technician B says it is easiest to remove the head bolts with an air gun when they are still warm. Who is correct?
   A. Technician A only
   B. Technician B only
   C. Both Technician A and Technician B
   D. Neither Technician A nor Technician B
6. Technician A says to spray penetrating oil on the exhaust manifold studs. Technician B says that it may be necessary to use heat to get the nuts off. Who is correct?
   A. Technician A only
   B. Technician B only
   C. Both Technician A and Technician B
   D. Neither Technician A nor Technician B
7. Technician A says that you can pry most harmonic balancers off with two big pry bars. Technician B says that you can damage the crankshaft if you do not protect the threads while using a puller. Who is correct?
   A. Technician A only
   B. Technician B only
   C. Both Technician A and Technician B
   D. Neither Technician A nor Technician B
8. Technician A says to rotate the engine to TDC #1 before removing the timing mechanism. Technician B says to make a mental or written note of the location of the timing marks. Who is correct?
   A. Technician A only
   B. Technician B only
   C. Both Technician A and Technician B
   D. Neither Technician A nor Technician B
9. Technician A says the pistons should come out of the top of the block. Technician B says to drive on the edge of the piston skirt with a punch to remove the pistons. Who is correct?
   A. Technician A only
   B. Technician B only
   C. Both Technician A and Technician B
   D. Neither Technician A nor Technician B
10. Technician A says that core plugs are generally replaced during a high-mileage overhaul. Technician B says that oil galleries should be cleaned with a long stiff-bristle brush. Who is correct?
    A. Technician A only
    B. Technician B only
    C. Both Technician A and Technician B
    D. Neither Technician A nor Technician B

Chapter 21 Engine Block Construction

Introduction

The cylinder block and the components within it make up the bottom end of the engine, also called a **short block**. Understanding the design and construction of the block, pistons, rings, connecting rods, and bearings will help you make good decisions about how you will repair an engine. Engine materials and machining have changed significantly in the past 20 years. Clearances have decreased, ring and piston choices abound, and block reconditioning is often limited. A silicon-impregnated cast-aluminum alloy block with integral bores cannot be refinished, for example. You must have access to thorough service information to properly service a modern engine **(Figure 1)**. It will provide you with information on the materials used for the block, pistons, and rings, along with the proper repair procedures.

You may be able to improve the performance and durability of an older engine by using newly designed replacement parts with improved engineering, materials, and machining.

BLOCK CONSTRUCTION

The block is the foundation of the engine. It houses all the rotating and reciprocating parts of the bottom end and provides lubrication and cooling passages. The engine block is cast; hot, liquid metal is poured into a mold and allowed to cool and solidify. This forms the structure of the block. Sand is used inside the mold to provide water jackets around the bores. Holes are drilled in the outside of the block to remove the sand after casting. These holes are then sealed with core plugs.

Blocks have traditionally been made of gray cast iron because of its strength and resistance to warpage. After it is cast it is easily machined to provide rigid bores for the pistons, crankshaft, and lifters and camshaft if applicable. The block is also machined on the top, bottom, and ends to provide good sealing surfaces for the head, oil pan, transaxle, and timing cover **(Figure 2)**.

Many newer designs use a cast-aluminum alloy for the block. This offers a significant weight reduction that helps manufacturers meet fuel economy standards without sacrificing performance. Cast-aluminum alloy blocks are more prone to cracking and warping problems when they are overheated, however. The cylinder wall may have replaceable or cast-in-place steel liners. When they are cast in place they are called dry liners because no coolant hits the surface directly. Replaceable liners are called wet sleeves; coolant circulates all around the sleeve. The sleeve is sealed by an o-ring on the bottom and by the head gasket and head on the top. Some aluminum blocks are fitted with cast-iron cylinder heads and/or thickly cast oil pans to provide added rigidity to the block. Other newer engines are all aluminum.

In some cases a silicon-impregnated aluminum alloy is used in the casting. After the bores are machined a special honing process removes the aluminum and uncovers a layer of hard silicon to form the thin wear wall of the bore. These cylinders cannot be bored oversize unless a sleeve is installed.

Other aluminum blocks use a special lost-foam casting process. The casting molds are made in a foam pattern that allows for more complex shapes. These aluminum blocks

GENERAL MOTORS ENGINES
3.0L, 3.8L and 3.8L "3800" V6 (Continued)

Engine Specifications (Continued)

Crankshaft Main and Connecting Rod Bearings

Engine	Main Bearings				Connecting Rod Bearings		
	Journal Diam. In. (mm)	Clearance In. (mm)	Thrust Bearing	Crankshaft End Play In. (mm)	Journal Diam. In. (mm)	Clearance In. (mm)	Side Play In. (mm)
All Models	[1] 2.4988-2.4998 (63.469-63.494)	.0003-.0018 (.008-.005)	2	.003-.009 (.08-.23)	[1] 2.2457-2.2499 (57.117-57.147)	.0003-.0028 (.008-.071)	.003-.015 (.076-.38)

[1] Maximum taper is .0003" (.008mm).

Pistons, Pins and Rings

Engine	Pistons	Pins		Rings		
	Clearance In. (mm)	Piston Fit In. (mm)	Rod Fit In. (mm)	Ring No.	End Gap In. (mm)	Side Clearance In. (mm)
All Models	[1] .0013-.0035 (.033-.089)	.0004-.0007 (.010-.018)	.00075-.00125 (.019-.032)	1	.010-.020 (.254-.508)	.003-.005 (.08-.13)
				2	.010-.020 (.254-.508)	.003-.005 (.08-.13)
				3	.015-.035 (.381-.889)	.0035 (.09)

[1] Measured at bottom of piston skirt. Clearance for 3.8L turbo is .001-.003" (.03-.08 mm).

Valve Springs

Engine	Free Length In. (mm)	Pressure Lbs. @ In. (Kg @ mm)	
		Valve Closed	Valve Open
3.8L (VIN C)	2.03 51.6	100-110@1.73 (45-49@44)	214-136@1.30 (97-61@33)
3.0L & 3.8L (VIN 3)	2.03 51.6	85-95@1.73 (39-42@44)	175-195@1.34 (79.1-88.2@34.04)
3.8L (VIN 7)	2.03 (51.6)	74-82@173 (33-37@44)	175-195@1.34 (79.1-88.2@34.04)

Camshaft

Engine	Journal Diam. In. (mm)	Clearance In. (mm)	Lobe Lift In. (mm)
3.0L	1.785-1.786 (45.34-45.36)	.0005-.0025 (.013-.064)	Int. .210 (5.334) Exh. .240 (6.096)
3.8L (VIN C)	1.785-1.786 (45.34-45.36)	.0005-.0025 (.013-.064)	[1] .272 (6.909)
3.8L (VIN 3)	1.785-1.786 (45.34-45.36)	.0005-.0025 (.013-.064)	[1] .245 (6.223)
3.8L (VIN 7)	1.785-1.786 (45.34-45.36)	.0005-.0025 (.013-.064)	....

[1] Specification applies to both intake and exhaust

Caution: Following specifications apply only to 3.0L(VIN L) 3.8L (VIN 3), 3.8L (VIN 7) and 3.8 "3800" (VIN C) engines.

TIGHTENING SPECIFICATIONS

Application	Ft. Lbs. (N.m)
Camshaft Sprocket Bolts	20 (27)
Balance Shaft Retainer Bolts	27 (37)
Balance Shaft Gear Bolt	45 (61)
Connecting Rod Bolts	45 (61)
Cylinder Head Bolts	[1] 60 (81)
Exhaust Manifold Bolts	37 (50)
Flywheel-to-Crankshaft Bolts	60 (81)
Front Engine Cover Bolts	22 (30)
Harmonic Balancer Bolt	219 (298)
Intake Manifold Bolts	2
Main Bearing Cap Bolts	100 (136)
Oil Pan Bolts	14 (19)
Outlet Exhaust Elbow-to-Turbo Housing	13 (17)
Pulley-to-HarmonicBalancer Bolts	20 (27)
Outlet Exhaust	
Right Side Exhaust Manifold-to-Turbo Housing	20 (27)
Rocker Arm Pedestal Bolts	37 (51)
Timing Chain Damper Bolt	14 (19)
Water Pump Bolts	13 (18)

[1] Maximum torque is given. Follow specified procedure and sequence.
[2] Tighten bolts to 80 INCH lbs. (9 N.m.)

Figure 1. This service information includes critical measurement and torque specifications. You cannot service an engine properly without engine specific information.

Figure 2. A bare V-8 block.

(and heads) almost look as though they are made of silver Styrofoam. The manufacturers claim increased reliability with lower machining costs.

Cylinder Bores

The cylinder bores may either be integral or sleeved. Integral bores are machined out of the casting to fit the pistons. They are bored first and then honed to size to provide the proper surface finish for the rings. Bores are **plateau honed** from the factory to provide good cylinder-wall lubrication with minimal wear. A rough stone hone is used to provide a crosshatch pattern in the cylinder. The diagonal scratches in the wall hold a little oil to lubricate the pistons and rings. The cylinder is then finish honed with a soft stone to cut the rough edges off the grooves in the

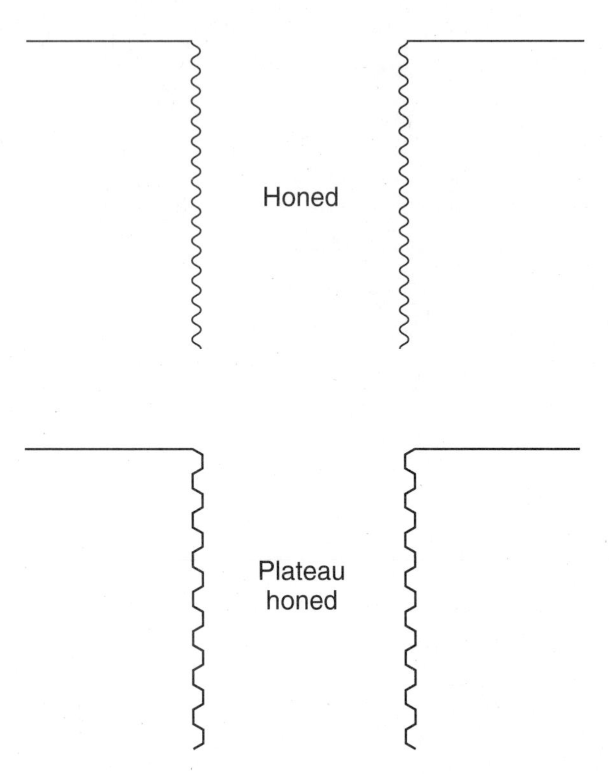

Figure 3. This exaggerated view shows the rough honing, which is then cut down by plateau honing. This process holds oil for the rings and pistons but causes less wear to the rings than simple honing.

cylinder wall. The result is a smooth plateau for the rings and pistons to contact **(Figure 3)**.

Clearances between the pistons and cylinder bores are commonly very small on modern engines, as little as .0005 in.; the machining processes are precise. Many aluminum blocks will use cast-iron sleeves either cast or dropped into the block. The block is supported at the top and bottom during the machining processes to prevent twisting. The lower portion of the block has webbing to help keep the cylinders straight during engine operation. The cylinder head helps secure the top half of the block.

Main Bore

The main bore holds the crankshaft. The bore is supported by the webbing in the lower part of the block casting. Some blocks use a cast-iron girdle that bolts onto the **main caps** to increase the rigidity of the bore. Main bearing caps are cast separately and then machined to bolt on the bottom of the block. The main bore is then line- or align-bored to form the round hole between the bottom of the block and the main caps. The main caps are bolted in place while a long boring bar passes through the end of the block. Because the boring bar will never turn exactly true,

Figure 4. Bearings are fit into the main bore to protect the crankshaft.

particularly toward the end of its reach, the bores will be round only if the same cap is installed in the same direction for each bore.

The main caps are held in place by two, four, or six bolts. A two-bolt main cap is the most common, but many newer lightweight engines or performance engines use a four-bolt main. The bolts may be next to each other, be parallel, or come into the cap from the bottom and the sides, cross bolted.

Main **bearing inserts** snap into the halves of the main caps. They provide a soft surface to support, lubricate, and protect the crankshaft. They also prevent the main bore from wear; the crankshaft does not contact the main bores **(Figure 4)**.

Modular Blocks

A few manufacturers are now designing blocks in two pieces. The upper half of the block holds the pistons and the upper half of the main bore. A separate casting is made to form the lower half of the main bores in one strong piece. The modules of the block are machined flat to bolt together, and then the main bore is line bored. This design changes the typical method of measuring main bearing clearances; each bearing is select fit to the bore. When replacing these bearings you will consult a table provided by the manufacturer to select the correct size. **(Figure 5)**.

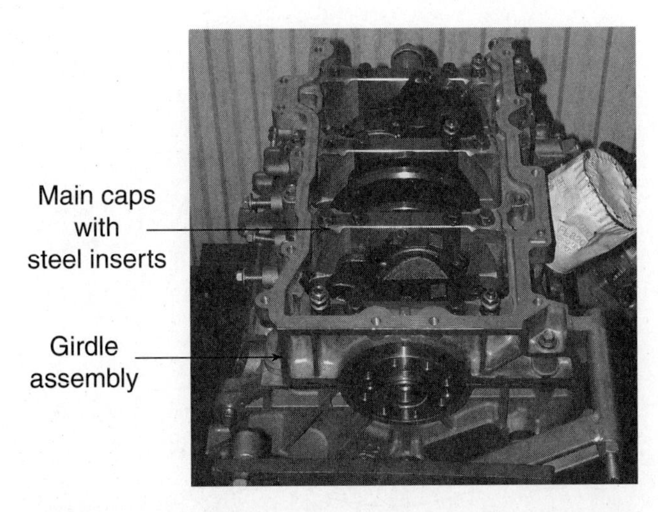

Figure 5. This all aluminum "modular" engine uses a main girdle; all the main caps are formed in one casting and bored out to fit the crankshaft.

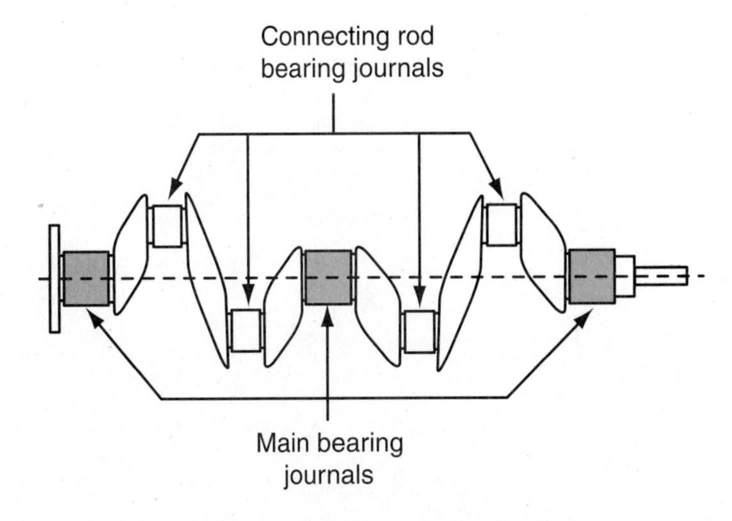

Figure 6. The rods journals are offset from the mains. The length of the offset times two is the engine's stroke.

CRANKSHAFT

The crankshaft withstands tons of force when the piston pushes the connecting rod down on the power stroke. It converts that reciprocating motion of the piston into useful rotary motion. Crankshafts may be made of cast iron or forged steel. Cast-iron cranks are quite common on production vehicles. Forged crankshafts are stronger and are often found on high-performance vehicles, engines with forced induction systems, and diesel engines. The crankshaft is bolted into the main bore on its main journals. These are machined on the centerline of the crank so it spins true. The connecting rod journals or pins are offset from the centerline **(Figure 6)**. The connecting rod clamps around these journals with a bearing insert in between them to protect the crankshaft and connecting rod. The distance they are offset determines the stroke of the engine. Rod journal offset times two equals the stroke. Crankshaft journals are machined and polished smooth to minimize bearing wear and friction; the soft bearing inserts protect the fine finish of the journals.

Crankshafts have many different configurations, depending on the engine. They will usually be ground so that the cylinders fire in equal intervals during one full cycle. A four-cylinder engine produces a power stroke every 180°; its rod journals are 180° off each other. A typical four-cylinder engine with a firing order of 1-3-4-2 will fire cylinder #1 at 180° of crank rotation, fire #3 at 360°, fire #4 at 540°, and complete the full cycle at 720° with the firing of cylinder #2. This puts cylinders #1 and #4 and cylinders #3 and #2 positioned at the same point in the cylinder during rotation. The connecting rod journals from an end view form a straight line above and below the main journals.

A V6 engine typically produces a power stroke every 120° of crank rotation. This is called an even-firing V6. The journals on the crankshaft are **splayed**. On one throw two journals are split and separated by 30°, in this case. Other V6 engines use one throw with two equal journal surfaces. This means that two rods will rotate together. The throws are machined at 120° intervals around the crankshaft. This produces an odd-firing V6; cylinders fire at 90°, then after 150° of rotation, then again after 90°, and on through the two full revolutions of the crankshaft that create a cycle **(Figure 7)**.

An even-firing V8 crankshaft has four throws, each with two journals **(Figure 8)**. The crankshaft produces power strokes every 90° of crank rotation. This is what makes a V8 such a smooth-running engine; the crankshaft is rotated by a connecting rod every 90°. This leaves little time for the crankshaft to speed up or slow down.

Crankshaft Oil Holes

The crankshaft is drilled to provide oil passages that feed each rod and main journal. The oil is held under pressure by the tight fit between the journal and the bearing. The bearing has an oil hole in it that lines up with the drilled hole in the crankshaft.

Crankshaft Counterweights

The crankshaft is cast or forged with large counterweights that oppose the rod journals. These are designed to even out the speed changes as the crank rotates through firing impulses. You will often see holes drilled in the counterweights. These are drilled when the crankshaft is balanced as it spins with rods and pistons attached. They also lighten the crank; you may see lightening holes on the sides of the throws as well.

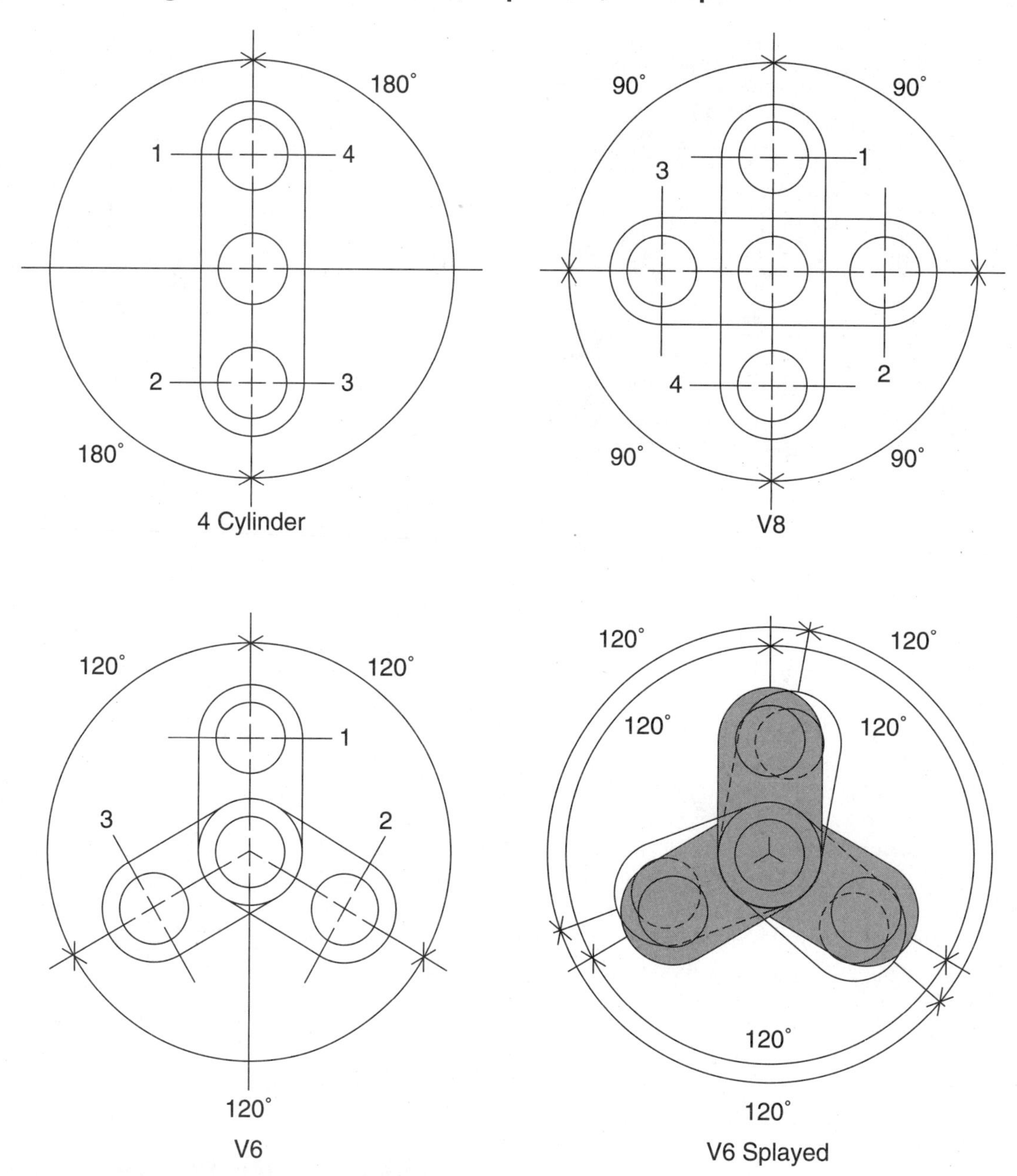

Figure 7. Typical firing intervals produced by different crankshafts.

Figure 8. Two rods share one journal on this V-8 crankshaft. Notice the oil holes to feed each of the rod bearings.

CAMSHAFT

The camshaft may be fitted in the block **(Figure 9)**. It rides on bearings placed in the camshaft bore. A thrust plate typically limits the camshaft endplay. We will discuss camshaft design and characteristics in Chapter 24,"Cylinder Head Construction."

BEARINGS

Bearings (bearing inserts) are used to protect the journals of the crankshaft. A similar type of bearing is used on the camshaft when it is in the block. The camshaft bearings

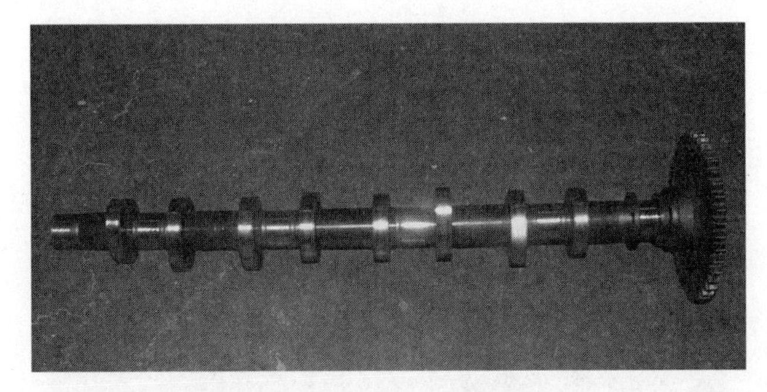

Figure 9. The camshaft may be housed in the block or in the head.

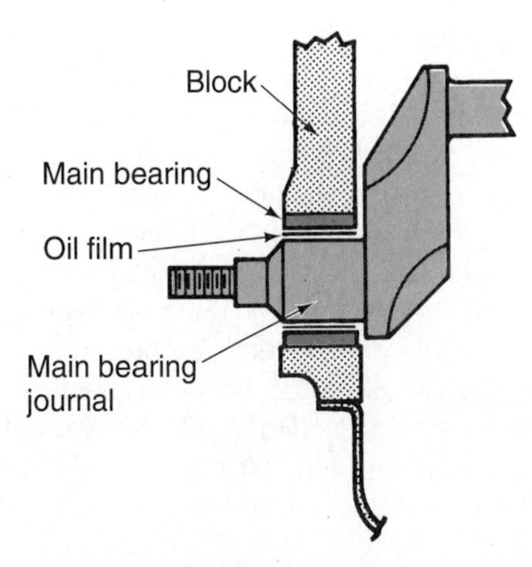

Figure 10. The crankshaft should ride on a film of oil about .001 in. thick. That is enough to reduce friction and protect the crank.

are one piece rather than two, but their materials and characteristics are nearly the same as those used to protect the crankshaft. The engine bearings are designed to wear before the crankshaft. The bearings snap into the main caps and rod caps and are then bolted around the crankshaft journals. The clearance between the bearing and the journal is critical. On modern engines a clearance of .0008–.002 in. is often specified; always check the allowable clearance for a particular engine. The clearance allows a thin film of oil to lubricate the bearing and the crankshaft. That oil also absorbs the pounding of the crank as the power stroke forces it downward in the main bores. The crankshaft should spin on a film of oil and not contact the bearing **(Figure 10)**. The oil pump sends oil under pressure through the holes drilled in the crankshaft, through an oil hole in the bearing, and into this small clearance. The correct bearing clearance allows a calibrated amount of oil to leak out past the side of the bearings, so the oil is constantly refreshed. When the bearing clearances become too great more oil leaks out, and engine oil pressure decreases. This loss of oil pressure results in a decrease in the oil film strength. The crankshaft will knock in its bore and cause rapid wear and distortion of the bearings and crankshaft. If the engine is not repaired promptly the damage to the crankshaft and bearings can become so severe that the bearings spin with the crankshaft in their bores. When bearings spin in their bores they can destroy the main bores, connecting rod bores, or camshaft bores and even cause the engine to seize. You will usually replace an engine with this type of severe damage to the block and crankshaft.

Bearings do wear out after high mileage and exposure to contaminated oil. When fitting new bearings you will check the clearance to be sure it is within the specified range. When a crankshaft is worn lightly and polished to be usable, bearings that are .001 in. and .002 in. undersize are often available. If the crankshaft has been ground to repair serious journal wear undersize bearings can be fitted. These bearings are generally available in .010 in., .020 in., and .030 in. undersizes. Undersize means that the inside diameter of the bearings is smaller to fit the reduced diameter of the crankshaft journals.

Bearing Characteristics

Engine bearings are made in halves that snap into the main and connecting rod bores and caps **(Figure 11)**. The back of the bearings is steel to provide strength to carry the tremendous loads imposed by the crankshaft. On top of the load-bearing shell are layers of soft metal to protect the crankshaft. The outer layer is typically a soft alloy of lead or tin babbitt, or of aluminum. Many bearings have a layer of copper below these alloys. These soft metals provide the bearings with two important characteristics, **embeddability** and **conformability**. When the oil carries impurities into the bearing clearances the soft alloy will allow the particles to become embedded in the bearing.

Figure 11. This is one half of a set of new aluminum coated main bearings.

This protects the crankshaft from being scored. The alloy also allows the bearing to conform to irregularities in the surface and shape of the crankshaft journals.

Bearings have a locating lug that fits into the bore to ensure proper positioning of the bearing. Bearings have oil holes and/or grooves to allow oil to circulate around the journal. One half of the main bearing has an oil hole that must line up with the oil hole in the block. **Bearing spread** ensures a positive lock in the bore and prevents the bearing from spinning. The bearing insert is not exactly round like the bore; it is spread wider **(Figure 12)**. This provides tension between the bearing and the bore; you will notice the spread when you try to push the inserts into place. The bearing is also slightly larger, .0005–.001 in., than the bore. This characteristic provides **bearing crush (Figure 13)**; when you install the bearings and torque the cap the bearing will be forced to crush slightly to fit.

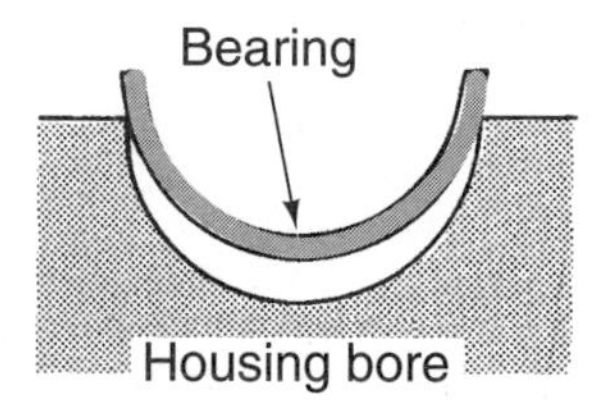

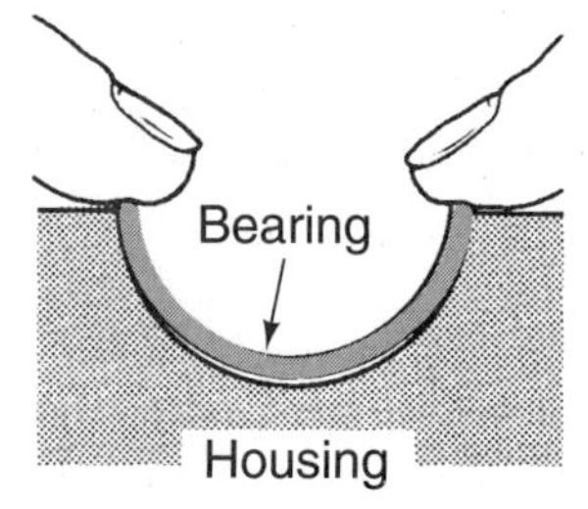

Figure 12. The bearing is spread wider than the bore. When it is pushed into the bore it makes solid contact all around.

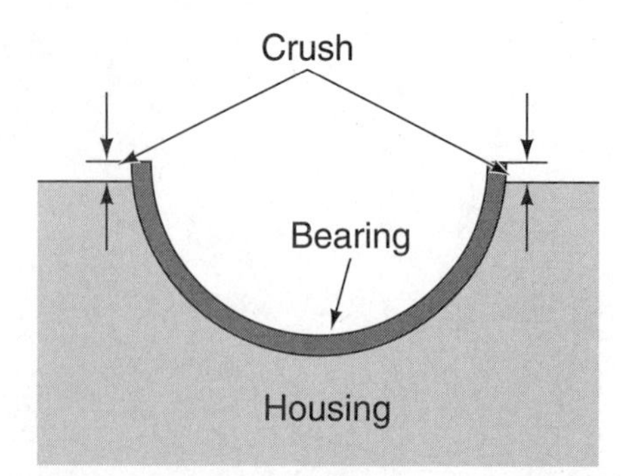

Figure 13. The bearing is slightly bigger than the bore. When the cap is torqued on, it crushes the bearing slightly to prevent it from spinning in the bore.

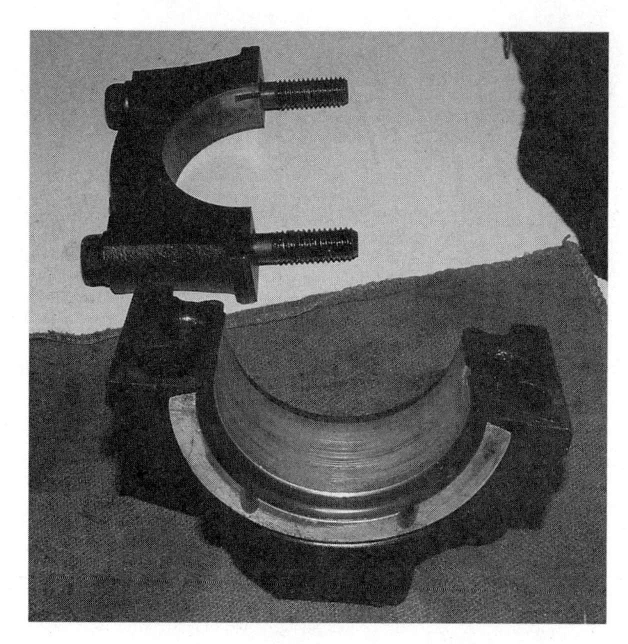

Figure 14. This thrust bearing has a soft aluminum-coated side flange to cushion crankshaft movement back and forth in the block.

Thrust Bearings

The crankshaft is fitted with a thrust bearing to protect the crank from slamming against the block as it moves forward and backward in its bore **(Figure 14)**. Under acceleration the crankshaft will be forced toward the back of the block. A thrust bearing allows a thin film of oil between a soft bearing and the crankshaft to absorb this force. The thrust bearing is often one of the main bearings, with flanges on the sides to control crank endplay. It may be positioned at an end of the crank or in the middle. Some thrust bearings are not part of a main bearing. They may be like washer halves and slide into a position between the crank and block.

PISTONS

The piston forms the bottom half of the combustion chamber. The shape of its head is carefully designed to provide the desired compression ratio, assist with airflow, and control emissions. Pistons must be strong enough to withstand the heat and pressure of combustion. They must be light enough to accelerate and decelerate rapidly. A piston may stop and start more than 150 times per second! Most pistons are made of aluminum. The piston must also fit the cylinder precisely to ensure proper sealing and long life.

The head, or crown, of the piston is the top of the piston. It may be flat, dished, or domed, depending on the compression ratio requirements of the engine. When modifying an engine you may change the type of piston head. If you add a turbocharger you may need to change the pistons to a dished design to lower compression enough to prevent detonation. The top of the piston may also have irregular contrasts to provide **turbulence** of the airflow to improve combustion. Many pistons have small cutouts in the top to make room for the valves; this is called valve relief **(Figure 15)**. There are typically three ring grooves to support the piston rings. The top two grooves hold the compression rings, and the lowest groove holds the oil control rings. The areas around the grooves are called the ring lands. The body of the piston is called the skirt; this helps keep the piston square in the cylinder and prevents rocking **(Figure 16)**. To reduce friction pistons are made with shorter skirts and use a slipper design to reduce piston weight. A slipper skirt refers to one that does not go all the way around the piston at one length. One small area is longer to provide stabilization. The piston is joined to the connecting rod by the wrist or **piston pin**. The piston pin bore is reinforced by pin bosses to support the huge forces transmitted to the connecting rod through the pin. The piston pin may be pressed into the piston, or it may be full floating. A full-floating pin is held in place by snaprings.

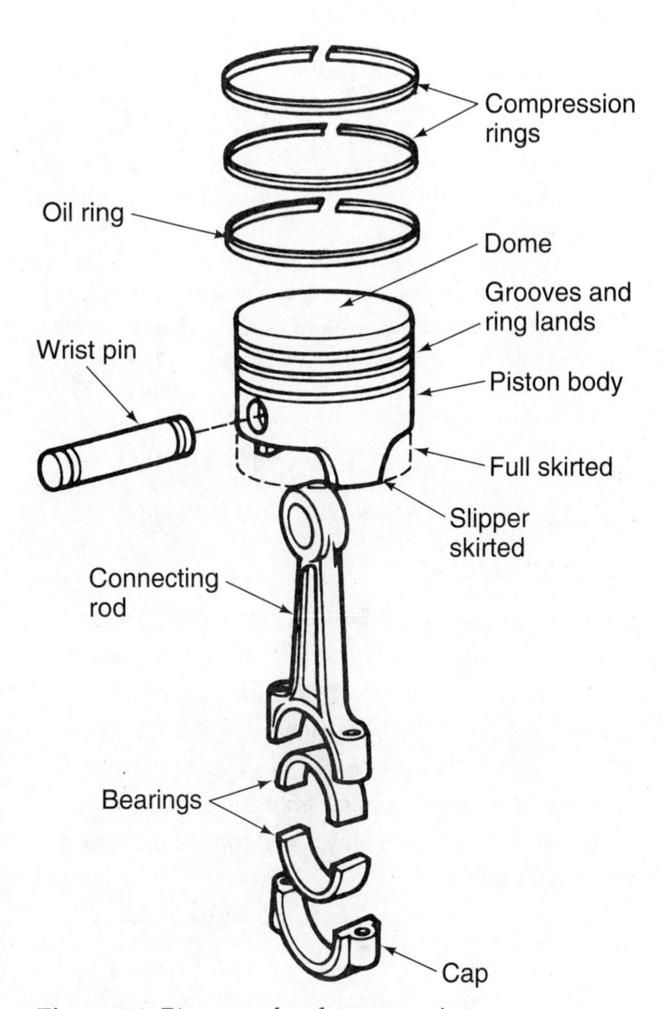

Figure 16. Piston and rod construction.

Piston Construction

Many engines use cast-aluminum pistons. These are lightweight and have good thermal expansion characteristics. They are strong enough for most applications. On some newer engines the top ring land may be anodized, manufactured using a special process, for added strength. The top ring and land suffer the most from the punishing forces of combustion.

Several newer engines use a **hypereutectic piston**. These pistons are significantly stronger than cast pistons and eliminate the need for extra support on the top land. Traditional cast-eutectic alloy pistons have 11–12 percent silicon in their aluminum alloy to improve strength and control expansion. A hypereutectic piston increases the silicon content to about 15–22 percent silicon. This brings the strength closer to that of a forged piston but provides better expansion control. Hypereutectic pistons are a popular replacement choice, particularly if the engine is modified.

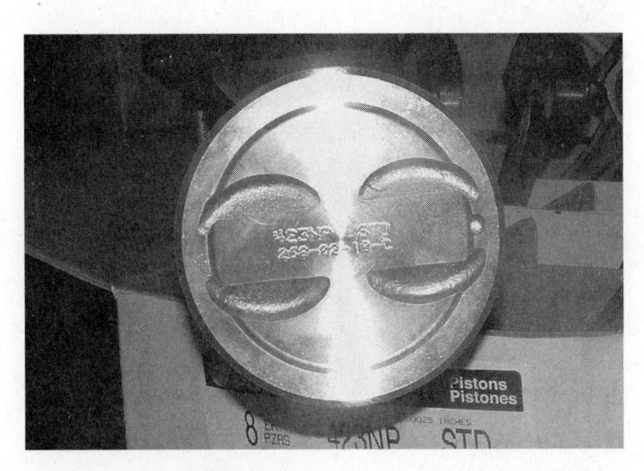

Figure 15. This piston has cut outs or valve relief to protect the valves from knocking the top of the piston.

Some high-performance engines, engines with forced induction systems, and diesel engines use forged-aluminum pistons. These, so far, are the strongest pistons used in production vehicles. They are not an ideal solution for today's engines. They require greater piston clearances because they expand at a greater rate than cast pistons. This allows blowby and noise when the engine is cold.

A graphite moly-disulfide coating is added to many modern piston skirts to reduce **scuffing** and increase durability. With closer piston-to-cylinder tolerances these coatings are helpful in reducing friction. Many parts manufacturers offer replacement pistons with coatings. They are slightly more expensive but will increase the longevity of the piston and reduce cylinder wear.

Expansion Controls

Most pistons are **cam ground** in a slight oval shape to improve piston fit when fully warmed up. The piston diameter is narrower on the pin side. As the piston expands the pin side gets hotter and expands more. A cam-ground piston is designed to become round and fit the cylinder closely at normal operating temperatures.

Pistons are also tapered from the top to the bottom. The top of the piston runs much hotter than the skirt. To allow room for the excess expansion the piston is manufactured smaller at the top. The top is designed to fit the cylinder perfectly when the piston is fully warm **(Figure 17)**. To measure piston diameter use the manufacturer's guidelines; measuring at the wrong height could give you improper readings. Most pistons are measured 90° off the pin and at the centerline or just below.

If you live in a cold climate you can notice the effects of these expansion controls. When the engine is just started and there is frost on the windshield it is normal to see new cars with some blue smoke coming from the tailpipe. If you follow them on the highway you will see the smoke soon disappear.

Piston Pin Offset

The piston pin is generally offset from the centerline toward the **major thrust side**. The major thrust side is the side of the cylinder wall the piston is pushed against on the power stroke. The minor thrust side is the side the piston rides against as it pushes upward on the compression stroke. The purpose of the offset is to reduce the amount of piston slap as the piston is forced from the minor to the major thrust side just after it passes top dead center. When the piston is pushed up on the compression stroke it rocks slightly so that the lower skirt on the major thrust side contacts the cylinder wall **(Figure 18)**. As the piston crosses TDC that contact allows the rest of the skirt to ease its way over to the major thrust side **(Figure 19)**. Without that tilt of the piston, the skirt would ride up the minor side, reach TDC, and be slammed against the other side by combustion pressure. If the piston pin is offset the piston will have a front identifying mark. If you were to install the piston in the cylinder backward you would notice a loud knocking noise as the piston skirt slapped the wall. When replacing pistons be sure to note which side of the connecting rod is toward the front. If you put the rod on backward and then line up the markings on the rod and cap during installation the cap will be on backward.

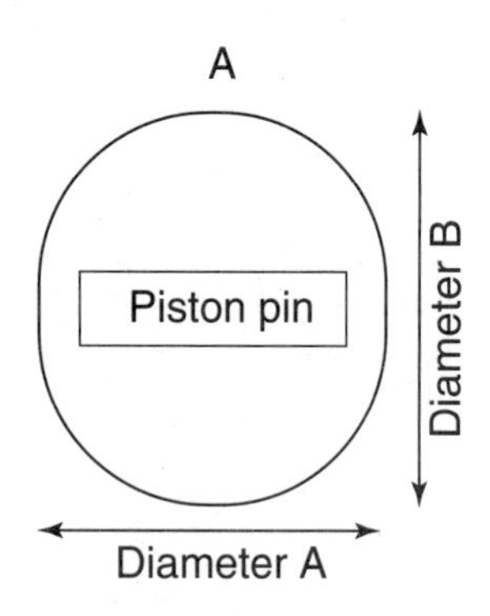

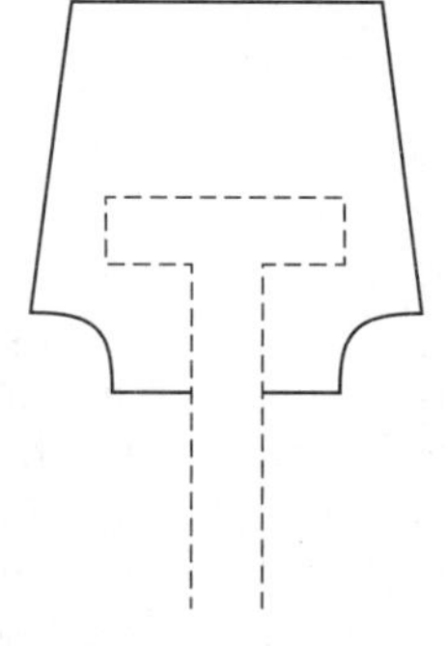

Figure 17. The piston is cam ground (A) and tapered (B) to provide good cylinder fit when running at normal operating temperature.

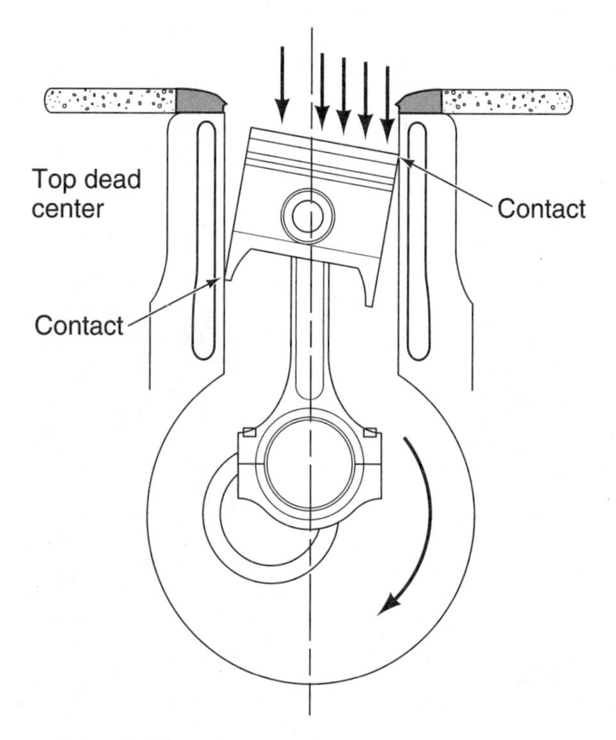

Figure 18. The pin offset causes the piston to tilt in the bore at TDC, allowing contact on both thrust sides of the cylinder.

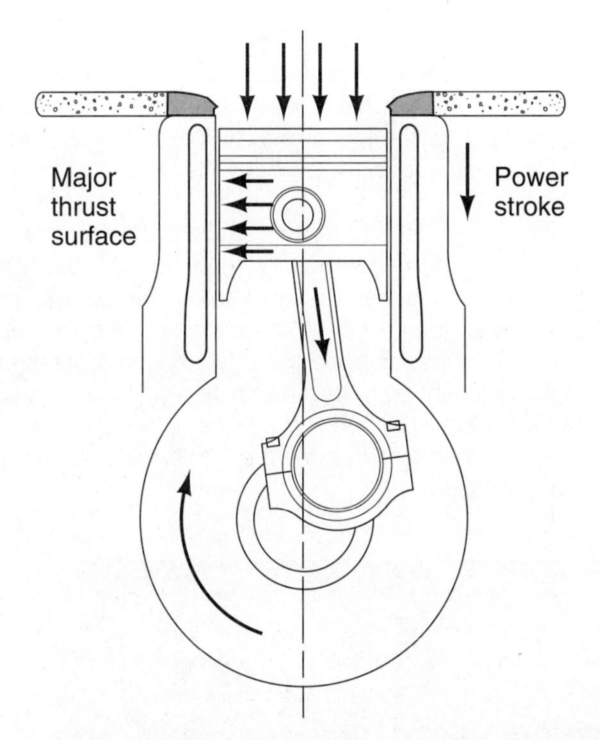

Figure 19. The pounding to the major thrust side caused by combustion is cushioned by the early contact at TDC.

A young man brought his newly restored car in with a knocking noise. He had just overhauled and modified the engine. It ran well and had good oil pressure, but there was a distinct knocking coming from the top of the block. After the head was removed we noticed the piston markings; they were all facing the back of the engine. The noise went away completely after the engine was reassembled with the pistons installed properly.

PISTON RINGS

The piston rings must seal the gap between the piston and the cylinder wall. They will keep the compression and combustion gases in the combustion chamber so that the engine can produce full power. Combustion gas that leaks by the rings is called blowby. A little blowby is inevitable, but excess blowby will reduce power, clog up the oil rings, and contaminate the oil, accelerating engine wear.

Rings have evolved over the years, so they are thinner than their earlier counterparts. This reduces engine friction and reciprocating weight. Newer rings are also made of harder materials for longer wear. These improvements cause lower ring tension on the walls. Older, softer rings could conform to out-of-round and tapered cylinder walls and do a satisfactory job of sealing them. Today's harder, thinner rings require smooth, round cylinders with the correct surface finish to function as designed.

The top ring is the primary compression ring. It does the lion's share of preventing combustion gases from being blown by the piston. The second ring should catch any gas that leaks past the top ring. While it is called the second compression ring it is also designed to assist with oil control. The bottom ring is the oil ring; it generally comprises two thin oil-scraping rings separated by an expander. The expander keeps the thin scrapers in position and provides holes for the oil to drain back into the piston and down to the oil pan. Rings have different designs, but all serve these basic purposes. Many compression rings have a chamfer or taper on one edge to improve sealing. The rings are carefully designed to match the operating characteristics of the engine. These rings are directional; when fitting new rings look closely for a top marking. This may be a small notch, dot, or hash, or an actual top marking **(Figure 20)**.

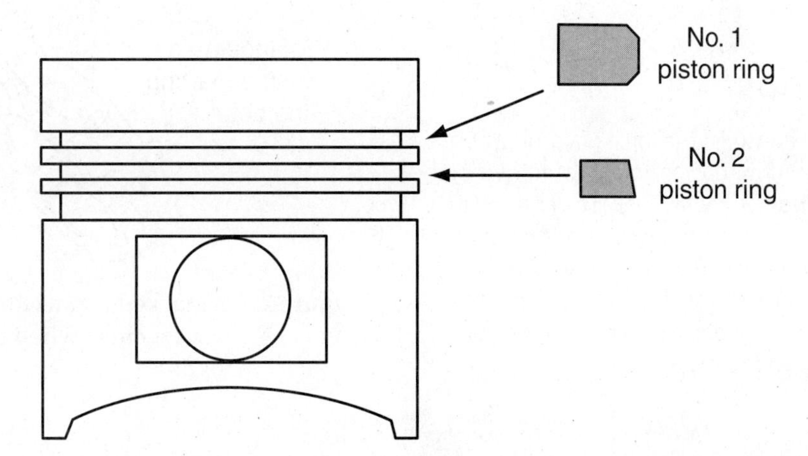

Figure 20. Many rings are designed to work with a particular side up. Look for top markings on rings that are not square cut.

Ring Materials

Rings have long been made from cast iron. These rings are a popular choice when overhauling an engine. The material is strong but soft. It can conform best to irregular cylinder surfaces and shapes. When reringing a worn engine without boring the cylinders cast rings are usually recommended. When selecting a replacement ring for a fully overhauled engine be sure that the quality meets or exceeds that of the original equipment.

Other rings are made of steel, or coated with chrome or moly. Many late-model engines are fitted with steel top rings for their increased durability. The top ring suffers the most; it is exposed to the combustion flame and the highest pressures. If an engine is originally equipped with a steel top ring it should be replaced with the same steel ring. Many modern turbocharged, supercharged, high-output engines and engines with the top ring positioned very close to the piston head require steel top rings.

Moly rings are cast rings with a groove cut in the face that is filled with molybdenum. Moly rings provide excellent scuff resistance and superior durability. The moly holds a little oil to lubricate the ring and reduce the friction against the wall. This is an excellent replacement choice for a freshly bored engine, particularly if has been modified.

Chrome-coated rings are another popular replacement option. They are stronger and can last 50 percent longer than plain cast-iron rings. They do not resist scuffing as well as moly rings, but they also do not absorb contaminants the way moly rings can. When an engine is operated in dusty or sandy conditions consider chrome rings. They will not seal an out-of-round cylinder, however, so they should be used only in freshly bored and properly finished cylinders.

Cast-iron rings seat best with a rougher cylinder finish. Chrome rings are harder to break in and also require a slightly rougher surface finish. The cylinders may be honed with 180- to 220-grit stones. The information provided with the rings should offer the optimal surface finish. Moly rings will seal better against a smoother wall; 220- to 280-grit stones may be required.

CONNECTING RODS

The connecting rod transmits the force of combustion to the crankshaft. Connecting rods are made of forged or cast steel or sintered powdered metal. They are designed in an I-beam construction to resist bending under extreme forces. The big end of the connecting rod houses the bore that the crankshaft rides in. The bore is machined with the cap in place. The cap and rod should be marked before disassembly so that the cap will be reassembled in the proper direction. The big end of the connecting rod is often reconditioned during an engine rebuild to make the bore round again; heavy use stretches it. The little end has the wrist pin bore. Sometimes the pin is press fit into the bore. Other times the pin rotates in the bore, and the rod is fitted with a bushing.

In a newer method of rod construction the rod is made of one piece of sintered metal **(Figure 21)**. The big end bore is formed, and then the cap is scored and cracked off. This forms an irregular parting surface between the cap and the rod. When the two pieces are bolted together it forms a very tight bond because there is so much surface area for gripping. The bore will be exactly the same shape as when it was formed because the cap fits precisely on the rod **(Figure 22)**.

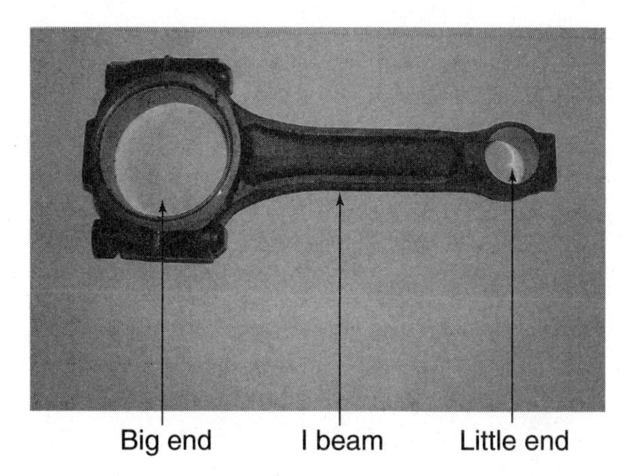

Figure 21. A typical connecting rod.

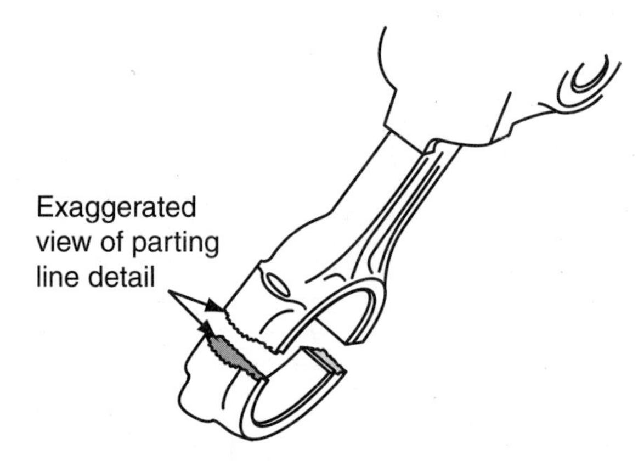

Figure 22. A cracked connecting rod cap will form a perfect circle when it is bolted back together.

HARMONIC BALANCER

A two-piece harmonic balancer is used on the front end of the crankshaft. It is used to absorb the torsional vibrations of the crankshaft created as the crankshaft speeds up and slows down due to firing impulses on different cylinders. The inner piece is bolted to the crankshaft. An inertia ring is bonded with rubber to the inner ring **(Figure 23)**. As the crankshaft winds and unwinds, the twisting between the rings absorbs some of the vibrations. If the bonding fails the two rings will separate, and you will need to replace the harmonic balancer. A faulty harmonic balancer can cause engine vibration, resulting in a broken crankshaft.

Figure 23. The rubber bonds the two-piece balancer together.

FLYWHEEL

The flywheel is a large round disc of heavy mass attached to the rear end of the crankshaft to smooth the rotation of the crankshaft. It helps the crankshaft speed remain fairly constant between firing pulses by storing inertial energy. The outer ring, the ring gear, is pressed on to provide teeth for the starter pinion gear to turn the engine over. Flywheels are used only on engines mounted to a manual transmission. An automatic transmission is mounted to the engine through a **torque converter**. This heavy fluid coupling does an excellent job of absorbing speed changes.

Summary

- The bottom end of the engine is made up of the block, crankshaft, pistons, and connecting rods.
- The cylinder block may be cast iron or aluminum.
- Cylinders are typically integral on a cast-iron block and are steel sleeves on aluminum blocks.
- The block's main bore holds the crankshaft.
- The main bore is line bored with the caps in place. Main caps must be installed in their original location and in the same direction.
- Crankshafts may be made of cast iron or forged steel; forged cranks are stronger.
- The crankshaft spins on its main journals. The connecting rod journals are offset from the centerline, which determines the stroke.
- Soft-faced main and connecting rod bearings protect the crankshaft journals from wear.
- Pistons transfer the power of combustion to the connecting rod.
- Pistons may be cast, forged, or hypereutectic aluminum.
- To provide a precise fit when warm, pistons are cam ground and tapered.
- The piston pin is offset toward the major thrust side to minimize piston slap on the power stroke.
- Piston rings seal the combustion chamber and prevent oil consumption.
- Rings are often cast iron, steel, moly, or chrome.
- The connecting rod transmits combustion forces to spin the crankshaft.
- The harmonic balancer bolts to the front of the crankshaft to reduce torsional vibration.
- The flywheel or torque converter helps maintain crankshaft speed between power pulses.

Review Questions

1. List the main components in the engine block.
2. What are the advantages and disadvantages of an aluminum block?
3. What two components are used to help maintain a fairly constant crankshaft speed?
4. The piston pin is offset toward the __________ thrust side to reduce __________.
5. List four characteristics of engine bearings.
6. Technician A says that cast-iron blocks use integral cylinders. Technician B says that aluminum blocks may use steel sleeves to form the cylinders. Who is correct?
   A. Technician A only
   B. Technician B only
   C. Both Technician A and Technician B
   D. Neither Technician A nor Technician B
7. Each of the following is a common type of piston *except:*
   A. Forged steel
   B. Forged aluminum
   C. Cast aluminum
   D. Hypereutectic
8. Technician A says that pistons are not round when they are cold. Technician B says the piston head is a larger diameter than the piston skirt. Who is correct?
   A. Technician A only
   B. Technician B only
   C. Both Technician A and Technician B
   D. Neither Technician A nor Technician B
9. Technician A says that main bearing clearance is typically .007 in. to .009 in. Technician B says that if the clearance is too great oil pressure will decrease. Who is correct?
   A. Technician A only
   B. Technician B only
   C. Both Technician A and Technician B
   D. Neither Technician A nor Technician B
10. A harmonic balancer has separated. Technician A says to reseal the bond with Loctite. Technician B says that a faulty balancer can cause the crankshaft to break. Who is correct?
    A. Technician A only
    B. Technician B only
    C. Both Technician A and Technician B
    D. Neither Technician A nor Technician B

Block Measurement and Analysis

Introduction

To properly overhaul an engine or to determine whether a rebuilt crate engine should be installed, carefully inspect, measure, and analyze the bottom end. In many cases you will be able to see problems if you look closely. Some components require measurement; you would not be able to see out of round in a rod bore. Make sure the components you are measuring are perfectly clean. A small particle of dirt could easily throw your measurement off by a couple of thousandths of an inch. This error is enough to cause unneeded machining or replacement of components. Your precise and thorough assessment of the engine will produce a quality rebuild or repair. Record each of your measurements and observations; this is called blueprinting your engine. It ensures that you will have all the information you need clearly laid out for your analysis.

When you finish, look at the whole condition of the engine to determine the necessary repairs. New rings, bearings, and gaskets are always installed during an overhaul, but many times additional repairs are required to fully recondition the bottom end. We will focus on the repairs normally performed within the general automotive repair shop where the vast majority of automotive technicians work. Many machining procedures are typically completed at an engine machine shop. If you work in a production engine shop the machining equipment would be available in house. If you go to work at a machine shop or production engine shop you will need additional training on the setup and procedures for the many pieces of machining equipment. The number of available jobs for engine machinists is limited. These numbers are decreasing as the trend in the industry is toward installing replacement crate engines rather than overhauling high-mileage engines.

An engine rebuilder can make short work of the many measurements required to thoroughly analyze an engine. Those of us who rebuild engines from time to time in a general service shop will often measure three times to be sure our readings are accurate.

BLOCK CLEANING

The block must be cleaned thoroughly. Strip the block of all parts, and remove the core plugs and oil gallery plugs. Clean the block in a **hot tank** to remove all oil, sludge, carbon, and varnish **(Figure 1)**. Brush the oil galleries well to be sure they are perfectly clean. Inspect the cooling passages thoroughly, and remove all deposits that have built up in them. Lightly oil the block after cleaning to prevent rust from forming. Many general repair shops do not have this cleaning equipment available. You may need to send the engine out to a machine shop to have it cleaned. When this is the case clean the block well enough to perform your inspections and measurements. After you have determined all the necessary repairs, send the block to the machine shop to have it cleaned and have any needed machining done.

BLOCK INSPECTION AND CRACK DETECTION

Carefully inspect the block for damage. Look for cracks, especially in the cylinder walls and in the lifter valley. If an engine had a coolant consumption problem or if it was

Figure 1. The block is throughly cleaned in a hot tank before inspection and machining.

overheated use a special procedure to check for cracks that cannot be detected visually. Cast-iron blocks can be checked by **magnafluxing.** To magnaflux the block you apply magnets at either end of the casting and spread a solution on the block that contains metal filings. If there is a crack the filings will line up on either side of it, attracted by the opposite charges across the crack. To check for cracks on an aluminum block pour a special dye penetrant over the block. If there is a crack the dye will settle into it and highlight the crack under a black light so you can see it. You may have to send the block out to the machine shop to have these procedures performed **(Figure 2)**.

Figure 2. The cracked block of this racing engine will be pinned and reused.

BLOCK DECK FLATNESS

Clean the top of the block thoroughly so you can measure the deck flatness. Lay a straightedge across the block lengthwise and diagonally. Try to fit feeler gauges under the straightedge **(Figure 3)**. The thickest blade that fits indicates the deck warpage. Check the specifications for your block; many manufacturers specify less than .004 in. warpage. If the block is warped the head gasket may not seal properly. The block can be "decked" or machined true using a special surfacing machine found in a machine shop or a production engine shop. Manufacturers specify different surface finishes to meet the requirements of the head gasket for proper sealing.

Figure 3. Clean the deck thoroughly and then check for warpage. If the deck is not flat, the head gasket will not seal.

MAIN BORE MEASUREMENTS

During your inspection of the main bearings you look for unusual wear patterns. If the bearings are worn differently front to back or show signs of wear on the sides of the bearing rather than on the top and bottom, you should check the condition of the crankshaft and the block main bore. Also, if the engine is aluminum, has been overheated, or has suffered a bent connecting rod, check the main bores for alignment and out of round. To confirm proper alignment of the main bore remove the bearing inserts, and torque the main caps into their proper positions. Lay a straightedge in the bore, and try to fit a .0015 in. feeler gauge under the straightedge. Place the straightedge in a few different locations in the bore. If the feeler gauge slides under the straightedge the main bore is out of alignment **(Figure 4)**. The block must be line (or align) bored to repair it, and oversize bearings must be installed. Main bore misalignment promotes rapid bearing wear and can even cause the crankshaft to seize. If the bore misalignment is less than .001 in. the block is aligned satisfactorily according to most manufacturers. As always you should check the particular specifications for the engine on which you are working.

With the caps still torqued on check the bores for stretch. The constant downward forces on the crankshaft can cause the main bores to stretch out of round. Use a telescoping gauge or inside micrometer to measure the diameter of the bore at the top and bottom, and then compare the measurement with the diameter of the bore from side to side **(Figure 5)**. If the diameters differ by more than .001 in. the bores are too far out of round to provide proper oil clearances. Knocking, premature bearing wear, and low oil pressure could result. The block must be line bored to repair main bore out of round. Then oversize bearings must be installed. An engine that requires line boring is a good candidate for replacement.

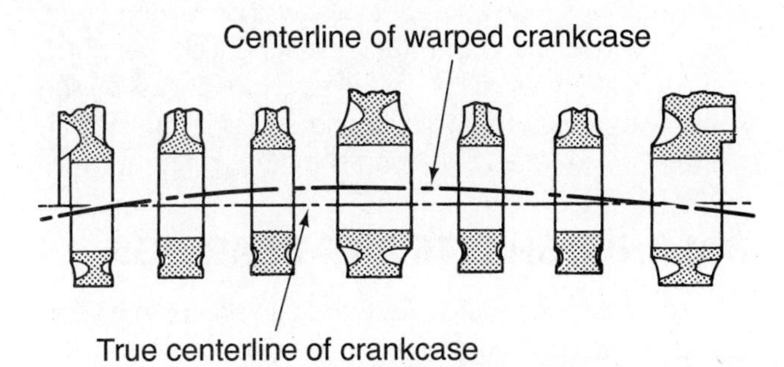

Figure 4. If the crankcase is misaligned, the block must be line bored and oversize bearings installed.

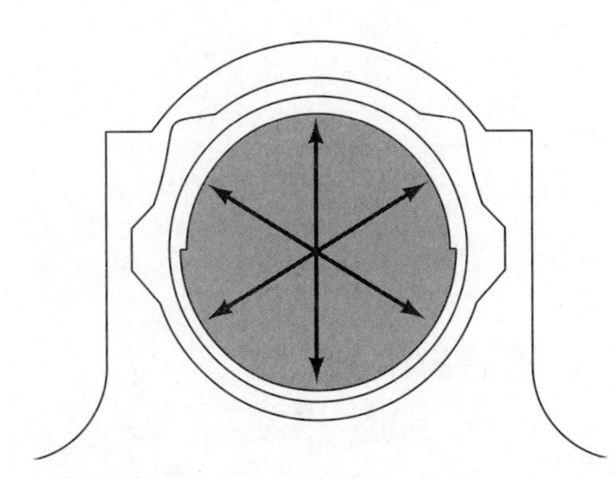

Figure 5. Measure the main bore for stretch along the top and bottom diameter. Line bore the block to repair out of round greater than .001 in.

LIFTER BORES

Visually inspect the lifter bores for scoring or wear. Clean the bores with a stiff brush to remove any varnish. Measure the diameter of the bores and compare the result with specifications. If necessary, the lifter bores can be honed to fit oversize lifters.

MAIN, ROD, AND CAM BEARING ANALYSIS

Engine bearings are always replaced during an engine overhaul. Bearings will wear as mileage is put on an engine. Dirt in the oil will increase the rate of wear dramatically. How they wear can tell you a lot about the condition of related components. **Figure 6** shows bearings with varying degrees of wear. These are normally worn bearings. Bearings will often wear through the top layer of babbitt and expose the copper underlayer. Severely worn bearings will have the copper exposed all the way around the bearing surface. Scoring will occur from oil contamination, infrequent oil changes, excessive blowby, or component wear. Main bearings will wear more on the lower halves because the downward force of the crank is exerted there. The rod bearings will wear more on the upper halves as the rod is pushed down against the crank. A bearing that has spun in its bore will be loose in its cap **(Figure 7)**. Both the inner and outer sides of the bearing will be scored. You will be able to see scoring in the bore as well. This requires repair to the bore; do not simply attempt to install a new bearing if the old one has spun. A repeat failure is almost inevitable. This fault with the block could also make replacement an effective alternative to overhaul. Also make sure that the oil supply holes in the block and crankshaft are clear. Oil starvation is a common cause of spun bearings. Severely worn bearings will usually damage the crankshaft journals; inspect them carefully.

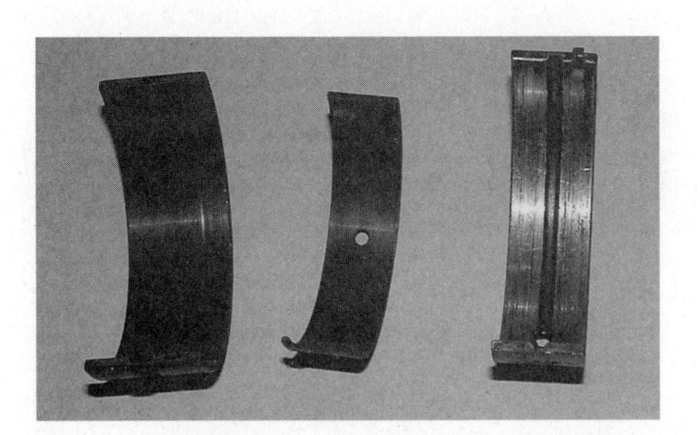

Figure 6. The bearings on the left are worn down to the copper. The ones on the right show moderate scoring.

Figure 7. The bearing spun on this journal and destroyed a specialty racing crank.

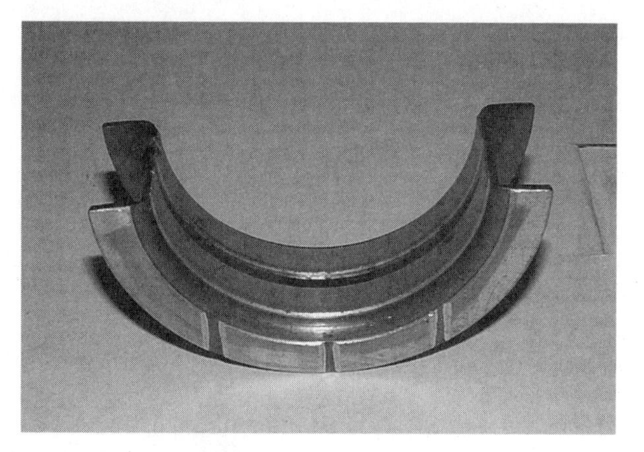

Figure 8. This thrust bearing is not badly worn, but the bearings are always replaced during an engine overhaul.

Line up the bearings in their proper order to examine them. They should be worn across the bearing. If a bearing is worn on just one edge it points to a problem with journal taper. If the tapered wear is on a rod bearing it could also be caused by a bent connecting rod. Look at the wear of the main bearings from one end of the crank to the other. The wear should occur evenly from the front to the back. The main bearings will normally wear more on the bottom bearing half. If the bearings are worn significantly on the tops at the end and on the bottoms in the center it is likely that the crankshaft is bent and/or the main bore is out of alignment. If the front bearings are worn on the top and the rear bearings are worn progressively toward the bottom it is possible that the accessory drive belt was overtightened. Wear on the sides of the bearing halves near the parting lines is a sign of an out-of-round bore. As the bore stretches downward with the force of the crank, the sides are pulled in and cause drag on the sides of the bearing. Inspect cam bearings for severe scoring or for signs that the cam is bent. If a bearing is worn severely or has spun in its bore be sure to check the oil supply passages.

The thrust bearing will also show signs of wear through normal use **(Figure 8)**. An excessively worn thrust bearing can cause damage to the crankshaft and block. Check the areas on the crank and block where the thrust bearing rides. If you can see wear in these areas repairs will be necessary. Thrust bearing failure can be caused by a defective clutch **pressure plate**, harsh shifting, or riding the clutch when the engine is coupled to a manual transmission. On an engine mated to an automatic transmission, fluid pressure in the torque converter normally pushes the crank toward the front of the engine. If the pressure is excessive the torque converter can swell or balloon under loads and damage the thrust bearing. Check the clutch pressure plate, transmission fluid pressure, or discuss driving habits with the customer if you find excessive thrust bearing wear on an otherwise sound engine.

BALANCE SHAFTS AND BEARINGS

When balance shafts are used in an engine the bearings must be checked. Some engines use traditional bearing inserts like cam bearings to support the shafts. These should be replaced during an engine overhaul. Other engines may use ball or roller bearings to support the balance shafts. These may not require replacement. When they are cleaned and oiled check them for smooth rotation and an absence of play or wear.

Check the balance shaft journals for damage. If the bearings are badly worn you will likely find scoring on the journals. If this is not corrected it will cause premature wear of the new bearings. It is typical to replace, not refinish, a worn balance shaft unless the scoring is so light that you can successfully polish the journals with emery cloth.

CYLINDER WEAR

The cylinders must be properly sized, round, and not tapered in order for new rings to seal properly. You can often gauge the degree of wear by the depth of the ring ridge. To determine whether the cylinders will need to be bored you will measure the amount of cylinder wear. If the cylinder walls are scored deeply they must be bored. Cylinders do not wear evenly. The cylinders wear more on the thrust diameter than on the nonthrust sides. This makes the cylinders out of round. The higher heat and pressure against the piston when it is at the top of the cylinder causes more wear at the top than at the bottom. The cylinders become tapered wider at the top and nearer their original dimension at the bottom of ring travel.

Cylinder Diameter and Out of Round

Measure cylinder diameter in the center of the cylinder with a telescoping gauge and micrometer or inside micrometer. To measure cylinder out of round use a dial bore gauge or a telescoping gauge and micrometer. Measure the cylinder near the top of ring travel on the thrust sides and then on the nonthrust sides **(Figure 9)**. Compare the readings on the dial of a bore gauge; the difference is out of round. When using a telescoping gauge and micrometer subtract the diameter of the nonthrust side from the thrust side diameter. This gives you the measurement of out of round. Look at the following example:

	Cylinder #1	Cylinder #2	Cylinder #3	Cylinder #4
Thrust Diameter	3.656"	3.656"	3.656"	3.657"
Nonthrust Diameter	−3.651"	−3.652"	−3.650"	−3.651"
Out of Round	.005"	.004"	.006"	.006"

The original diameter of the cylinders is 3.650. Compare your measurements with the specified diameter; the engine could have been bored oversize. You would need to know this when ordering parts. This engine has cylinder out of round between .004 in. and .006 in. On an older engine this may be just barely acceptable. Many newer engines specify a maximum of .002 in. or .003 in. of out of round. Some newer engines allow for absolutely no out of round or taper. It is important to check the specifications for the engine you are working on to make a good repair decision. If the out of round is beyond the allowable limit the engine should be bored to produce a powerful and long-lasting engine. When new rings are installed in a cylinder with too much out of round blowby, compression loss and oil consumption can result. When an engine requires boring new oversized pistons will have to be fitted. Given today's trend toward replacement these findings may dictate engine replacement rather than overhaul.

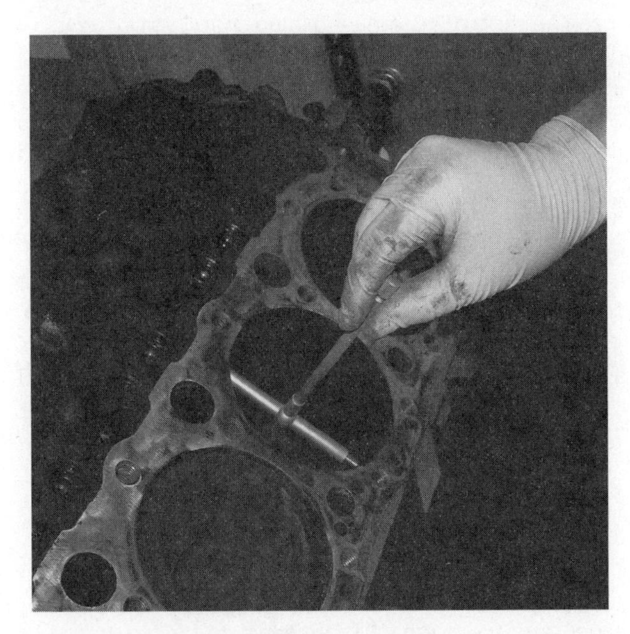

Figure 9. Measuring bore diameter on the thrust sides with a telescoping gauge and micrometer.

Cylinder Taper

The cylinders become tapered as mileage increases on an engine or if the engine is allowed to run below normal operating temperature for extended periods. Something as simple as a stuck-open thermostat can cause the cylinder to become tapered beyond allowable limits. To measure taper use a dial bore gauge **(Figure 10)**. Place the gauge at the bottom of ring travel in the cylinder, and zero the dial. Steadily drag the bore gauge up to the top of ring travel, and read the amount of taper on the dial. Repeat the procedure a few times to be certain you are getting accurate readings.

Check the specifications for the maximum allowable taper on your engine. Many manufacturers specify less than .005 in. of taper. Typically, the newer the engine is, the lower the allowable taper will be. If an engine is assembled with too much cylinder taper blowby, compression loss, and oil consumption may result. In addition, rings may break if the cylinders are tapered excessively. At high rpms the ring may

Figure 10. Use a dial bore gauge to measure cylinder taper. The cylinder will wear more at the top, showing a larger diameter.

not have time to contract to the smaller dimension as it nears the bottom of the cylinder. Instead it may score the cylinder and snap, often breaking a piston ring land.

CYLINDER REPAIRS

If the cylinders are slightly worn but within wear limits the manufacturer often specifies that you brush hone the cylinders to deglaze the walls **(Figure 11)**. A brush hone is used on the end of a drill and run up and down in the cylinder. This will break the glaze and put a proper crosshatch in the surface. The crosshatch helps seat the new rings and holds some oil to lubricate the walls and the rings. The crosshatch should usually be at a 20°–40° angle off the horizontal plane of the deck **(Figure 12)**. Run the hone only until a distinct crosshatch pattern is visible; you do not want to remove excessive material from the cylinders. Brush honing will not correct problems of taper or out of round. Some manufacturers do not suggest brush honing; follow the manufacturer's recommendations closely.

Clean the cylinders thoroughly with hot soapy water and a brush after brush honing. Then dry and lubricate the cylinder walls to prevent rust. It is critical that you get all the loose material off the cylinder walls and out of the block. If the block is not thoroughly cleaned the new bearings will be contaminated with the small metal shavings. It can also cause ring and cylinder wall damage.

Figure 11. Use a brush hone to deglaze a cylinder that measures within specifications for out of round and taper.

Figure 12. The brush hone produced a fine crosshatch that will help hold oil for the rings and pistons and speed up the ring break-in period.

When the cylinders are worn beyond specifications they must be repaired. The most common way to repair the cylinders is to bore them .020 in., .030 in., .040 in., or .060 in. oversize and install new oversize pistons. Be sure that the correct-size pistons are available before specifying a bore size. Every block has a bore limit; if the block has previously been bored be sure to check the limit. Some new engines do not allow boring. When the cylinders are worn or damaged the block must be replaced. To prepare for boring a machine shop or engine rebuilding shop installs torque or deck plates on the top of the engine to simulate the load of a torqued cylinder head. Then the boring bar is carefully centered in the bore. The cutting diameter is adjusted to the proper dimension, and the engine is bored. When an engine is to be bored .030 in. oversize the cylinders are actually bored to .028 in. The last .002 in. is taken out of the cylinder with stiff hones to achieve the proper surface finish **(Figure 13)**. Boring restores the cylinders to excellent condition and ensures long engine life.

When one cylinder is badly damaged but the others are in fine condition the affected cylinder may be bored greatly oversize. Then a sleeve is press fit into the new bore. This would be a desirable repair option on a low-mileage engine that had one connecting rod let loose, for example. On an engine originally equipped with wet sleeves they are replaced when worn rather than machined.

On many occasions you will have to make a judgment call about boring. Boring an engine will add significant expense to the overhaul. The machining process may cost between $100 and $300, and then new pistons are required. This may make engine replacement a good option. You should clearly explain the repair options to the customer

Figure 13. This hone will finish a cylinder after boring. The stones can be changed to provide the proper surface finish; these are 400 grit stones for use with moly rings.

when the measurements are close. If an engine is barely out of specification the customer may not want to spend much money on the overhaul. If the engine is going back into a 10-year-old car that the customer intends to drive only for another year or two it might be practical to brush hone the cylinders and install new cast-iron rings. The engine may smoke a little over time, but it will likely perform satisfactorily for the relatively short period the customer wants the car. You must thoroughly explain the consequences of a compromised repair choice to the customer. It is wise to document your discussion on the repair order in case there is a concern later on. The best answer for a worn engine would be to bore it and install oversize pistons. Then you could reassure the customer that the engine will provide trouble-free service for an extended period. If the engine is more than a few thousandths out of specification do not offer a cheaper repair option to the customer; the work will not result in a satisfactory outcome.

PISTON INSPECTION

Inspect the piston for excessive scuffing. Some light scuffing is normal and does not require piston replacement **(Figure 14)**. Look at the scuffing carefully; it should be straight up and down on the piston. Irregular or diagonal

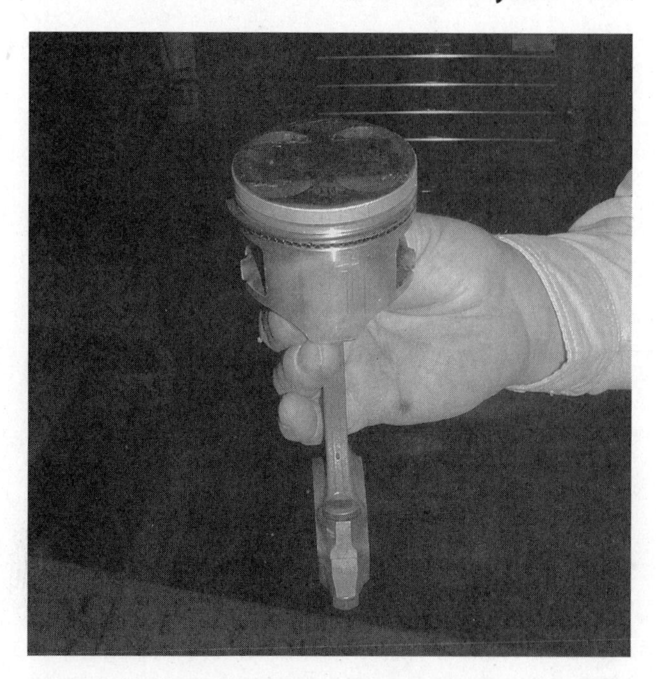

Figure 14. This piston has some moderate vertical scuffing on the major thrust side. Its diameter is still within specification; both the piston and rod can be reused.

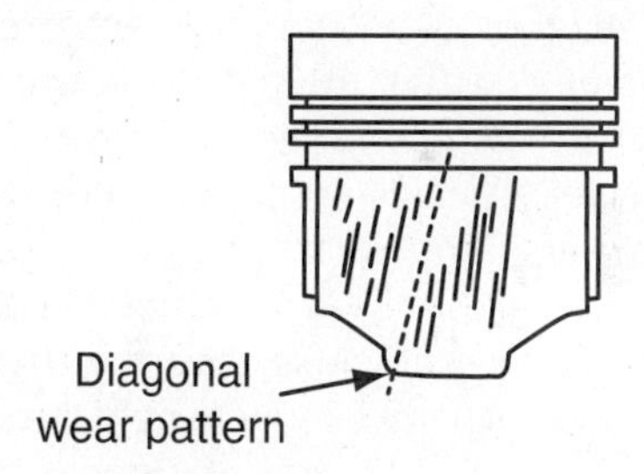

Figure 15. Diagonal scuffing on the piston skirt indicates a bent rod. A machine shop can verify this on a rod aligner.

scuff marks indicate a bent connecting rod **(Figure 15)**. A bend may be difficult to see on the connecting rod itself.

Measure the piston diameter using a micrometer at the point specified in the service information. Subtract the piston diameter from the cylinder diameter to calculate piston clearance. Compare the measurement with the specification. If piston clearance is too great compression loss, blowby, and oil consumption may occur. Replace worn pistons, or bore worn cylinders to correct piston fit issues. When installing new pistons check that the piston clearance matches the specifications. Pistons with too little clearance will score the cylinder walls and may break.

Look at the pin boss area for cracks. Check the fit of the piston pin by pulling and pushing the piston up and down

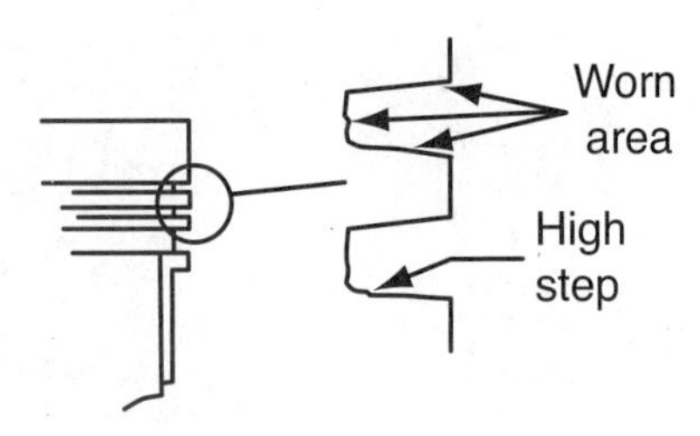

Figure 16. The ring grooves can wear in a bell-mouth fashion. New rings will not seal the cylinder properly and can break from the excess movement in the groove.

on the rod. Any detectable movement indicates a worn pin or bore. If there is wear remove the piston pin, measure the pin bore, and inspect the pin. It may be possible to bore out the pin hole and install bigger pins, but usually the pistons are replaced. If the bore is fitted with a bushing to support the pin the bushing can be replaced.

If the pistons will be reused remove the rings and clean the grooves. It is best to use a ring expander to remove the rings to avoid scratching the side of the piston. A ring groove cleaner can be used to clean the carbon from the grooves. If the proper size bit is not available snap an old ring and use it to scrape the groove clean. Inspect the grooves for wear. They may wear in a bell-mouth fashion; the back of the groove retains its original diameter, but the opening of the groove has worn significantly from the ring rocking up and down as the piston changes direction. Worn grooves necessitate piston replacement. Inspect the ring lands for wear, damage, or cracks **(Figure 16)**.

To remove a piston from the rod you will need a press and some special tools to hold the piston. On pistons with full-floating wrist pins simply remove the snaprings and push the pins out by hand. Be sure to note the direction of the piston on the rod using the piston front markings and the rod number markings. The new piston should be installed in the same orientation on the rod on press fit pins. A press tool must be used to secure the piston in the press on the boss area without contacting the top of the piston. When the piston is supported properly press the pin fully out of the bore. Inspect the pin closely for signs of wear, and measure its diameter. The pin is case hardened, so it is usually the pin bore that wears. Many technicians and machinists will heat the rods and freeze the pins to ease the press installation procedure. Be very careful to start the pin squarely in the bore or you will damage the new piston.

CONNECTING ROD INSPECTION

Inspect the connecting rod for cracks, straightness, and signs of damage. Look closely at the scuffing on the pistons and the wear on the rod bearings to determine whether the connecting rod may be bent. Wear on the outside edges of the bearings may indicate a bent rod. To verify a suspicion of a bent rod check it on a rod-aligning fixture. These are found at machine shops and engine rebuilding shops. Compare the costs of machining versus replacement when deciding how to correct connecting rod problems.

Check the big end of the bore for stretch if the bearings show wear on the edges near the parting lines or if it is a high-mileage engine. Measure the diameter of the bore vertically and then horizontally; compare the results to see if the big end bore is out of round. The rod should be replaced or reconditioned if the bore is more than .001 in. out of round. Rods can be resized at the machine shop to restore the big end bore **(Figure 17)**. A small amount of material is shaved from the parting faces to make the bore smaller. Then the bore is honed to the proper size. Sintered powder metal rods with cracked caps are typically replaced when damaged, but oversize bearing shells may be available to allow resizing.

The rod little end should be inspected when the pistons are removed with press-fit pins. Look for damage in the rod bore. Measure the diameter and compare it with specifications. If the bore is worn the rod can either be replaced or it can be bored out and have a sleeve installed. On rods with full-floating pins the pins should be disassembled and inspected during an overhaul. The bushings are commonly replaced during an overhaul. This is typically a machine-shop procedure. The old bushing is pressed out,

Figure 17. This rod reconditioning tool can resize the big or little end of the rod so it can be reused.

and a new one is pressed in. Then the bushing is rolled to tension it within the bore and honed out to size.

CRANKSHAFT INSPECTION AND MEASUREMENT

In many cases you will be able to condemn a crankshaft just by looking at it. If the journals are scored enough to catch a fingernail the crankshaft must be reground or replaced. Most crankshafts used on common production engines are replaced rather than reground due to the labor-intensive process of setting up the crank for grinding. Check each journal all the way around for scoring. Very light scoring may be polished with emery cloth.

Inspect the fillet area next to the journals for cracks in the crankshaft. This is where the edge of the journal is radius cut to allow oil to leak out the sides and prevent varnish buildup from restricting crank movement. You rap the counterweights lightly with a hammer and listen for the ring of the crankshaft. A dull thud indicates a crack in the crank. If the crank shows signs of overheating, the harmonic balancer is separated, or if the engine suffered a catastrophic failure have the crankshaft magnafluxed to be certain it is not cracked **(Figure 18)**.

Use your bearing inspections to help uncover problems with the crankshaft. The main bearings should be worn more on the bottoms. The wear should be centered in the bearing. If the main bearings are worn on the top halves in the front and rear and the bottom halves in the middle, for example, suspect a bent crankshaft. Bearings that are worn on the edges suggest a tapered journal. Whenever these indications are found measure the crankshaft carefully and thoroughly.

Figure 18. Despite significant journal damage, this $1,400 racing crank could have been repaired if a crack had not been detected during this magnafluxing procedure.

Crankshaft Straightness

To measure crankshaft straightness lay the crankshaft in vee blocks. Set up a dial indicator to contact one of the center main journals. Spin the crank while watching the dial indicator. Repeat at another journal toward one end. Deflection of the indicator over .001 in. dictates that the crankshaft be replaced or straightened. Only forged crankshafts can be straightened. A bent crankshaft can seize in the block or prematurely wear new bearings.

Another way to verify that the crankshaft is straight is to torque it into place with new oiled bearings. Rotate the crankshaft using an inch-pound torque-o-meter. The tension on the gauge should be consistent throughout rotation. The downside to this technique is that discovering another faulty part during the reassembly process could change your estimate and hold up your work while you wait for a new or straightened crank.

Crankshaft Journal Measurements

The crankshaft journals should be measured for size, out of round, and taper **(Figure 19)**. If any of these measurements indicates wear over .001 in. the crankshaft must be replaced or reground. Use a micrometer to measure a rod and a main journal to be sure the crank has not previously been ground undersize. If it has you will have to be careful to order the correct undersize bearings. To measure journal out of round use a micrometer to measure the journal diameter in two spots 90° apart on the journal. The difference is out of round. Taper is found by measuring the diameter at both ends of the journal and comparing the measurements.

Figure 19. Measure the crankshaft journals to be sure the crank is within specifications and can safely be reused.

Summary

- Precise and thorough measurements and inspections are essential to a successful engine overhaul.
- Many general repair shops will send out much of the reconditioning of components to be performed at a machine shop.
- Core and gallery plugs should be removed and the block thoroughly cleaned.
- Inspect the block for damage. Have it checked for cracks if it was burning coolant or if it was overheated.
- Measure the deck of the block for warpage; generally it should be less than .004 in. to ensure head gasket sealing.
- Measure the main bores for alignment and out of round. Readings in excess of .001 in. require that the block be line bored or replaced.
- Carefully inspect the main, rod, and cam bearings for unusual wear that could point you to related component failure.
- Cylinders wear the most at the top and on the thrust sides.
- Cylinder diameter, taper, and out of round must be within specifications to provide proper ring sealing.
- Worn cylinders can be bored oversize to restore an engine to nearly new condition.
- Inspect the piston for scuffing, cracks at the pin boss, wear in the grooves, and damage to the lands.
- Check for proper piston clearance by subtracting the piston diameter from the cylinder diameter.
- Check the piston pin fit by rocking the piston up and down on the rod; any detectable movement indicates a problem.
- Analysis of the connecting rod includes verifying straightness, measuring the big end bore for stretch, and checking the pin bore for wear.
- Look at the fillet areas of the crankshaft for cracks.
- Inspect the journals for excessive scoring. Most crankshafts are replaced rather than reground.
- Lay the crankshaft in vee blocks, and set a dial indicator against the journals to measure the straightness of the crank.
- Measure the crankshaft main and rod journals for out of round and taper; readings over .001 in. necessitate repair or replacement.
- In many cases when significant engine machining would be required a replacement engine will be fitted.

Review Questions

1. What three measurements should be made on crankshaft journals?
2. Cylinders wear out of round on the __________ sides of the cylinder and become tapered toward the __________ of the cylinder.
3. What problems can occur if the engine block's deck is warped beyond the manufacturer's limit?
4. Describe the procedure for measuring piston clearance.
5. Is it true or false that all engines can be bored at least .010 in.?
6. Technician A says that the main bore can be only .010 in. out of alignment. Technician B says that a misaligned main bore can cause premature bearing failure. Who is correct?
   A. Technician A only
   B. Technician B only
   C. Both Technician A and Technician B
   D. Neither Technician A nor Technician B
7. A late-model engine has .007–.009 in. out of round and .006–.008 in. taper in the cylinders. Technician A says the engine should be bored or replaced. Technician B says brush honing and new rings would be a reasonable alternative. Who is correct?
   A. Technician A only
   B. Technician B only
   C. Both Technician A and Technician B
   D. Neither Technician A nor Technician B
8. Technician A says that a bent crankshaft will create irregular wear on the bearings. Technician B says that riding the clutch can cause thrust bearing failure. Who is correct?
   A. Technician A only
   B. Technician B only
   C. Both Technician A and Technician B
   D. Neither Technician A nor Technician B
9. Technician A says that excessive bearing clearances will cause knocking and low oil pressure. Technician B says that a stretched rod bore will cause wear on the sides of the bearing near the parting line. Who is correct?
   A. Technician A only
   B. Technician B only
   C. Both Technician A and Technician B
   D. Neither Technician A nor Technician B
10. Technician A says that if you can see scuffing on the pistons they need to be replaced. Technician B says that diagonal scuffing points to a bent connecting rod. Who is correct?
    A. Technician A only
    B. Technician B only
    C. Both Technician A and Technician B
    D. Neither Technician A nor Technician B

Block Repair and Assembly

Introduction

Once you have analyzed the engine blueprint you should make a list of all the needed parts and repairs. If the block or another component needs to be sent to a machine shop write a thorough list of the required repairs. Sometimes you will determine that the cost of all the repairs will be more than the cost of a short-block or tall-block assembly. Replacement engines are often more cost effective than overhaul when machining processes are required. They also come with a warranty. Consider parts and labor in your decision to repair an engine rather than replace it. Many parts suppliers can provide you with an engine overhaul kit that includes all gaskets and seals, rings, and bearings. A master kit is likely to include new pistons. Be sure to correctly identify the engine using the VIN and any identification codes stamped on the block. Use the manufacturer's service information to locate and interpret these engine codes. Check your blueprint to be sure you have addressed all the concerns and have the needed replacement parts available.

When you are ready to begin assembly of the bottom end make an extremely clean work area for yourself. The engine and new parts must be perfectly clean to achieve a long-lasting overhaul. Lay the parts out on a clean workbench to facilitate assembly. Make sure that you have cleaned bolts and threaded bores. They can be very lightly lubricated. If the manufacturer uses torque-to-yield bolts for the rod and main bolts they must be replaced; have the necessary parts available. Many technicians will create a checklist of procedures to perform so they are sure to do a complete job. One missed step can cause a catastrophic failure. A checklist is particularly advisable if you will be interrupted to perform other work during the assembly process.

Some production engine shops have one or more technicians whose specialty is engine assembly. One technician may analyze the engines, another will perform the machining, and the assembly specialist will build the blocks. Each technician must be an expert to ensure that each product is of high quality.

INSTALLING OIL GALLERY PLUGS AND CORE PLUGS

Install the oil gallery plugs if they have not already been installed. Make sure the threads in the block and on the plug are clean. Apply pipe sealant to the threads, and torque the plugs in to their specification. Check to be sure that each gallery hole is properly sealed.

Make sure there are no deposits or corrosion in the core plug bores. Any dirt in the bores or on the new plugs can cause leakage. Apply a hardening sealant as specified to the outside diameter of the core plug. Use a core plug tool or bearing driver to install the core plugs **(Figure 1)**. The driving face should be just slightly smaller than the edges of the plug. Do not hit the core plug in the center area; this will collapse it and distort the sealing edge. Using a punch to tap the plug in will also deform the core plug and cause leaks **(Figure 2)**.

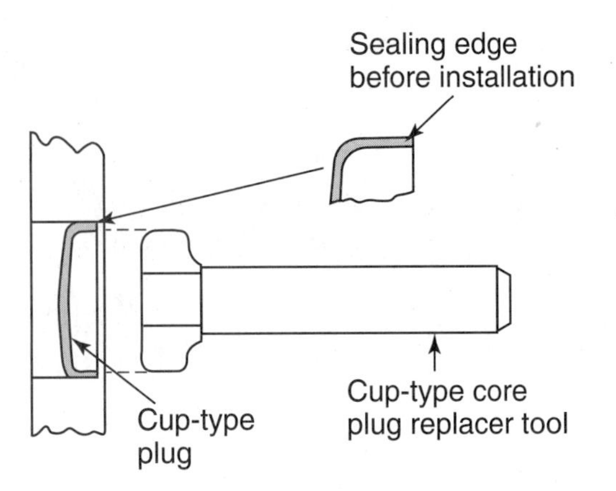

Figure 1. Use the proper tool to install core plugs so you do not damage them.

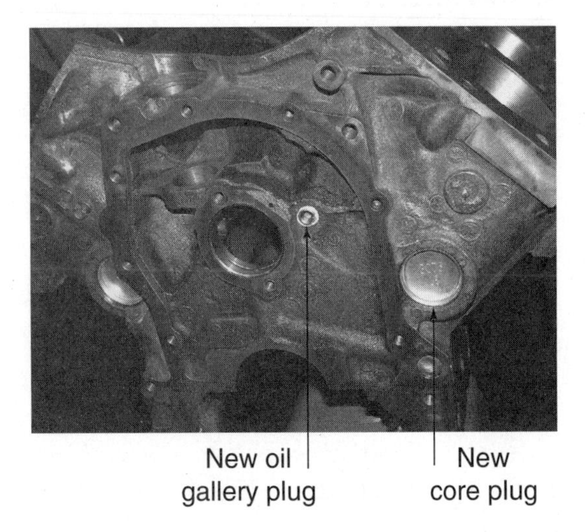

Figure 2. With new core plugs and oil gallery plugs installed, the block is clean and ready for reassembly.

INSTALLING THE CRANKSHAFT AND BEARINGS

The importance of cleanliness cannot be overstated when discussing installation of the bearings and crankshaft. Wash your hands, and lay the main bearing halves out on your clean bench. Note the difference between the upper and lower insert halves. The upper bearing half will have an oil hole. Locate the thrust bearing, and determine its proper location in the block. Place the main cap bolts on the bench, clean and lightly oiled. Use new bolts if the manufacturer specifies them. Clean the main bores with a lint-free rag.

Pick up the bearings, touching only the ends. Do not put anything on the backs of the bearings; be sure the bore is clean. Push the dry upper bearing halves into place in the bores. Line up the main caps in their proper order and direction. Push the lower bearing halves into place. Gently place

Figure 3. Lay a strip of plastigauge across each main journal to check the bearing clearances.

the clean and dry crankshaft into place. Do not rotate the crank. Lay a strip of **plastigauge** across the main journals. Plastigauge is an extruded piece of plastic manufactured to a precise thickness **(Figure 3)**.

Place the main caps over the journals, and tighten them using the proper procedure and sequence to prevent twisting the block or distorting the bores. Most main caps are tightened in stages, starting with half the torque, then three-quarters torque, then full torque. Reverse the tightening sequence, and remove the main caps. Compare the squished plastigauge with the paper scale provided with the plastigauge **(Figure 4)**. This gives you the actual main bearing clearances. Make sure they are within the specified range, typically .0008 in. to .002 in. If the clearances are not correct you may have mismeasured the crankshaft journals or received the wrong bearings. Carefully check your work to correct the problem. Do not ignore improper clearances; they can make your "new" engine scrap metal in a short time.

Figure 4. Compare the squished plastigauge to the paper scale to find the bearing clearance.

When the clearances are correct scrape the plastigauge off the crank journal with your fingernail. Do not touch the bearing insert. Lube the bearings well with oil or engine assembly lube, and torque the caps into place using the proper stages and sequence. Torque the main cap bolts twice when they are tightened to a final torque value. If the final stage of tightening is performed by twisting the bolt a number of degrees you cannot double-check your torque. Pay close attention to your work so that you are sure you have tightened each of the bolts.

With the new bearings installed and the caps fully tightened check for smooth rotation of the crankshaft. You may learn to do this by feel, but it is best to start by checking rotation with an inch-pound torque-o-meter. Rotate it by hand first to learn what a smooth crank feels like. Then check that the drag on the crank is only a few inch-pounds and that it is consistent throughout rotation. If rotation is not smooth either the crankshaft is bent or the main bore is misaligned.

Checking Crankshaft Endplay

Next you must check that the crankshaft endplay is within specifications. A typical specification might be between .002 in. and .004 in., but many newer engines specify less than that. Check the specification for the engine you are working on. Place a dial indicator on the block, and push the measuring tip squarely against the end of the crankshaft **(Figure 5)**. Rock the crankshaft forward and backward in the block using a pry bar. Note the movement of the indicator, and compare the reading with specifications.

An alternate method of checking crankshaft endplay is performed using feeler gauges. Push the engine all the way forward, and slide feeler gauges into the space between the thrust washer and the machined surface of the block. Push the crankshaft all the way to the back of the block, and measure the clearance again. The difference between the two measurements is the crankshaft endplay.

If the crankshaft endplay is out of specification it may be possible to locate a different-sized thrust bearing or washer. If the endplay is left too tight the crank can bind and cause serious damage to the block. When the endplay is too great the crankshaft will knock as it walks back and forth in its bore. This will prematurely wear the thrust bearing and could cause engine failure if the crankshaft knocks through the end of the block.

Figure 5. Check to be sure that the crankshaft endplay is within specifications. Too much endplay will cause a knocking noise on acceleration; too little could cause the crank to seize.

INSTALLING THE CAMSHAFT AND BEARINGS

When the engine has an in-the-block cam new camshaft bearings should be installed. Often the machine shop will perform this operation after cleaning the block. Installation requires a special cam bearing tool **(Figure 6)**. You only get one chance to install the camshaft bearings; be sure of your work before you drive them in. Set the bearings squarely into the bore with the chamfered edge toward the bore. Be certain that the oil hole in the bearing will line up with the oil hole drilled in the block. Hold the cam bearing driver tightly against the bearing ends. Use a hammer to knock the bearings into place. They are interference fit to prevent spinning, so it takes some force to drive them home. When they are installed check the alignment of each oil hole with an inspection mirror. A partially blocked passage can destroy the new bearing and the camshaft. Lubricate the bearings with oil or assembly lube.

Use a cam holder or install the camshaft sprocket to give you some leverage to hold the camshaft. Guide the

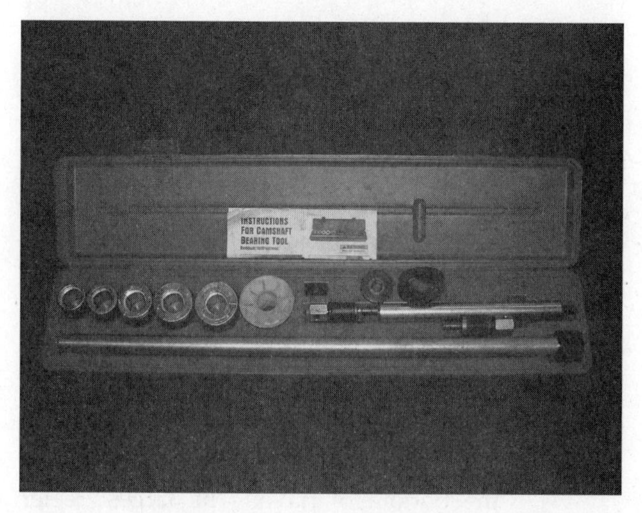

Figure 6. The only way to install cam bearings is to use a special cam bearing installation tool.

camshaft into place very carefully; avoid knocking the lobes on the edge of the block. Install the cam thrust plate at the end of the camshaft. Rotate the camshaft to be sure it spins freely and smoothly in the new bearings.

INSTALLING RINGS

Rings must be fitted to the cylinders and checked for proper side clearance. It is possible to receive the wrong set of rings; do not install them without checking their fit. Like the rest of the engine components the rings must be installed clean and well lubricated to break in properly and provide good service. The gaps should be properly spread to prevent blowby or oil consumption.

Ring End Gap

Ring end gap specifications vary widely. The top ring generally has a different gap than the second ring and the oil rings. Check the specifications, and compare each measurement with them. To measure ring end gap place a ring into the cylinder. Push it down toward the bottom of ring travel, and square it in the cylinder by using a piston to push it. Measure the gap with a feeler gauge **(Figure 7)**. If the gap is too small or too large you have the wrong set of rings. If the gap is too small sometimes you can grind or file the end of the ring until the gap is within the specified range. You cannot file a chrome ring because the coating will chip off. Repeat the measurement procedure for each of the rings in one cylinder. If the rings are all well within specification and the cylinder bores have been carefully measured you need not check each individual ring for the engine.

Figure 7. Measure ring end gap with the ring near the bottom of travel. The largest feeler gauge that fits the gap must be within the specified range.

Figure 8. The proper ring side clearance allows the ring to seat properly in the piston.

Ring Side Clearance

To check **ring side clearance** place the ring into its proper groove. Find the feeler gauge that just fits between the top of the ring and the ring land; this is your ring side clearance **(Figure 8)**. Typical specifications for side clearances are between .002 in. and .005 in. Check the actual specification on your engine to be sure. Measure the clearance at each of the rings. Rings with too much clearance will rock in the groove and allow blowby and oil to leak past. They will wear the ring grooves prematurely and could snap. The excessive clearance could be caused by using the wrong ring or from having excessive piston ring groove wear. Correct the problem before assembling the engine further. Too little clearance will cause the ring to bind and break as it gets hot. Be sure the ring groove is clean and that you have the proper rings if the side clearance is too small.

Some performance engine builders will file each ring to create the desired gap, usually toward the lower end of the range to provide good sealing and maximum wear.

Installing Rings

Use a ring expander to install the rings into their proper grooves **(Figure 9)**. Twisting rings on can snap them or stretch them depending on the material. Double-check that the rings are installed facing the correct direction. Rings installed upside down will not seal properly. Rotate

Figure 9. Use a ring expander so you do not twist the new ring or score the piston.

the rings so that neither of the gaps is on top of the other nor directly on the pin bore or the center of the thrust face of the piston. Space each gap at least 90° off the one above it. Manufacturers may have specific ring spacing instructions; look for them, and follow their recommendations.

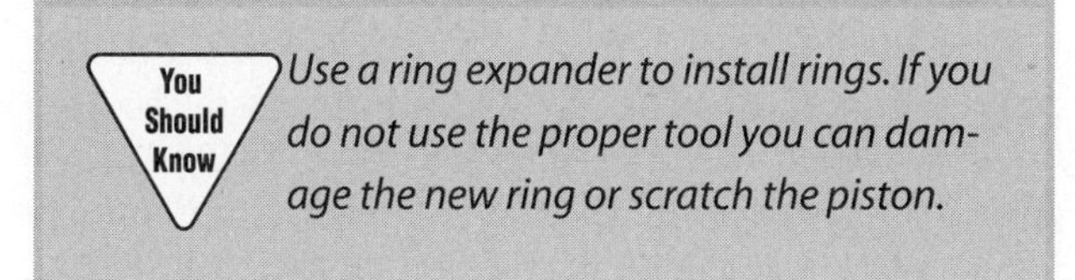

Use a ring expander to install rings. If you do not use the proper tool you can damage the new ring or scratch the piston.

INSTALLING PISTONS

With the rings properly placed on the clean pistons you are ready to install the pistons into their bores. Line the pistons up with their caps in the correct order. Lay out the new or cleaned rod bolts on your workbench. Snap the bearings into the connecting rods and caps, and leave them dry. Lubricate the clean cylinder walls. Clean and dry the crankshaft rod journals. Oil the pistons and rings well, or dip them into a coffee can full of clean oil. This part of the job will be messy. Place a drain pan or rags below the engine to catch the excess oil.

Locate the correct piston for the cylinder. Notice the front marking on the piston. Put a vacuum hose on the rod bolts to protect the crankshaft and cylinder walls during installation. Rotate the crank so the journal is straight down in the cylinder. Install a ring compressor around the piston to squeeze the rings flush with the piston. Be sure that a ring gap is not in the seam of the compressor, or the edge will stick out. There is a bottom side to most ring compressors; be sure it is installed properly on the piston.

Place the piston and rod into the cylinder. Tap the top of the piston with the wooden or plastic shaft of a hammer

Figure 10. With the ring compressor fully seating the rings in their grooves, tap the pistons into place. If you encounter resistance, stop and check that a ring has not popped out of the compressor; you can easily break one.

(Figure 10). Go slowly, and do not use much force. Guide the rod so that it goes around the journal properly. If you feel resistance check the ring compressor to be sure it is seated firmly on the block and that no rings have popped out. Start over if you can see the ring exposed. Do not be tempted to hit the piston head harder; you will break the new ring. Drive the piston until the rod is on the crankshaft journal.

Checking Rod Bearing Clearance

Flip the engine over, and select the correct rod cap. Lay a strip of plastigauge lengthwise across the dry rod journal. Install the new bearing in the cap; do not oil it. Install the cap in the proper direction so the markings line up. Torque the cap bolts to specification. If the cap bolts require that you finish to a number of degrees past the final torque specification omit that step to prevent stretching the bolts. Just tighten them to the final foot-pounds of torque. Remove the cap, and compare the squished plastigauge with the paper scale. Be certain that the rod bearing clearance is within specification. Lubricate the fronts of the bearings thoroughly, reinstall the rod cap, and torque it fully to specification **(Figure 11)**. Repeat the process with each of the pistons and rods.

Rotate the crankshaft after each installation to be sure it does not bind or require excessive force to turn it. Note that with each piston installed the crankshaft will become incrementally harder to rotate.

Checking Rod Side Clearance

With the rods properly torqued onto the crank you need to check for proper **rod side clearance** on vee engines. Spread the rods away from each other using a pry

Figure 11. Be certain to torque each of the rod bolts properly; one loose rod bolt will ruin an engine.

bar. Check the clearance between the rods using a feeler gauge **(Figure 12)**. Typical side clearance ranges between .005 in. and .020 in. Too little rod side clearance usually indicates that the rods have been installed improperly. Check the forward marks on the pistons and the markings on the cap. Without adequate side clearance the rods can seize as they warm up. Too much clearance is caused by excessive wear on the sides of the rods; this requires rod replacement.

TIMING MECHANISM

If the engine has an overhead camshaft wait until the head or heads are on to assemble the front end of the engine. On engines with the camshaft in the block, assemble the timing mechanism and its cover next. Rotate the engine until the piston in cylinder #1 is at TDC. Locate the timing mark on the camshaft sprocket, and turn the cam until it is in the proper position. You can find the proper orientation of the valve timing mark drawn in the service information. Do not guess about the correct alignment! Place the chain around the sprocket, and place the sprocket or gear on the crankshaft. Locate the timing marks on the crankshaft and camshaft sprockets or gears, and line them up precisely as shown in the service information **(Figure 13)**. Rotate the engine two full revolutions, and be sure the timing marks realign perfectly. If the sprocket or gear is just one tooth off the improper camshaft timing can cause significant drivability and emissions concerns. Torque the crankshaft and camshaft sprockets or gears onto their appropriate shafts.

Drive the front main seal into its bore. Use a seal installer or a tool that will contact the seal all the way around the circumference. Do not use a punch to install the seal. Lubricate the radial lip with oil.

Make sure that the timing cover and gasket surface on the block are completely clean and free of old gasket material. Check the service information to see which if any sealant should be used on the gasket. Some engines use RTV sealant to form a gasket. Other manufacturers may specify the use of RTV or aviation sealant to assist gasket sealing. Do not use any sealant if none is specified. Place the gasket in position using the appropriate sealant. Thread the timing cover bolts into place. Be sure that the correct-length bolt is used in each bore. Torque the cover to the

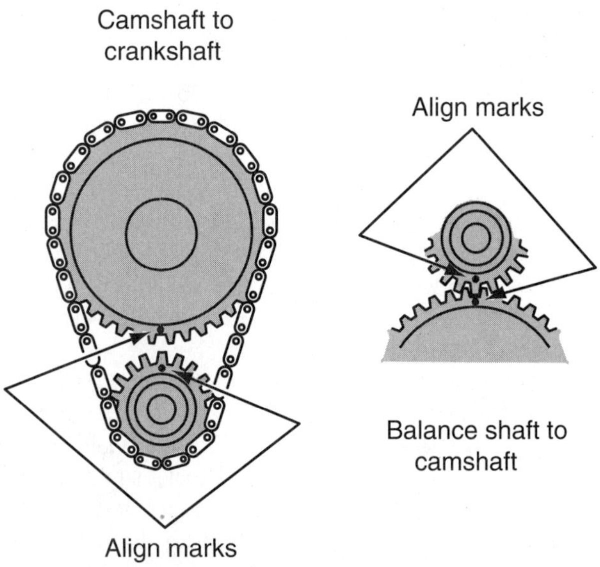

Figure 12. Use a feeler gauge to measure the clearance between two rods sharing a journal.

Camshaft to crankshaft

Align marks

Balance shaft to camshaft

Align marks

Figure 13. Check the service information to be sure you know the proper alignment of the timing marks.

proper specification, working in a diagonal pattern around the cover.

INSTALLING THE HARMONIC BALANCER

The best way to install a harmonic balancer is to use a special tool that draws the balancer onto the crankshaft. Make sure the key is properly positioned in the keyway. When the tool is not available the balancer can be driven on using a dead-blow hammer. Be careful to hit the damper on the center ring, not the outside ring, or you could break the bond and separate the rings. Torque the balancer bolt to specification.

INSTALLING THE OIL PUMP

If you are using a new oil pump check it against the old one for proper application. You may have to remove the strainer from the old pump and install it on the new pump. Be sure it is perfectly clean. Some technicians tack weld the strainer onto the pump pick-up tube to be sure it is securely held. Lubricate the oil pump thoroughly with oil. Spin the pump to fill the pump body with oil. Make sure the mating surfaces are clean. Mount the oil pump, and torque it to specification. On a crankshaft-mounted oil pump be sure the new o-ring seats properly in the groove.

INSTALLING THE OIL PAN

Make sure the oil pan and the gasket surface on the block are clean. Follow the manufacturer's recommendations for sealing the oil pan. Use the proper sealant when indicated and none if not specified **(Figure 14)**. Place the gasket in position, and thread the oil pan bolts into their bores. Tighten the bolts to the proper torque specification in the proper sequence. If no tightening sequence is provided work your way across the oil pan diagonally. Do not tighten the bolts in a sequential circle around the pan. Be careful not to overtighten the oil pan bolts; that is a common cause of oil leakage.

Figure 14. This oil pan gasket is installed without any added sealant.

You Should Know

Applying too much RTV sealant to the oil pan can allow the excess into the crankcase and clog the pickup screen. Use a 1/8 in. bead of sealant; more is not better.

Summary

- When assembling the engine the components must be completely clean to ensure that the overhaul is long lasting.
- Use a checklist to be sure that you perform all the necessary repairs and checks.
- Install the oil gallery plugs and new core plugs.
- Lay the dry crankshaft into its bore, place a piece of plastigauge across each of the journals, and torque the main caps. Remove the caps, and measure the oil clearance as read on the plastigauge scale. Typical clearance specifications are .0008–.002 in., but always refer to the manufacturer's specifications.
- Torque the main caps into place using the proper tightening stages and sequence.
- Check crankshaft endplay using a dial indicator or feeler gauges.
- Carefully install camshaft bearings using a special bearing installer. Be sure to check the alignment of the oil holes.
- Slide the new rings into the proper grooves, and check side clearance using a feeler gauge. Typical clearance is .002–.005 in.
- Push the new rings toward the bottom of the cylinder with a piston, and measure the end gap using feeler gauges. Replace the rings, or file them to achieve the proper end gap.

- Space the gaps around the ring so that none line up or fall directly on the pin or the center of the thrust face. Follow the manufacturer's specified spacing instructions if applicable.
- Install the pistons using a ring compressor and tapping them lightly into place.
- Check the rod bearing clearances using plastigauge. Torque on the main caps using the proper torque specification.
- On a vee engine check the rod side clearance between the two rods.
- On an engine with an in-the-block cam mount the timing mechanism being careful to set the valve timing precisely as directed. Torque the timing cover in place.
- Lubricate the oil pump thoroughly, and fill it with oil. Mount the pump, and torque it to specifications.
- Use the correct sealant on the oil pan, and torque the bolts down only to the specified torque. Use an appropriate tightening sequence.

Review Questions

1. Typical main bearing clearance specifications are ____________ to ____________ in.
2. Describe the proper procedure for checking main bearing clearances.
3. What problems can occur if the crankshaft endplay is not within specifications?
4. When installing camshaft bearings be sure that the ____________ line up perfectly.
5. Is it true or false that you should check ring end gap with the new rings on the piston?
6. Technician A says that if the bearing clearances are too great knocking and low oil pressure will result. Technician B says if the clearances are too small check to be sure the bearings are correct for the application. Who is correct?
   A. Technician A only
   B. Technician B only
   C. Both Technician A and Technician B
   D. Neither Technician A nor Technician B
7. Technician A says the main caps should be tightened to the final torque specification and then rotated another 90°. Technician B says that manufacturers may specify using new main cap bolts. Who is correct?
   A. Technician A only
   B. Technician B only
   C. Both Technician A and Technician B
   D. Neither Technician A nor Technician B
8. Technician A says that proper ring side clearance is typically .002 in. to .005 in. Technician B says that many rings have a top marking and must be installed correctly to provide proper sealing. Who is correct?
   A. Technician A only
   B. Technician B only
   C. Both Technician A and Technician B
   D. Neither Technician A nor Technician B
9. Technician A says that rod side clearance should be checked on vee engines. Technician B says that too much rod side clearance will cause low oil pressure. Who is correct?
   A. Technician A only
   B. Technician B only
   C. Both Technician A and Technician B
   D. Neither Technician A nor Technician B
10. Technician A says the oil pump should be filled with oil when installed. Technician B says the oil pan bolts should be tightened snugly working in a circle around the pan. Who is correct?
    A. Technician A only
    B. Technician B only
    C. Both Technician A and Technician B
    D. Neither Technician A nor Technician B

Cylinder Head Construction, Inspection, and Repair

Interesting Fact

When performing a valve job (refinishing the valves and seats), some enthusiasts try to eke out a little more power by cutting five angles into the valve seat to "round" the corner and help airflow into and out of the cylinders.

SECTION OBJECTIVES

After you have read, studied, and practiced the contents of this section you should be able to:

- Understand the advantages and disadvantages of different combustion chamber designs.
- Know the parts of a valve: the head, margin, face, fillet, neck, stem, keeper grooves, and tip.
- Understand the operation of lifters and know the different designs.
- Understand camshaft lift, duration, and overlap.
- Replace valve seals and springs without removing the cylinder head.
- Remove the cylinder head for head gasket replacement or head reconditioning.
- Measure and replace valve guides that have clearance beyond the specified amount.
- Check the valves for bends or burning.
- Consider all the possible reasons why a valve has burned so that you can correct the cause of failure.
- Measure the valve spring free length, squareness, and tension.
- Inspect the lifters, camshaft, pushrods, and rocker arms to determine whether to reuse or replace them.
- Check a cylinder head for cracks, warpage, and passage condition.
- Resurface the valves (if allowed) to remove all pitting and cupping.
- Recognize when valve seats are badly damaged or deeply recessed and require replacement.
- Use a carbide, diamond, or stone cutter to refinish the valve seat angle.
- Check for proper valve seating.
- Install the cylinder head and torque it into place using the proper sequence and procedure.
- Decide whether to reuse or replace head bolts.

Chapter 24 Cylinder Head Construction

Introduction

The top end of the engine consists of the cylinder head, valves, springs, lifters, and, very often, one or more camshafts. Every component that works to open and close the valves is part of the valvetrain. Problems in the cylinder head and associated components can cause noise, drivability problems, and serious engine damage. The cylinder head is just as critical as the engine block in supporting powerful combustion. Air must flow efficiently into and out of the cylinder head to supply maximum power. The head seals the top of the block and creates the combustion chamber. The combustion chamber is carefully designed to create good air-fuel mixing while still getting the most air charge in. Engines may use two, three, four, or five valves per cylinder to maximize airflow. Lighter valves allow higher engine speeds. The valves seal perfectly on narrow, precisely machined seats. They must dissipate the tremendous heat of combustion in milliseconds. Carbon buildup is cleaned away as the valves snap closed on their seats. The springs must overcome tremendous momentum when the engine is spinning near the red line to hold the valves against the profile of the camshaft. One stray movement, a stretched component, or a snapped lock can reduce the whole assembly to a wrecked heap of parts. In this chapter we will look at how the top end works to provide good engine operation; this will help you understand how problems occur and how they can be repaired.

CYLINDER HEAD

The cylinder head houses the valves, valve seats, valve guides, springs, and combustion chamber. In general, we will refer to cylinder heads with overhead camshafts, noting when other heads are being discussed. You already know that some engines have the camshafts in the block. More passenger car engines now use overhead camshafts to simplify valve opening **(Figure 1)**. The head may be made of

> **Interesting Fact**
>
> *A combustion chamber that houses the intake and exhaust valves on opposite sides of the chamber offers cross-flow through the combustion chamber during valve overlap. This helps cylinder filling; the added volumetric efficiency means increased power.*

Figure 1. This is a typical aluminum DOHC cylinder head.

cast iron or aluminum. Aluminum is now far more popular, although some engines using aluminum blocks will seal the chambers with cast-iron heads. The head has oil passages to allow oil flow up to the camshaft, rockers, and guides. Return ports allow the oil to flow back to the crankcase. Coolant flows through small passages around the valve seats. The seats are very hot spots in the engine because they absorb the heat of the valves. It is critical that the cooling passages be cleaned when the head is serviced, especially when valve failures have occurred. The surface of the head must be machined flat and to the proper surface finish to allow the head gasket to seal properly.

Combustion Chambers

The cylinder head forms the upper half of the combustion chamber. Its design is typically a **wedge**, **pent roof**, or a **hemispherical** (hemi) design, though each type has variations. The different combustion chambers have advantages and disadvantages. The hemispherical head is theoretically one half of a sphere. The now-common pent roof design is similar and shares many of the benefits of a hemi head **(Figure 2)**. The combustion area is wide open with no hidden areas. The valves are canted each to one side so that airflow occurs across the chamber during the period of valve overlap. The spark plug is centrally located in the chamber to help the flame front reach all areas of the chamber quickly **(Figure 3)**. The ratio of surface area of the chamber walls to the volume the combustion chamber can hold is very low. This low **surface to volume (S/V) ratio** has a positive influence on emissions. The pent roof design has a slightly higher S/V ratio than the hemi head,

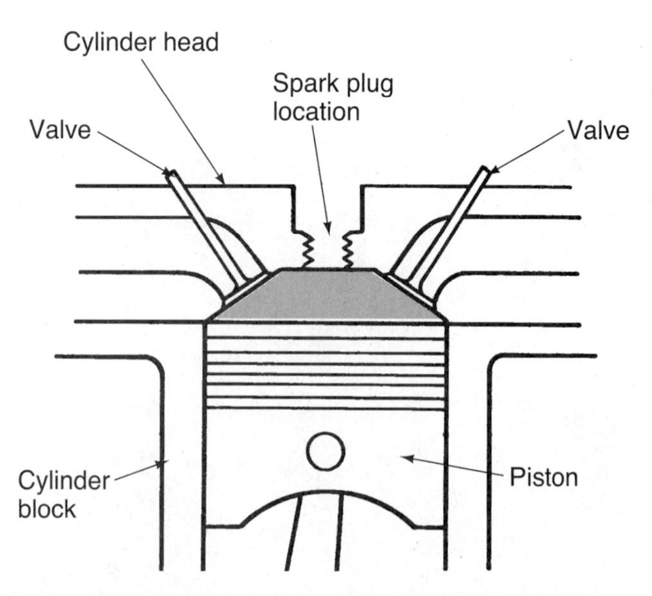

Figure 2. This pent roof combustion chamber places the spark plug centrally and the valves opposed for good airflow.

Figure 3. This modified hemi combustion chamber uses four valves per cylinder. Notice the central location of the spark plug.

but it is still significantly lower than the wedge combustion chamber. When the fuel enters the chamber it will naturally be drawn toward the cooler cylinder walls rather than hang in the hot, highly compressed air. Because there is less surface area and no cooler hidden pockets this is less likely to become a significant problem. When, in contrast, raw fuel clings to the cool surface area of a wedge combustion chamber a higher percentage of that fuel is emitted unburned as hydrocarbons (HCs). The hemi and pent roof heads also allow high volumetric efficiency, particularly at higher rpms. The chamber is wide open and allows uninterrupted airflow. This low-turbulence design has a downside; during low-rpm operation the air-fuel mixture does not mix as well and can lead to slightly rougher running. The pent roof design offers somewhat of a compromise; it has slightly higher turbulence because of the slanted edges of the chamber. When the piston comes up on the compression stroke the air from the corners is pushed toward the center, improving air movement. Pent roof combustion chambers are a very common production design used on multivalve cylinder heads, especially those with dual overhead camshafts.

The wedge combustion chamber is shaped like its name, as shown in **Figure 4**. The tight part of the wedge at the end of the chamber is what gives this design its high S/V ratio. This small corner of the wedge serves as both a **squish** and a **quench** area. As the piston moves upward on the compression stroke with a full chamber, the air in the corner is compressed into the rest of the air stream, creating excellent turbulence and air-fuel mixing. Once the flame front starts, the end gases in the corner are cooled by the large quenching (cooling) surface area. This helps to prevent detonation. Some of the fuel is held against that cool wall, however, and contributes to higher HC emissions. It is harder to fill the wedge combustion chamber with air;

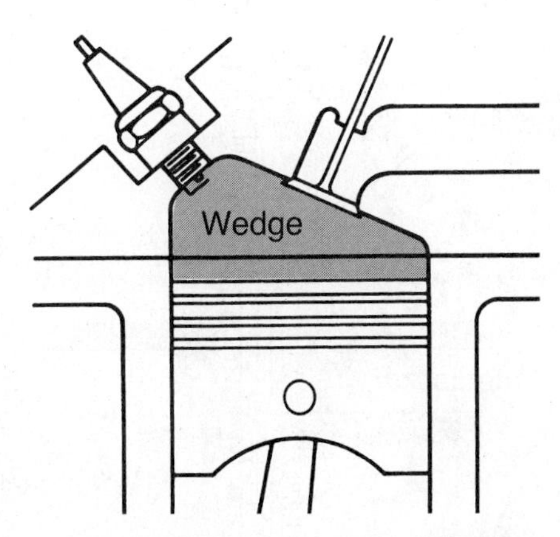

Figure 4. The small corner at the right of this wedge combustion chamber is the squish and quench area.

there is no natural airflow toward the wedge. This design has been used successfully for many years and provides a lower-cost alternative to the more complex valvetrain setup of a hemi or pent roof head.

Intake and Exhaust Ports

The intake and exhaust ports are carefully machined to match the flow of air from the intake manifold and out to the exhaust manifold. Some cylinders have Siamese ports, meaning that two cylinders share the same opening. This design is not ideal but may be used to save space or cost. Intake ports may be designed with a hump in them to increase air velocity and thus flow. Having the intake ports on one side of the head and the exhausts on the other is called a cross-flow design. This adds to the volumetric efficiency of the head, especially during valve overlap. As the intake valve opens the incoming air is drawn into the air moving across the chamber from the exhaust flow **(Figure 5)**. Noncross-flow designs have a much harder time fully filling the combustion chamber with air. Exhaust airflow out of the chamber may actually provide resistance to incoming air as it is pushing out in the opposite direction of the incoming air. The proximity of the hot exhaust gases to the intake charge also has a tendency to heat (and therefore reduce the density of) the intake charge.

Interesting Fact

Many enthusiasts port and polish their heads during a performance overhaul. A die grinder can be used to polish the surface of the passages and enlarge the actual ports to better match the opening of the manifold. This is called match porting. Manufacturers have spent countless hours designing the flow through these ports, however, so take care not to change the basic flow design.

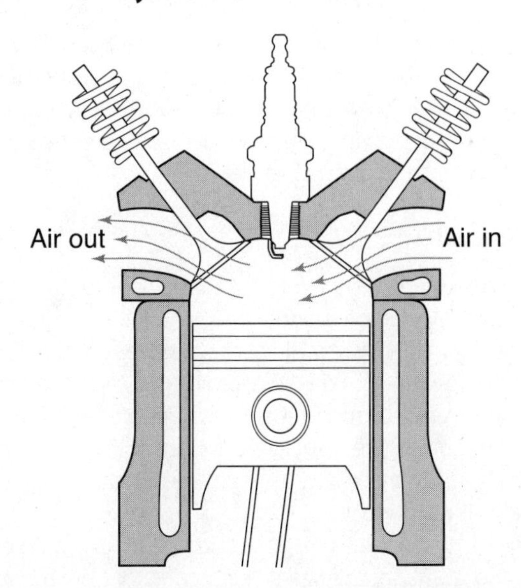

Figure 5. The momentum of airflow helps draw in a fresh charge; in a cross-flow chamber this is particularly effective.

Camshaft Bore

The camshaft or camshafts on an overhead-cam engine have a bore machined into the head to house the camshaft. The top halves of the bores are caps or a cap housing that bolts to the head **(Figure 6)**. On some engines the bores may be fixed, and you can slide the camshaft into place from either the front or rear of the head. Overhead camshafts sometimes use bearings. More often the integral

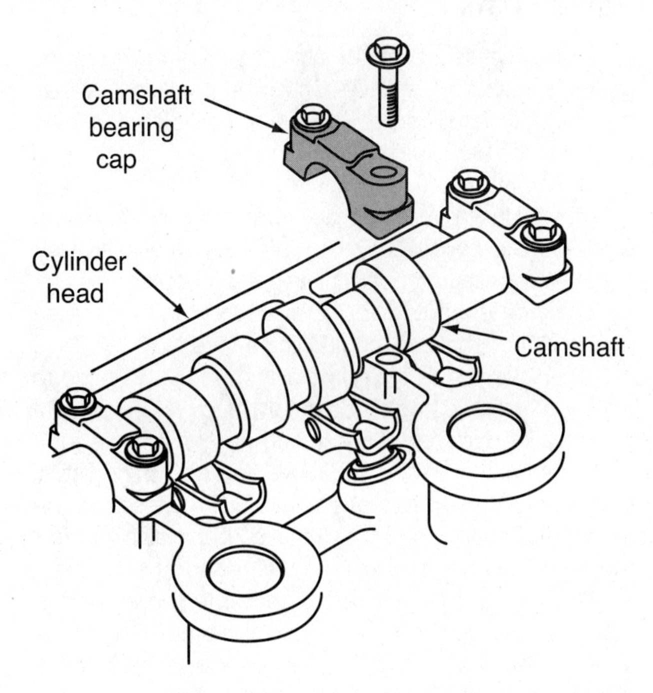

Figure 6. These soft aluminum caps serve as the cam bearing surface.

bearing surface is the soft aluminum of the head. Oil passages lead into these bores to provide essential lubrication. If the cam bores are damaged the head usually requires replacement, though in some cases the bores are machined oversize and inserts can be installed.

Valve Guides

Valve guides may be integral to the head, or they may be pressed in. Cast-iron cylinder heads typically use integral valve guides. Holes are drilled down through the head where the valves will ride. On aluminum heads the holes are bored oversized, and guide inserts are pressed into the head. The guides are usually cast iron or bronze. While bronze is a softer material it has a natural lubricating quality that helps it resist wear.

The guide ensures that the valve rides straight up and down and contacts the valve seat accurately. When valve guides wear, the valve can wobble in the head, and the valve-face-to-seat sealing is compromised. The wobbling of the stem also reduces valve seal contact and distorts the seals. Guide wear allows leakage past the valves and seals and reduces compression and power. The guides are machined to provide just the right amount of clearance between the guide and the stem of the valve. The clearance is typically .001 in. to .003 in., though some manufacturers may specify more clearance. If the fit is too loose the valve will wobble, and if it is too tight the valve can stick. The small clearance allows a thin film of oil to lubricate the valve and guide. The guides have coolant passages around them and help the valves dissipate their heat.

Valve Seats

The **valve seats** are cut into the head to provide the sealing surface for the valves. On cast-iron heads they may be integral, carefully drilled holes in the head casting. They are induction hardened to withstand the repeated beating they take without any lubrication. On aluminum heads they are pressed in and may be made from induction-hardened steel, a nickel-cobalt alloy, or a high chrome-steel alloy. Many performance modifications press in stellite seats for added hardness and durability. Valve seats can be replaced during cylinder head reconditioning as needed.

The valve seat is typically machined to a 45° angle to closely match the angle of the valve face and provide a good sealing surface. Valve seats and valves may have additional transitional angles cut above and below the actual seat contact area to improve airflow. The valve seat is surrounded by cooling passages **(Figure 7)**. The valve transfers much of its heat to the seat during the short intervals it is closed. The width of the seat is critical to proper performance and longevity. The seat is usually about 1/16 in. wide for an intake valve and 3/32 in. wide for an exhaust valve. Always check the manufacturer's specifications for the desired width on a particular engine. The exhaust valve often has a wider seat because it has to absorb more heat; the valve opens at the end of the power stroke. The seat must not be too wide, or it will not cool adequately when the valve is open and the valve will wear faster. A narrower seat helps to chip away at the carbon that has a tendency to build up on the valve seats. Carbon can hold a valve open and lead to valve leaks and burning.

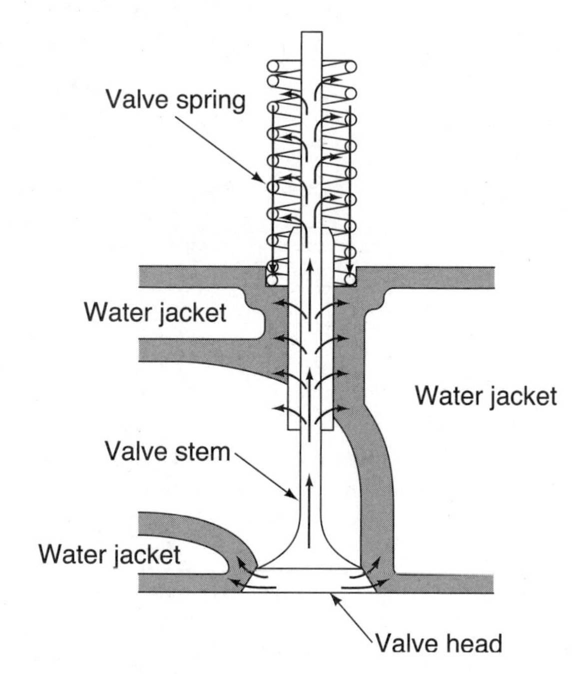

Figure 7. The valve dissipates its heat through the face and stem to the surrounding water jackets.

> **You Should Know** *Older cast-iron cylinder heads used integral cast-iron seats without any additional manufacturing processes necessary. The lead in the fuel served as an excellent lubricant to cushion the blow between the seat and the valve face. When updating a 1960s or early 1970s cylinder head, hardened seats should be installed.*

VALVES

Valves open to allow air to flow and close to seal in compression and combustion. Production valves are usually made of a stainless steel alloy. The camshaft opens the valves while the valve springs close the valves. The seal between the valve face and the valve seat in the head is essential to proper engine performance. The large round area of the valve is called the head **(Figure 8)**. The valve face is a machined surface that contacts the valve seat to ensure

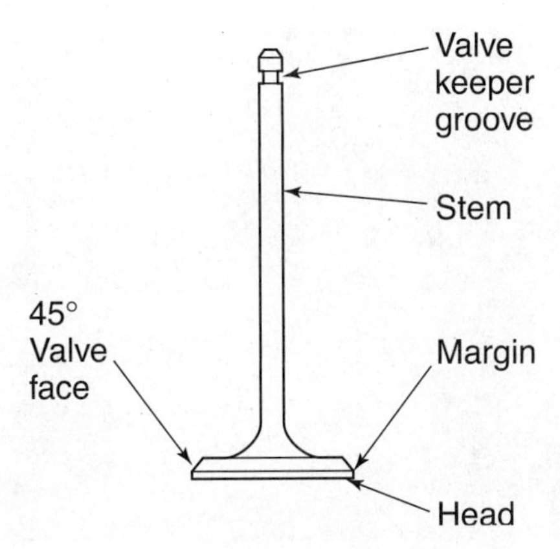

Figure 8. You should be familiar with the parts of the valve.

good sealing of the combustion chamber. The thickness of the head above the face of the valve is the **margin**. The margin adds substance to the valve and helps to prevent the valve face from burning or distorting. If the valve face were cut to a thin point meeting the valve head the resulting edge would quickly warp and burn. The valve fillet is the area below the valve face. It may be cut differently to give adequate valve strength while minimizing obstruction of airflow. The valve neck connects the head of the valve to the stem. The valve stem is machined smooth and to a precise diameter to fit the valve guide. Toward the top of the stem are keeper grooves. Valve **keepers** are fitted into the grooves to hold the spring retainer on the valve. The retainer holds the valve spring in place. The **valve tip** is where the **valve follower** or rocker arm rides to open the valve. Some valve tips have a stellite or hardened alloy coating for added strength and wear resistance.

Intake valves are often larger than exhaust valves. The intake valve has to flow air into the engine with only the pressure differential between atmospheric pressure and manifold vacuum. Exhaust gases may still be under pressures nearing 200 psi when the exhaust valves open; less space is needed to allow the same amount of air to flow. When an engine uses three or five valves per cylinder any extra valve is an intake valve for this same reason. A four-valve engine uses two intake valves and two exhaust valves. Sometimes the valve sizes on multiple-valve heads are the same. Aside from the ability to flow more air, multiple valves allow the size of each to be smaller. This increases the possible engine rpm without valve damage; the heavier the valve, the slower it must open and close to prevent damage to the valve or to prevent valve float. Valve float occurs when the springs can no longer hold the valve closing consistent with the profile of the cam lobe. Instead of closing swiftly with the drop of the camshaft lobe the valve floats over the ramp and hangs open at higher rpms, when the momentum is greater. This causes engine misfire.

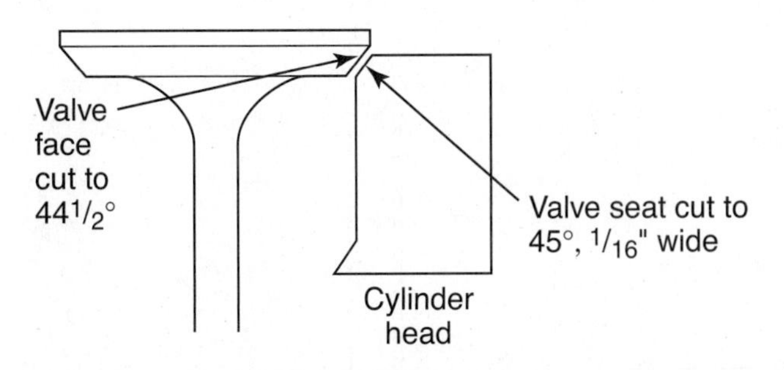

Figure 9. This valve was cut to 44 1/2° to create a 1/2° interference angle.

Valve Face

The seating surface of the valve, the face, is typically cut at a 45° angle. Sometimes, when performing a valve job (reconditioning the valves and seats) the manufacturer will specify an **interference angle** between the valve face and the valve seat. It is typically a 1/2 to 1° difference between the angle of the face and the seat **(Figure 9)**. For example, the valve face may be cut at a 44° angle, while the seat is cut at a 45° angle. This allows for quicker break-in of the valve-face-to-seat seal; the pressure per square inch on the valve is much greater when the contact area is so small. A few manufacturers use a 30° angle on the valve seat and face. Be sure to refer to the service information specific to the vehicle you are working on.

Sodium-Filled Valves

Turbocharged, supercharged, or other high-performance engines may use sodium-filled valves for their excellent heat dissipation quality and decreased weight. A sodium-filled valve has a hollow stem that is partially filled with metallic sodium **(Figure 10)**. When the valve opens the sodium splashes down toward the head and picks up heat. When the valve closes the sodium moves up the stem, taking heat away from the hottest area of the valve, the head. The heat is transferred from the stem to the guide and then dissipated in the coolant passages around the guide. Sodium-filled valves should not be machined; if you were to cut through the valve the sodium could explode. These valves should be replaced when refurbishing the head.

Titanium Valves

Many production valves have some titanium in the alloy to lighten the valve. Titanium valves are commonly used in high-performance engines and racing engines. They dissipate heat better than a standard valve and are significantly lighter. The reduced weight is particularly well suited for racing applications, where engine rpm limits are pushed far beyond production red lines. This reduces the tendency for valve float at higher rpms. Titanium valves

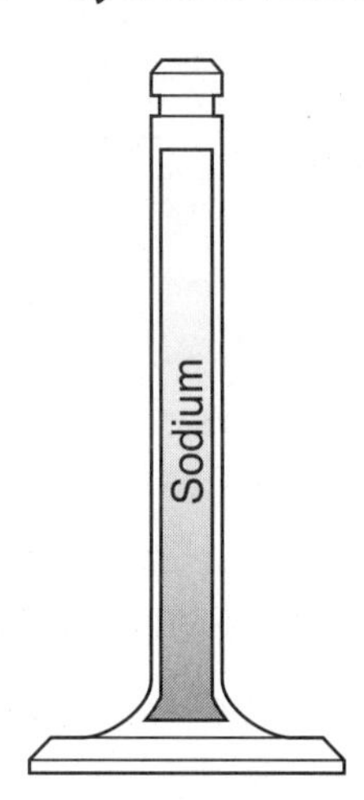

Figure 10. The hollow valve is partially filled with sodium. When the valve closes the sodium splashes up to help take heat away from the head and into the stem.

Figure 11. The positive-lock seal on the left shows the retaining ring and spring fit; the umbrella seal on the right fits loosely over the guide.

may be used as replacement valves, though they are much more expensive. Unless the whole engine is being modified for performance stock valves should provide adequate performance.

VALVE SEALS

Valve seals keep oil from draining down through the space between the valve stem and the valve guide. Intake valves are particularly prone to oil consumption through the guides because when the intake valve is open vacuum is pulling at the oil near the top of the valve. Valve seals for the intake and exhaust valves are commonly of different designs and materials. When valve seals fail the engine burns oil particularly after startup, and at idle and deceleration when vacuum is high. Valve seals can be replaced without removing the cylinder head.

There are three basic designs of valve seals: positive lock, umbrella type, and simple o-ring seals. The positive-lock valve seal locks onto the base of the head at the top of the guide **(Figure 11)**. These seals may be made of synthetic rubber or hard plastic. They have different locking rings around them to hold them tightly in place. Umbrella seals or oil shedders are just round cups that fit over the valve stem. They do not lock into place; they just sit over the top of the guide. O-ring seals are more commonly used on exhaust valves, and they ride just below the keepers. They effectively form an umbrella seal with the retainers sitting above them.

VALVE SPRINGS

Valve springs close the valves and hold them along the profile of the camshaft lobes. A valve spring may be a simple single coil, or it may be a double- or even triple-coil spring. Additional coils increase spring stiffness. Other springs use one coil spring with a flat wound vibration damper inside to reduce harmonic vibrations. Some springs have a variable rate; their stiffness increases the further they are opened. The coils toward the head are wound tighter and closer together **(Figure 12)**. The springs on the intake valve may be different from those on the exhaust valve. Make a note of spring location and orientation when disassembling the cylinder head.

Valve Spring Retainers and Rotators

Valve springs are held in place by retainers locked onto the valves with keepers **(Figure 13)**. The keepers lock onto the valve stem and are tapered so the retainer opening cannot fit over the keepers. The retainer may be a simple piece of steel to hold the spring in place. Other designs use a rotator built into the retainer. A rotator is used to turn the valve on its seat. It looks thicker than an ordinary retainer

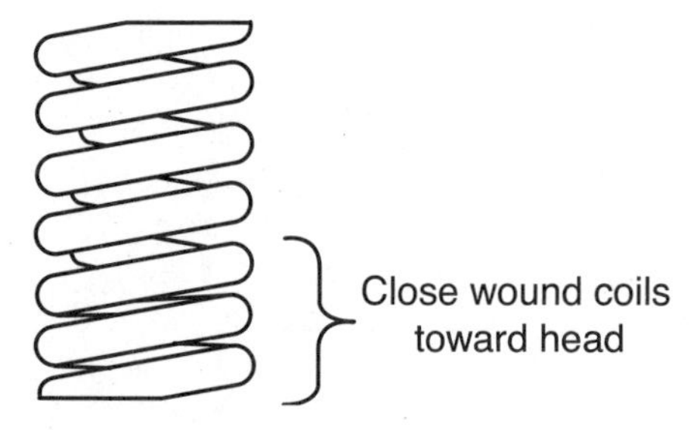

Figure 12. The tightly wound coils at the bottom of this variable-rate spring make it stiffer the more it is compressed.

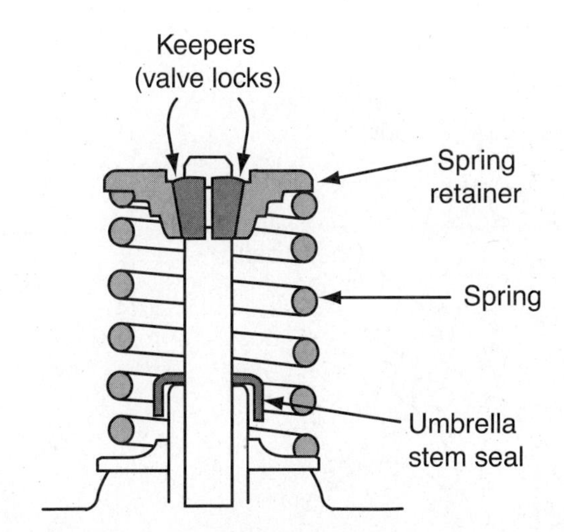

Figure 13. The valve keepers and spring retainer hold the spring in place.

and is either ball or spring type. The ball type has small ball bearings that rotate on a ramped track in the assembly to turn the valve. The spring type has an unequal-height spring to achieve the same effect. Often they are used only on the exhaust valves. The rotation helps clean carbon away from the valve and seat and eliminates hot spots on the valve **(Figure 14)**.

LIFTERS AND HYDRAULIC VALVE LASH TENSIONERS

Valve lifters and hydraulic valve lash tensioners are used between the camshaft and the valve as part of the valvetrain. Lifters can be hydraulic, solid, or roller lifters or simple followers. Each type of lifter has certain advantages and disadvantages; the manufacturer uses what it thinks is

Figure 14. The exhaust valves on this engine are fitted with rotators to help clean the valve face and seat.

best for the engine design. Hydraulic valve lash tensioners are used on many newer overhead-camshaft engines to eliminate the need for periodic valve lash adjustments.

Hydraulic Lifters

Hydraulic lifters use oil pressure inside the body of the lifter to keep the valvetrain in constant contact with the valve. This means there is always zero lash (clearance) between the valve, valvetrain, and camshaft. **Figure 15** shows a typical in-the-block camshaft valvetrain using a hydraulic lifter for automatic adjustment. The lifter is slightly convex (rounded outward) on its bottom side. The lifter bore is also slightly off center from the camshaft. This causes the lifter to rotate in its bore, minimizing wear.

The hydraulic lifter works by covering and uncovering the oil pressure passage in the lifter bore. When the lifter is riding on the base circle of the cam oil pressure is let in to fill the lifter body and the plunger. This moves the plunger outward to keep light tension on the valvetrain. A check valve allows oil to flow from the lifter body to the plunger but not from the plunger to the body. Once the lifter body is moved up in its bore by the camshaft lobe oil can no longer reach the lifter body. The added tension on the plunger closes the check valve. The lifter is now full of trapped pressurized oil; it forms a nearly solid connection between the camshaft and the valvetrain that forces the valve open **(Figure 16)**. A hydraulic lifter can also have a roller on its bottom to reduce the friction between the lifter and the camshaft **(Figure 17)**. This reduces wear on both components.

Hydraulic lifters have the advantage of not requiring periodic adjustment. You may have to center the plunger in the lifter body when installing them, but that adjustment

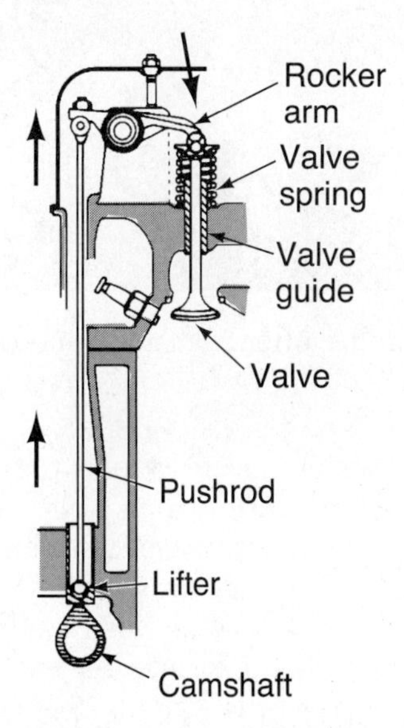

Figure 15. Hydraulic valve lifters may be used on in-block camshaft valvetrains or OHC valvetrains.

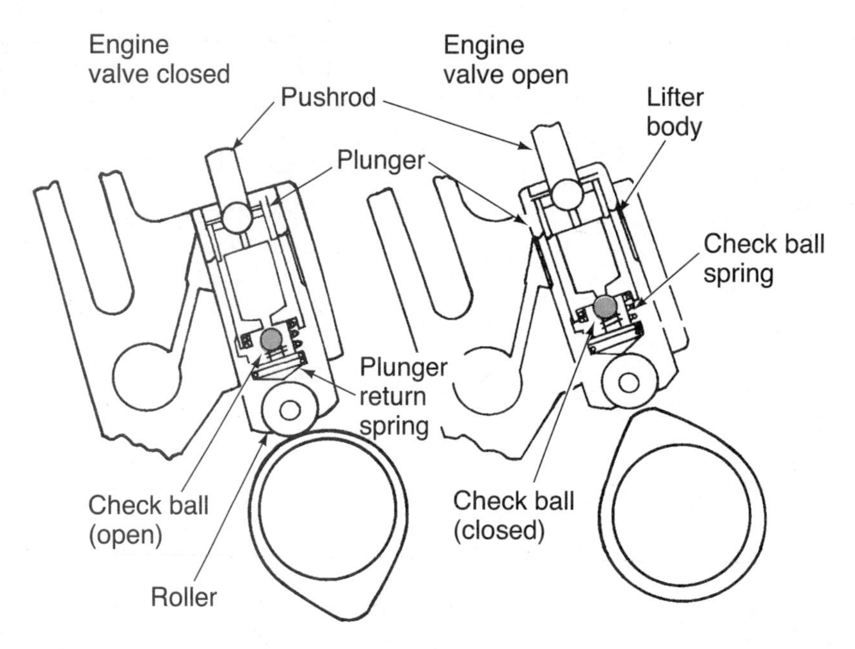

Figure 16. The plunger moves up and down in the lifter body to maintain zero clearance in the valvetrain.

Figure 17. A roller lifter from a late-model engine.

should last the life of the lifter or until the rocker arm is removed. We will discuss valve and lifter adjustment in Chapter 31. The fact that the hydraulic lifter maintains zero lash means that the valve is always lifted as much as it can be. When there is clearance between the valve and the camshaft the amount of clearance becomes lost lift. One disadvantage of hydraulic lifters is that they can "pump up" at high rpms. The lifter becomes effectively longer during the valve open period and cannot displace the oil fast enough to shorten up. It causes the valves to float when there is not enough time for the oil to move out of the body after it has passed over the camshaft lobe.

You can hear faulty hydraulic lifters; they make quite a clatter. Their noise is evident when you first start up the engine. The clattering should disappear within a few seconds as oil fills the body. When lifters begin to leak it will take longer and longer for them to fill after startup. Eventually they will become unable to hold oil. This will cause noisy valvetrain clatter and loss of power as the valves do not open fully if at all.

Hydraulic Valve Lash Tensioners

Hydraulic valve lash tensioners operate in the same manner as hydraulic lifters. They are placed on one end of a rocker arm to create the proper tension on the rocker arm to maintain zero clearance between the rocker and the valve tip **(Figure 18)**. The tensioners shown receive oil pressure through a passage in the head. Some tensioners are fed through an oil pressure tube that runs along the camshaft bore.

There is no initial or periodic adjustment required on hydraulic valve lash tensioners. When physical wear is visible on the tensioners they should be replaced. They, like hydraulic lifters, can collapse and cause significant valvetrain clatter. The force required to depress the plunger on a hydraulic valve lash tensioner is much lower on a faulty one than on the rest.

Solid Lifters

Solid lifters are just that: a solid piece of steel that makes a link between the camshaft and the valve. These

Figure 18. The hydraulic valve lash tensioner keeps light tension between the rocker arm and valve tip. No periodic valve adjustment is needed.

too may have rollers on the bottom to reduce friction and wear. Solid lifters ride in their bores to push the valve open when on the cam lobe. When the lifter reaches the base circle of the cam the valve spring tension will close the valve. The lifter must be adjusted periodically to compensate for wear in the valve and valvetrain components. This is often recommended every 30,000 or 60,000 miles. Check the specific maintenance schedule. You will adjust the valve to have some lash between the rocker arm and the valve tip. This allows for a margin of error in valvetrain operation and also allows for the diminishing clearance as the valvetrain components heat up and expand. Many consumers see periodic valve adjustment as a disadvantage; it represents more costly maintenance.

Followers

Cam followers are often used on overhead camshafts where the cam sits directly over the valves. The follower, often called a tappet, is just a round cup that sits on top of the valve and provides a wider base for the camshaft to push on **(Figure 19)**. Often there are shims either under or on top of the followers that can be fitted to size to make the proper adjustment. These lifters also need periodic adjustment. There are a few manufacturers that use hydraulic followers. The hydraulic body sits under a cup (like a follower) and works just like a hydraulic lifter to constantly adjust valve clearance to zero.

PUSHRODS AND ROCKER ARMS

A **pushrod** is a simple hollow tube of carbon steel; some have hardened tips. Engines with cams in the block, overhead-valve engines, use pushrods to transmit the

Figure 19. This follower sits under the cam and has an adjusting shim on top for valve adjustments.

motion of the lifter up to the rocker arm. The hollow tube in the pushrod typically delivers oil from the lifter up to the rocker arm and valve stem tip. Different-length pushrods are available as replacements when the valvetrain geometry is significantly altered by heavy machining of the valves. This is not often required on newer gas engines because little material can be machined off the valves before they require replacement.

Rocker arms may be used with camshafts in the block or with overhead cams. They are used to transmit the motion of the camshaft on one end to the valve tip on the other end **(Figure 20)**. Rocker arms pivot on a stand to allow the movement of the camshaft to affect valve opening. Rocker arms are often used as a means to adjust valve lash, as we will cover in Chapter 31.

On many engines the rocker arm uses mechanical advantage to increase the valve **lift** as compared to the camshaft lift **(Figure 21)**. The ratio of the distance is calculated by comparing the distance from the rocker pivot to the point where the rocker hits the valve tip, to the distance from the rocker pivot to the point where the rocker contacts the pushrod. Divide the longer distance to the valve tip by the shorter distance to the pushrod to find the ratio. For example, if the distance to the pushrod is 0.75 in. and the distance to the valve tip is 1.25 in., divide 1.25/0.75. The answer is 1.67 in.; the rocker ratio is stated as 1.67 : 1. If the

When installing a "hot" performance camshaft to try to boost horsepower and torque many enthusiasts will go the extra mile and install roller lifters to reduce engine friction. It may also be necessary to replace the stock valve springs if the cam profile is more aggressive.

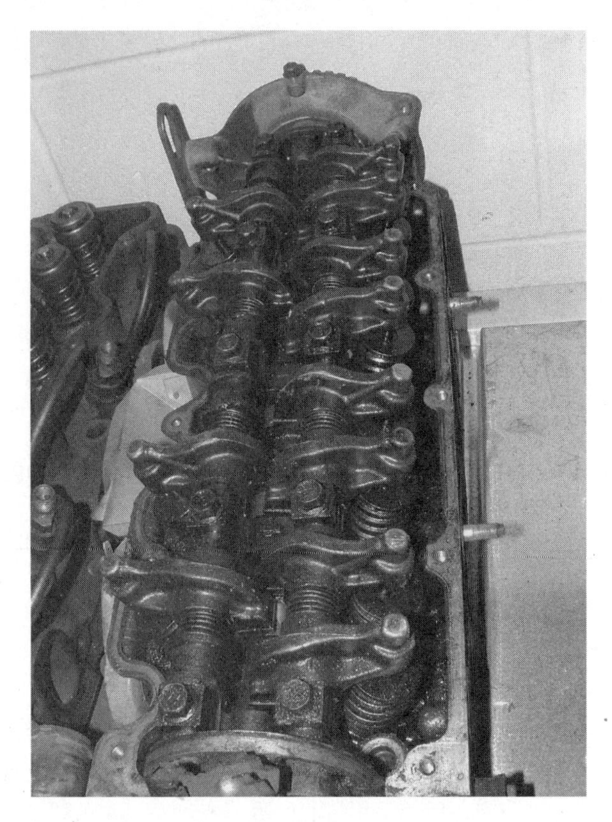

Figure 20. This OHC head uses rocker arms with integral hydraulic lifters to open the valves.

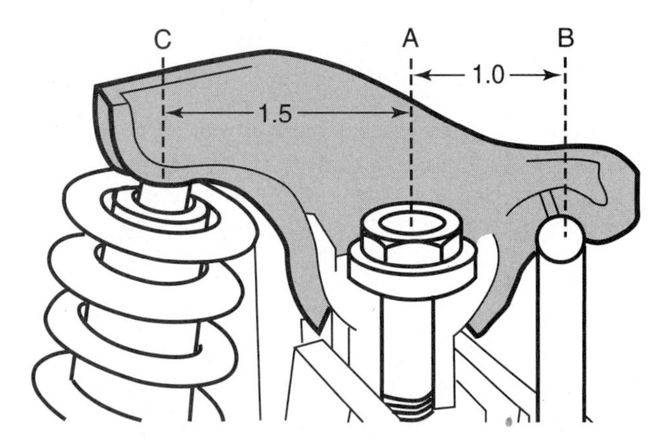

Figure 21. This drawing shows a typical rocker ratio used to provide additional valve lift beyond the camshaft lift.

cam shaft lift is 0.3 in. multiply that by the rocker ratio to find the valve lift: 0.3 in. x 1.67 in. = 0.5 in. of valve lift. **Figure 21** shows a simple ratio of 1.5 : 1.

CAMSHAFTS

The camshaft opens the valves for a certain **duration** (time), at a certain rate, and to a particular lift (height). The camshaft is timed to the crankshaft to be sure the valves open at the correct time during the strokes. It spins at half the speed of the crankshaft to allow each valve to open only once during a complete engine cycle. The camshaft may be driven by a gear, belt, or chain off the crankshaft. Camshafts are designed differently depending on the type of lifter to be used with them. There are roller lifter camshafts, hydraulic camshafts, and mechanical camshafts. The difference in the camshaft may be in the material used or in the shape of the lobe. Many camshafts are made of cast iron or cast steel, but they may be forged steel as well. Many newer camshafts are made from a steel tube, where the lobes are welded onto the camshaft. These camshafts are lighter and stronger than older-style cams **(Figure 22)**.

The camshaft rides in bearings or bearing surfaces on their journals. Oil pressure is received through ports in the cam bores. Like crankshaft main and rod bearings, proper cam bearing clearances are essential to good lubrication and adequate oil pressure. Camshaft bearings are one-piece inserts mounted in the block. When the cam is in the cylinder head the bearings may be one- or two-piece inserts, or the aluminum bore and caps in the head may serve as the bearing surface.

The camshaft has elongated lobes on the shaft that are designed to lift the valve the correct amount for the specified duration. The height of the lobe above the base circle determines the camshaft lift **(Figure 23)**. When rocker arms are not used the camshaft lift is the same as the valve lift. As we discussed the rocker arm may produce additional valve lift beyond the cam lift. A typical amount of valve lift is 0.4 in. to 0.55 in. The design of the lobe also determines how rapidly the valve opens. The steepness of the ramp dictates how quickly the valve opens and closes. A steeper ramp requires heavier springs to pull the valve closed quickly on the closing ramp. The camshaft duration is determined by the width of the top of the lobe. The wider the lobe, the longer the valve stays open. Camshaft duration is measured in degrees of *crankshaft* rotation. If the valve is open while the crankshaft rotates 178° the camshaft duration is 178°.

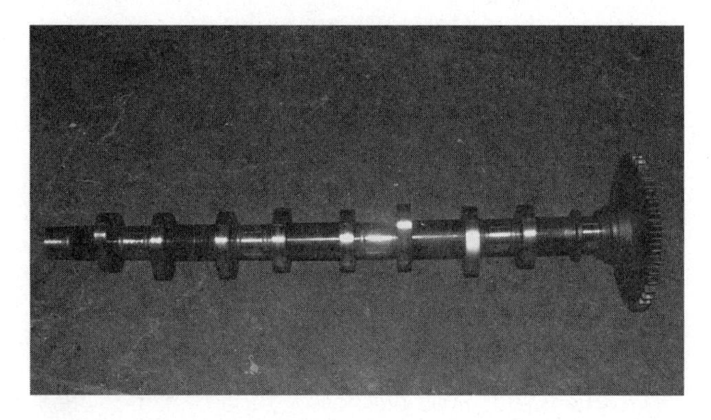

Figure 22. A lightweight late model camshaft used in a DOHC V6 engine.

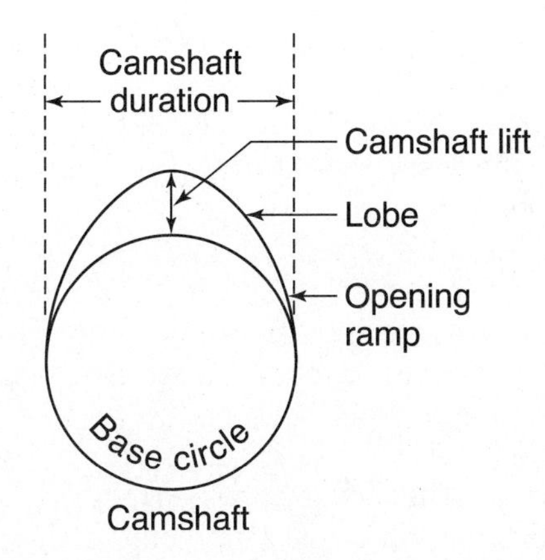

Figure 23. The height and width of the camshaft lobe determine the cam lift and duration, respectively.

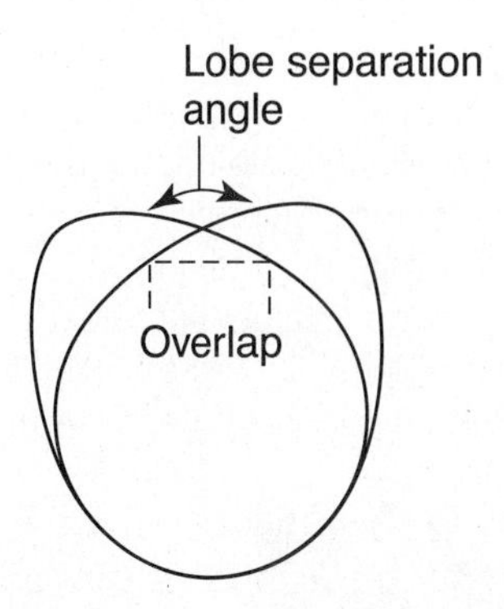

Figure 24. The time, in crankshaft degrees, that both valves are open is overlap. This is defined by the lobe separation angle.

Intake valve duration may differ from exhaust valve duration. When choosing a camshaft be wary of how the duration is measured. Some manufacturers advertise the duration once the valve is open .006 in., while others measure duration beginning when the valve is opened .050 in. This means that a cam with 220° of duration at .006 in. could have less usable duration than a cam with 200° of duration at .050 in.

Valve Overlap

At the end of the exhaust stroke and the beginning of the intake stroke both valves are open at the same time. This is called valve overlap. The amount of overlap is defined by the camshaft lobe separation angle **(Figure 24)**. The larger or wider the separation angle, the less overlap. A tighter or lower separation angle means that the valves will be open together for a longer time, providing more overlap. A wider separation angle, used on most passenger cars, is 112°–116°. A high-performance engine or performance camshaft may have a tighter angle of 106°–110°. Valve overlap helps the engine breathe better, particularly at higher rpms. The movement of air out of the exhaust valve helps draw the fresh air in through the intake valve. Too much overlap at a lower rpm, however, can allow some of the fresh charge to flow out of the exhaust valve and some exhaust to flow back into the intake. This is called reversion and leads to a rough idle, increased emissions, and reduced fuel economy. Production vehicles use a wider separation angle to prevent the negative effects of reversion. Some engines using variable valve-timing systems use two sets of camshaft lobes. At higher rpms the camshaft in use will provide greater overlap to improve volumetric efficiency and power.

Aftermarket performance camshafts often produce an uneven idle accompanied by low vacuum. This is the result of reversion. When top-end performance is what you are looking for long overlap is essential, but be aware that you will sacrifice some low-end torque to achieve it. Most camshaft upgrades are not legal for street use. They have a significant impact on emissions because of the long period of overlap. Any engine modification that increases the emissions of the vehicle is illegal for street use.

Summary

- The cylinder head houses the valves, guides, valve seats, combustion chamber, intake and exhaust ports, and on overhead-camshaft engines the camshaft bore.
- The hemi and pent roof combustion chambers offer a low S/V ratio for lower HC emissions and excellent volumetric efficiency.

- The wedge combustion chamber has excellent turbulence and resistance to detonation. Its high S/V ratio can increase HC emissions, and the turbulence can reduce volumetric efficiency.
- Valve guides may be integral on cast-iron heads or pressed-in inserts on aluminum heads.
- Worn valve guides allow the valves to wobble, preventing good valve seating and destroying the valve seals.
- Valve seats provide the sealing surface for the valve face. A proper seat is necessary to properly seal the combustion chamber and to dissipate the valve's heat.
- Most seats are cut at a 45° angle and about 1/16 in. wide on intake seats and 3/32 in. on exhaust seats. Newer engines with smaller valves may have narrower seats.
- The parts of a valve include the head, margin, face, fillet, neck, stem, keeper grooves, and tip.
- An interference angle of 1/2 to 1° difference between the angle of the valve seat and valve face may be used to speed up valve seating after grinding.
- Sodium-filled and titanium valves offer lighter weight and better heat dissipation than standard valves.
- Valve seals prevent oil from leaking through the guides into the combustion chamber.
- Valve springs are used to close the valves; they may be single, double, triple, or variable-rate coils.
- Valve retainers hold the valves and springs in place. They may incorporate a rotator, especially on the exhaust valves, to rotate the valves to help clean the seats and minimize hot spots.
- Valve lifters may be hydraulic, solid, roller, or followers. Hydraulic lifters automatically adjust for valvetrain wear.
- The rocker arm transfers the motion of the camshaft to the valve on some engines. A rocker arm ratio may be used to increase the valve lift.
- The camshaft opens the valves for a certain duration and to a particular lift at a rate designed into the ramp. Duration is measured in degrees of crankshaft rotation; lift is measured in inches or millimeters.
- Camshaft overlap is defined by the lobe separation angle. Overlap helps volumetric efficiency, but too much can cause reversion at lower rpms.

Review Questions

1. List the components that make up the valvetrain.
2. Describe the advantages of a pent roof combustion chamber.
3. A __________ S/V ratio will provide lower hydrocarbon emissions.
4. What problems will worn valve guides cause?
5. Why is the width of the valve seat important?
6. Technician A says that valve seats on a cast-iron head may be integral. Technician B says that aluminum heads have pressed-in seats. Who is correct?
   A. Technician A only
   B. Technician B only
   C. Both Technician A and Technician B
   D. Neither Technician A nor Technician B
7. Each of the following is part of the valve *except:*
   A. Fillet
   B. Margin
   C. Face
   D. Lip
8. Technician A says that rotators may be used on exhaust valves to help keep the seats free of carbon. Technician B says the valve springs open the valves. Who is correct?
   A. Technician A only
   B. Technician B only
   C. Both Technician A and Technician B
   D. Neither Technician A nor Technician B
9. Technician A says that camshaft duration is measured in degrees of camshaft rotation. Technician B says that camshaft lift is the height of the lobe above the base circle. Who is correct?
   A. Technician A only
   B. Technician B only
   C. Both Technician A and Technician B
   D. Neither Technician A nor Technician B
10. Technician A says that more overlap improves low-rpm performance. Technician B says that a tight lobe separation angle increases the amount of overlap. Who is correct?
    A. Technician A only
    B. Technician B only
    C. Both Technician A and Technician B
    D. Neither Technician A nor Technician B

Chapter 25 Cylinder Head Component Replacement and Disassembly

Introduction

The cylinder head should be removed for service when the head gasket fails, as part of a thorough engine overhaul, or when low compression readings indicate a burned valve. You must remove the cylinder head properly to prevent further damage. Valve seals, springs, rocker arms, overhead camshafts, and many valve lifters or followers can be replaced with the cylinder head installed. Good organization of components will help you during the reassembly process and ensure proper placement of parts that will be reused. Lifters, valves, rocker arms, and camshaft caps should all be replaced in their original location if you are going to reuse them.

When you disassemble the cylinder head for service it is critical that you inspect each component for signs of wear or damage. This may be the time that you find the important clue about what caused the failure. Overlooking that sign may cause a repeat failure. Keep parts well organized; it may help you solve the mystery of why a fault occurred, and it will certainly help you reassemble the cylinder head. It is also essential that some parts be returned to their original location; carefully mark and store valvetrain components. Reconditioning a cylinder head and its related components is commonly called performing a valve job. The first step in this service is careful disassembly and inspection. Keep notes about any unusual observations so that you do not miss anything during parts ordering and replacement. This is also essential to providing the customer with an accurate estimate and informed choices about repair options.

Interesting Fact

An experienced technician once made a very expensive mistake. He was so quick to get the head ready for removal that it was still quite hot. He carelessly removed the head bolts with an air gun, using no particular sequence. The cylinder head suffered an irreparable crack. Do not let your attempts to complete the job fast take over what you know about proper procedures.

VALVE SEAL REPLACEMENT

You can replace worn valve seals or springs without removing the cylinder head from the engine. Worn valve seals will cause oil consumption. Valve seals often wear as a result of worn valve guides **(Figure 1)**. The customer will likely notice a big cloud of blue smoke just after starting the vehicle. The oil on the top of the head leaks down past the valve seals as the vehicle sits; when it starts, the oil in the combustion chamber burns. Blue smoke is also often seen after a period of idling or deceleration. High engine vacuum pulls oil past the weak intake valve seals into the combustion chamber. The resulting blue smoke is a telltale sign of worn seals. This pattern of smoking is different from that of oil consumption caused by worn rings. Ring troubles generally lead to more consistent smoking that is worse when the engine is under a load.

Before replacing the valve seals it is a good idea to check for valve guide wear. Often the seals are damaged because the guides are allowing the valves to wobble back and forth in the seals. Check the engine vacuum at idle and at 2500 rpm. If the vacuum fluctuates at idle but smoothes out at higher rpms suspect a worn valve guide. The valve is

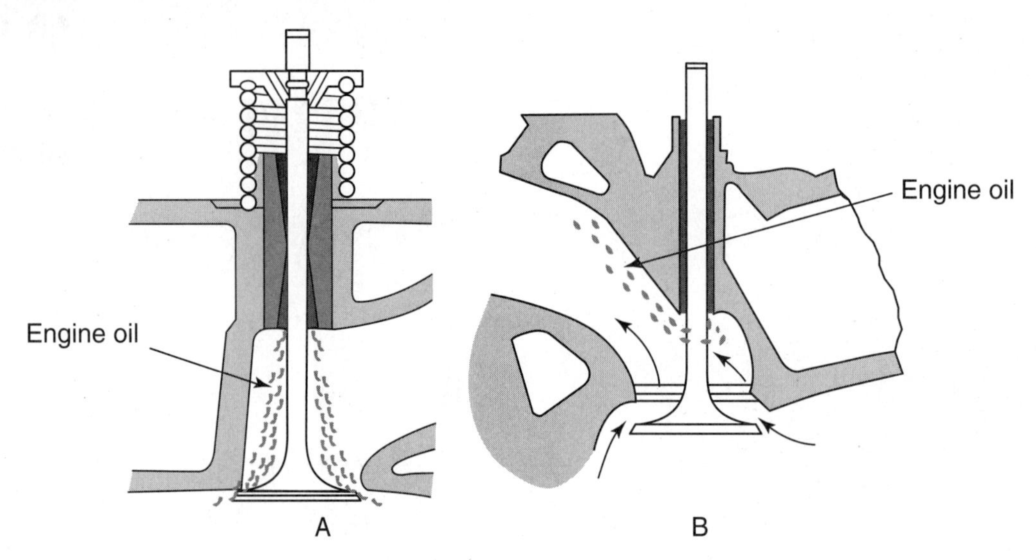

Figure 1. Worn valve seals allow oil to be pulled into the combustion chamber through the guides. It exits the exhaust port as blue smoke.

not being guided correctly onto its seat at lower rpms. You can also check running compression. The compression would be lower at idle, when the valve has more time to rock back and forth in the guide. Running compression would return to a nearly normal value at 2500 rpm.

To replace a valve seal with the head installed you must work one cylinder at a time. Remove the valve cover or covers first. The critical issue during this service is to prevent the valve from dropping into the engine when the keepers are removed. That mistake could require that you remove the cylinder head. Place the first cylinder at TDC on the compression stroke. Remove the camshaft, rocker arm, and lifter or follower as needed to gain access to the valve retainer. Use a cylinder leakage tester or regulated shop air to apply 100 psi of air pressure into the cylinder with the valves closed. This will keep the valves up in the head while you release the valve retainer and keepers. Place a socket over the valve retainer, and give it a light rap to break any varnish loose; this may make it much easier to compress the spring. Use a valve spring compressor to compress the spring, and remove the keepers **(Figure 2)**. Set the spring, retainer, and keepers safely aside. Note which end of the spring faces up; often tighter coils are placed at the bottom, toward the head. Use a screwdriver or seal puller to remove the old valve seal. Check for damage or excessive scoring on the valve stem that could ruin a new valve seal. Lubricate the inner lip of the new valve seal with oil, and install it onto the valve. Many positive-lock seals must be driven down onto the head to seat properly. Use a valve seal installation tool to fit the seal into its proper position. If the correct tool is not available you can use a deep socket that matches the diameter of the seal but will easily slide over the valve stem. Be sure the socket is deep enough to tap the seal down without hitting the valve tip. Knocking the

Figure 2. Compressed air holds the valve up while you compress the spring to remove the keepers.

tip could release the air pressure in the cylinder and cause you to drop a valve into the cylinder, necessitating head removal. Refit the valve spring, retainer, and keepers. When the valve is back in place take the socket and lightly rap again on the retainer to check that the keepers are fully seated. Repeat the procedure for any other valves on this cylinder. Then move to the next cylinder and repeat the process to replace the whole set of valve seals.

VALVE SPRING REPLACEMENT

Weak or broken valve springs can cause valve float and valve burning. Valve float occurs at higher rpms, when it is hardest for the spring to hold the valve tightly to the profile of the cam. If the spring is too weak or broken the valve has a tendency to float over the top of the camshaft lobe and not begin to close immediately. When the valve hangs open like this you can notice misfiring and power loss at higher rpms. The engine will usually run normally at lower rpms. If the valve spring is broken the valve may never close. This can cause a steady misfire and popping out of the intake or exhaust depending on which spring is at fault. A broken valve spring must be replaced immediately to prevent the valve from burning.

You can use a vacuum test to help diagnose weak valve springs. Engine vacuum should be normal at idle but will fluctuate significantly at 2500 rpm or higher. A broken valve spring will cause a steady drop in vacuum each time that cylinder is on the intake stroke. You will most likely hear valvetrain noise and popping in the intake or exhaust when a spring is broken.

The procedure to replace valve springs is the same as for replacing valve seals. As a matter of fact, technicians should replace the valve seals while they have the valves disassembled. Once the valve spring is removed carefully inspect it for cracks, proper free length, squareness, and opening tension. These procedures will be described fully in the next chapter. You want to be sure that you have correctly diagnosed the problem and that spring replacement will cure the symptom.

CYLINDER HEAD REMOVAL

Depending on the engine configuration removing the cylinder head or heads can be a short and simple job or a relatively long and complex undertaking. It is important to read through the service information for the vehicle you are working on to be sure that no critical or obscure steps are overlooked. The manufacturer's information will include shortcuts and techniques that may not be obvious if you try to figure out by yourself how to remove the head(s). While you are learning about a particular engine it makes sense to follow the service procedures step-by-step.

Some general procedures are common to most head removal jobs. These points include:

- Place the engine at TDC firing for cylinder #1.
- Disconnect the battery negative terminal.
- Drain the oil and the coolant.
- Remove the valve cover. Loosen the nuts or bolts in a diagonal pattern around the cover. Removing the bolts on a circular pattern can distort the valve cover **(Figure 3)**.
- Disconnect and label vacuum lines, hoses, electrical connectors, and spark plug wires if applicable.
- Remove the intake ducting from the throttle body.
- Relieve the fuel pressure, and disconnect the fuel rail or fuel line.
- Remove the intake manifold, following the specified loosening sequence **(Figure 4)**.

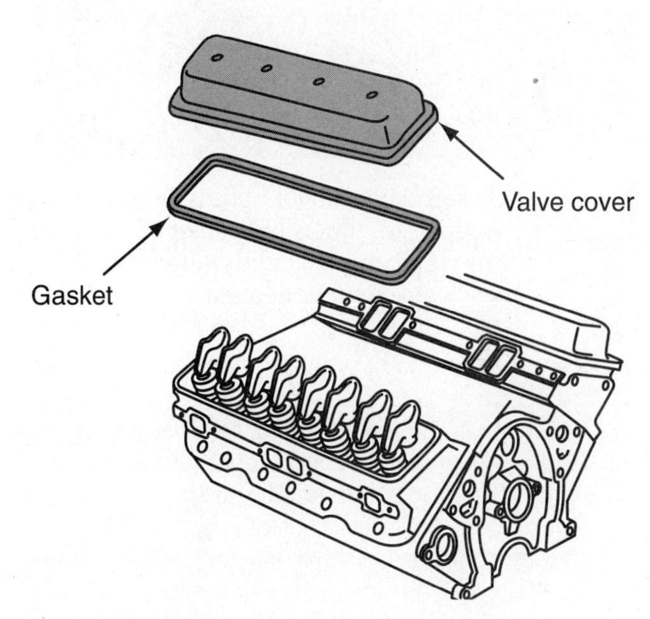

Figure 3. Remove the valve cover to access the cylinder head bolts.

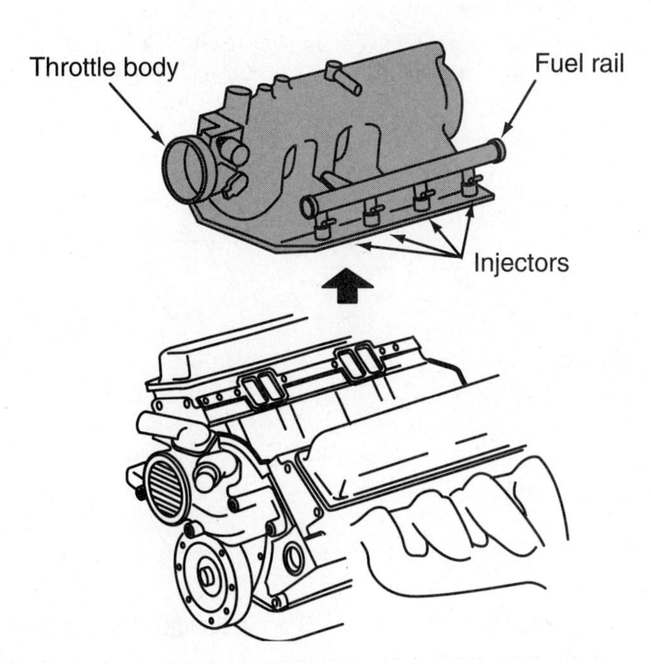

Figure 4. Use a diagonal pattern to loosen the intake.

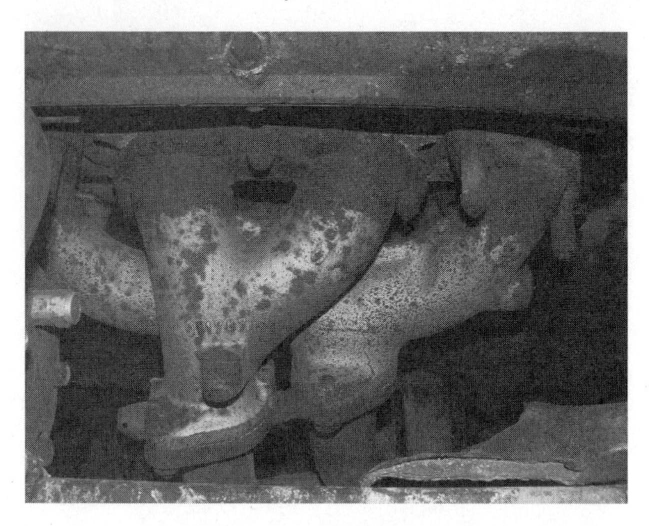

Figure 5. These rusty manifold studs will need to be replaced. It may be easier to loosen the front pipe and deal with the studs on the bench; heat will be necessary.

- Remove the exhaust manifold. On some engines it may be more practical to remove the cylinder head first and then work on loosening the exhaust manifold fasteners. They are often quite rusty and require penetrant or heat to loosen them **(Figure 5)**.
- Remove the timing chain tensioner on overhead-camshaft engines. On some engines it will be necessary to remove the timing cover to achieve this.
- Remove the rocker arms, if applicable, and mark their location. Sometimes you can string the rocker arms on a piece of mechanic's wire in the same order as the cylinders.
- Remove the camshaft sprocket and timing chain or belt. Note the positioning marks on the cam sprocket and head as well as on the crankshaft pulley or flywheel.
- Many engines require one or more special holding tools to retain the camshaft sprockets and keep the timing belt or chain correctly in place.
- Remove the pushrods and lifters or followers as applicable. Mark their position for proper reinstallation.
- Remove any brackets or components that are bolted to the cylinder head, marking their position to ease reassembly.
- Make sure the cylinder head is cool. Loosen the head bolts in the manufacturer's specified sequence. It is best to loosen each bolt one turn and then repeat the sequence to fully loosen each bolt. Aluminum cylinder heads are prone to warpage if they are removed hot or without using an appropriate loosening sequence. A typical sequence is shown in **Figure 6**.

HEAD GASKET REPLACEMENT

With the cylinder head removed carefully inspect the gasket for tears, cracks, and blowby past the fire ring. If the head gasket was leaking or the cylinder was burning coolant you want to clearly identify the cause to be certain that you repair it. Look for corrosion around the coolant passages on the head gasket and on the head. Inspect the head bolts for signs of stretching or thread damage. Replace any suspect bolts, or replace all bolts if specified by the manufacturer. Torque-to-yield bolts are replaced whenever they are removed. Check the threaded bore in the engine block. If any threads are damaged beyond repair you will need to install a helicoil. This is quite common on aluminum cylinder blocks. Obtain a helicoil kit that will replace the threads with the same size as the others. Use the drill provided with the kit to enlarge the threaded bore. Then thread the drilled hole to fit the helicoil. Use the helicoil tool to thread the helicoil into the new bore and lock it into place.

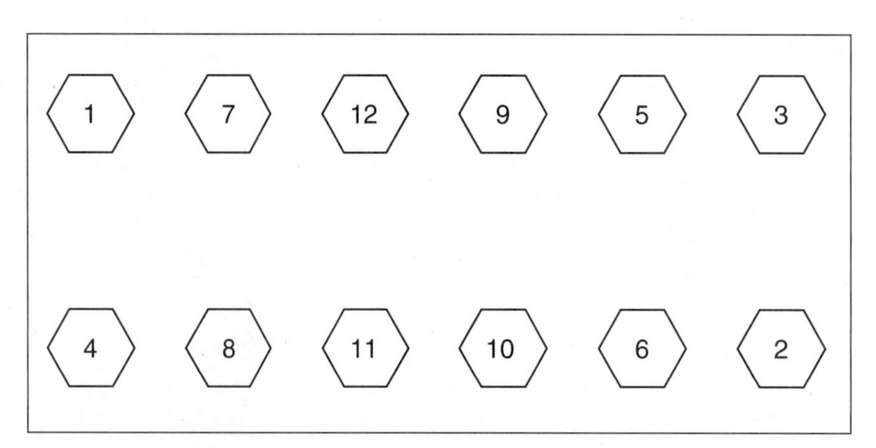

Figure 6. Loosen the cylinder head in the proper sequence. This is usually the reverse of the tightening process.

Clean the gasket surfaces on the block and head thoroughly. Particularly on aluminum components the manufacturer may allow only a gasket removing chemical and a plastic scraper. Abrasive discs or steel scrapers can damage the surface of the block or head. Check the head and block surfaces for warpage. Place a straightedge across the head, and try to fit feeler gauges underneath. Place the straightedge lengthwise in two places and diagonally across the head in both directions. The largest feeler gauge that fits under the straightedge in any of these positions indicates the amount of cylinder head warpage. Repeat the procedure for the engine deck. Compare the warpage with the maximum allowable specification. If there is excessive warpage the cylinder head or block will need to be machined flat.

Inspect the cylinder head thoroughly for cracks. Look closely between and around the valve seats; this is the most common area in which cracks occur. Check also between two cylinders and in the coolant passages. If the engine overheated you should definitely have the head pressure tested or magnafluxed to be sure there are no cracks undetectable by your eye. The condition of the coolant passages should be checked carefully. Clean all deposits from the cooling passages. This may require professional cleaning at the machine shop.

To reinstall the head and gasket make sure the surfaces are perfectly clean. Look at the gasket closely, and look for any top markings. Place the gasket on the block, and closely examine the coolant and oil passages to be sure they line up properly with the gasket. Do not use any sealant on the head gasket unless specifically requested by the manufacturer. Replace torque-to-yield bolts, or clean the threads of bolts that will be reused. Refer to the manufacturer's service information to determine whether the bolts should be lightly lubricated, sealed with thread sealant, or installed in a particular position if the bolts are different lengths. Place the cylinder head on the block and gasket, being careful not to disturb the gasket. Many engines use dowel pins in the block to help locate the head gasket and position the head. Start each of the head bolts by hand. Refer to the manufacturer's specifications, and follow the torque procedure precisely. They will provide the proper torque sequence and torque specification, including an additional turning angle if torque-to-yield bolts are used. Reassemble the engine by reversing the removal procedure. We will cover head installation more fully in Chapter 27.

CYLINDER HEAD DISASSEMBLY

When beginning the process of a valve job or thorough cylinder head reconditioning continue with head disassembly. Place the cylinder heads on stands to prevent bending open valves and to ease disassembly **(Figure 7)**. If any camshafts were not removed previously remove them now, reversing the tightening sequence to loosen the cap bolts or housing. The camshaft bearing cap halves must be reinstalled in the same location and direction; note the appropriate markings. Inspect the bearing surface for excessive wear or scoring. Make a note of any problems found; severe wear may require head replacement. Remove the rocker arms, if equipped, and mark them for reinstallation in their original location. Remove the valve lifters or followers as applicable. If the cam is in the block use a lifter removing tool to prevent damaging the lifters in case they can be reused **(Figure 8)**. These components must be replaced in

Figure 7. Use head stands to hold the head on a convenient base.

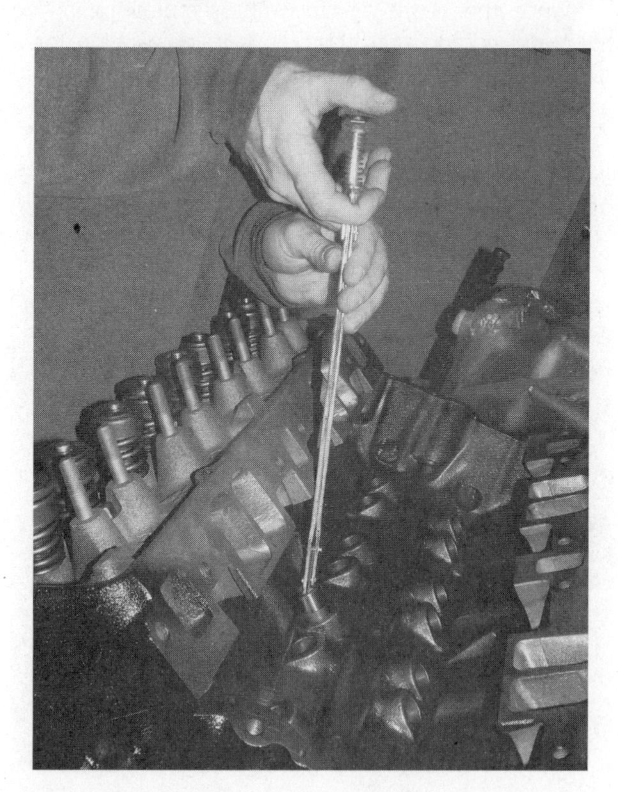

Figure 8. This lifter removal tool will help get lifters out without damaging them.

their original location, so organize and label the parts accordingly. Mechanical valve followers may have shims in a groove on the top or underneath the follower. Be sure to keep these shims with their followers. This will make valve adjustment much easier during the reassembly process.

In preparation for valve removal tap on the valve retainer of each valve to loosen any varnish. Be careful not to hit the valve tip. Using a socket that contacts the valve retainer is a good idea to prevent inadvertently damaging the valve tip. Use an appropriate valve spring compressor to compress the spring and lock it **(Figure 9)**. We use a valve spring compressor with a manual lever. **Figure 10** shows a hydraulic-assisted valve spring compressor commonly used in machine shops or production engine shops. Remove the valve keepers with a pocket screwdriver or a magnet. Keep them together in a safe place; you do not want to hold up a job because you lost one keeper. Remove the valve spring retainers; if there are some valve rotators they will be placed on the exhaust valves. Remove the tension on the springs slowly. Inspect the valve tips for mushrooming. If they are distorted wider than the stem they must be filed before removal or they can seriously damage the guide or even crack the head. Remove the valve and place it, the retainer, and the spring into a valve holder, or use a piece of cardboard to hold the components. Mark their positions accordingly; the valves should be returned to their original location. Exhaust springs may be different from the intake springs; be careful not to mix them up.

Once the heads are fully disassembled clean the head and components thoroughly in preparation for a careful inspection. On cast-iron heads clean the head gasket surface on the head and the block thoroughly. Use a scraper or razor blade nearly parallel to the head surface to scrape away any gasket residue or corrosion buildup. Some manufacturers allow the use of an abrasive disc or "cookie" on a die grinder as long as the material is suitable for the job. On many aluminum cylinder heads the only acceptable method of removing the old gasket material is to use an aerosol spray or brush-on gasket remover and a plastic scraper. If abrasive discs are allowed you must be sure to specify a soft, usually white, plastic disc with 120-grit abrasiveness that will not remove any material from the head. Spend just enough time in any spot to remove the old gasket. You must be careful not to deform the surface of the cylinder head or the block.

Figure 9. Compress the spring and use a magnet or a pocket screwdriver to remove the keepers.

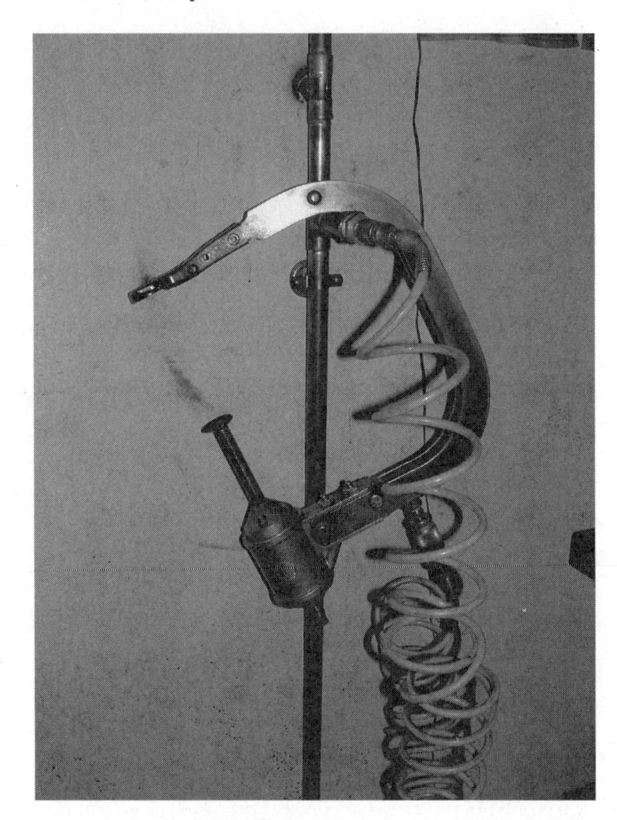

Figure 10. Our local machine shop uses this hydraulic spring compressor; the "head man" appreciates the help.

Remove any worn exhaust manifold studs. It is common, especially in rust-prone areas, to remove and replace all the exhaust manifold studs. Clean the intake and exhaust manifold surfaces using an appropriate technique. Remember that these surfaces must also remain flat for a new gasket to seal properly. Clean the threads for the exhaust manifold mounting, and replace the studs. Thoroughly inspect the intake and exhaust manifolds for cracks; either one is prone to cracking when the cylinder head has been overheated. Look carefully at the seams of a plastic or composite intake manifold.

Use a valve guide brush to clean the valve guides **(Figure 11)**. Pick a brush that fits just snugly; be careful not to ream material off the guide. Blow the loose carbon from the guides using shop air. Use a wire brush lightly on a drill to clean the valve seats and throat below the seats. This will allow for a good inspection of the seats and ports. Remov-

Figure 11. Use the proper valve guide brush to clean all the carbon from the guides.

ing the carbon will also help when reconditioning the seats with grinding stones. Be careful not to scratch the cylinder head surface with the wire brush.

Now you should be ready for a careful inspection of the cylinder head and its related components before reconditioning. Organize the parts and hardware you had to remove to get the cylinder head off, and set them aside for reassembly. Next your focus will be on identifying worn or damaged components and analyzing the cause of failure.

Summary

- Valve seals and springs can be replaced without removing the cylinder head.
- Apply pressurized air to a cylinder before removing valve keepers to prevent dropping the valve into the cylinder.
- Worn valve seals will cause oil consumption and blue smoke after startup and at idle and deceleration.
- Weak valve springs can cause valve float at higher rpms; this causes misfire as the valve hangs open when it should be closed.
- Broken springs can allow a valve to stay open constantly, causing a regular misfire and a burned valve.
- Remove the cylinder head for head gasket replacement or head reconditioning.
- Mark components as they are removed to facilitate reassembly.
- Keep lifters, followers and shims, rocker arms, pushrods, valves, springs, and retainers in order so they can be returned to their original locations.
- Be careful to follow an appropriate loosening sequence when removing head bolts, intake and exhaust manifold hardware, and camshaft caps or housings to prevent cracking or warping components.
- Inspect the camshaft bearing surfaces for excessive wear or scoring.
- Look closely at the pivot and contact points of the rocker arms for damage.
- Check the tips of the valves for mushrooming before removing the valves from the head; file them as needed to prevent damage to the guides or head.
- Use a spring compressor to release the spring tension and remove the keepers, valve retainers, springs, and valves.
- Clean the cylinder head surfaces and valve guides in preparation for a thorough inspection.

Review Questions

1. What can happen if the camshaft caps are not installed in their proper positions?
2. Briefly outline the procedure to replace valve seals with the cylinder head installed.
3. Why is it important to follow a loosening sequence when removing a cylinder head?
4. Technician A says that you should replace the valve seals when replacing valve springs. Technician B says that before replacing valve seals you should check the condition of the guides. Who is correct?
   A. Technician A only
   B. Technician B only
   C. Both Technician A and Technician B
   D. Neither Technician A nor Technician B
5. Technician A says that low cranking compression indicates worn valve guides. Technician B says that worn guides often cause fluctuating vacuum at low rpms. Who is correct?
   A. Technician A only
   B. Technician B only
   C. Both Technician A and Technician B
   D. Neither Technician A nor Technician B
6. Technician A says you should rap on the tip of the valve before compressing the spring. Technician B says you can use a magnet to remove the valve keepers and prevent them from dropping into the head. Who is correct?
   A. Technician A only
   B. Technician B only
   C. Both Technician A and Technician B
   D. Neither Technician A nor Technician B
7. Technician A says that weak valve springs cause a rough idle. Technician B says that weak valve springs can cause valve float. Who is correct?
   A. Technician A only
   B. Technician B only
   C. Both Technician A and Technician B
   D. Neither Technician A nor Technician B
8. Technician A says to remove the intake ducting and manifold when removing the cylinder head. Technician B says to remove the exhaust manifold and replace rusted studs. Who is correct?
   A. Technician A only
   B. Technician B only
   C. Both Technician A and Technician B
   D. Neither Technician A nor Technician B
9. Technician A says to keep rocker arms organized so they can be reinstalled in their original location. Technician B says to inspect their pivot and contact points for wear or cracks. Who is correct?
   A. Technician A only
   B. Technician B only
   C. Both Technician A and Technician B
   D. Neither Technician A nor Technician B
10. Technician A says you can use a plastic scraper to remove any gasket material from the head. Technician B says to use a reamer to clean the valve guides. Who is correct?
    A. Technician A only
    B. Technician B only
    C. Both Technician A and Technician B
    D. Neither Technician A nor Technician B

Chapter 26 Cylinder Head Component Inspection

Introduction

As part of performing a valve job or thorough cylinder head overhaul you will need to make some critical inspections and measurements on components to determine the actions needed. Be thorough in your inspections to prevent your repairs from failing. Overlooking a small crack in the head could cause a valve to burn or cause coolant consumption to persist. The head itself must be measured for flatness and checked carefully for cracks. You must assess valve guide wear prior to performing any valve work. The valves must be evaluated thoroughly to determine whether they can be refinished and reused. There are several methods used to gauge valve spring condition. Lifters or followers and rocker arms must be inspected closely before deciding to reuse them. Inspect and measure the camshaft journals and lobes.

In many cases your job will be to diagnose the need for cylinder head work, remove the cylinder head, and send it out to a machine shop for the repairs. Whether or not you do some of the work in-house it is essential to establish a good working relationship with a machinist who produces high-quality work. She may point out other repairs that are needed. Similarly, you should provide the machinist with as much information as possible so that she can perform a complete and successful repair. When you return the vehicle the customer relies on you to have resolved the issues satisfactorily. Your machinist becomes part of your service team.

Interesting Fact

Experienced engine rebuilders or machinists can gauge the condition of some components by visual inspection. This ability comes only after a lot of practice looking then measuring to verify conclusions.

CYLINDER HEAD INSPECTION AND MEASUREMENT

The cylinder head should be checked for flatness using a straightedge and feeler gauges. Check for warpage with the straightedge placed horizontally at the center of the head and also on each diagonal **(Figure 1)**. Most manufacturers do not allow more than .004 in. of warpage, but many specify even less; check the specifications for the engine you are working on. If a head is reinstalled on the block with excessive warpage the head gasket is likely to fail. When the cylinder head is sent out to a machine shop the machinists often automatically resurface the head to make it flat and achieve the proper surface finish.

On an overhead-cam engine, particularly when the head has been overheated or the surface is warped, check the alignment of the camshaft bores. Lay a straightedge in

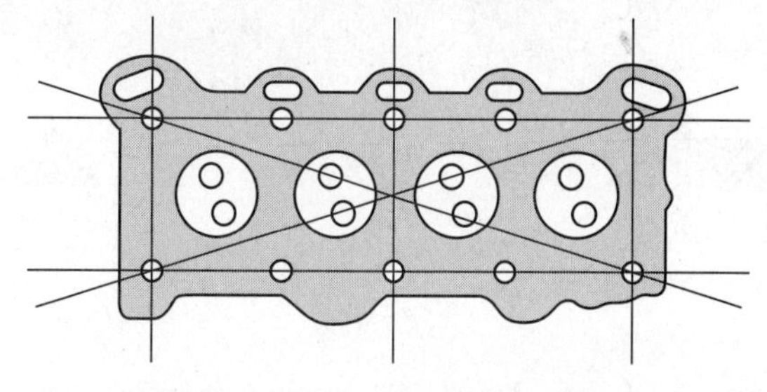

Figure 1. Measure the cylinder head in the planes shown to detect any excessive warpage.

the base of the bores, and use feeler gauges to make sure there is less than .001 in. of warpage in the bore. If you measure excessive warpage the head will have to be straightened or replaced. Straightening should be performed before the gasket surface is machined. Speak with a machinist for a recommendation for that specific cylinder head.

Interesting Fact

During what I thought was a routine head gasket replacement I sent the head out to have the sealing surface trued up; it had significant warpage. I never checked the camshaft bore alignment. I had the head back on the engine and nearly assembled before I tried to fit the camshaft. I could not even bolt the camshaft caps down because the bore was warped so badly. Luckily the machine shop was able to salvage the aluminum head by straightening it, or the customer would have been shocked by the bill!

Carefully inspect the cylinder head for cracks. Very often a head cracks between the valve seats or around the exhaust valve seats **(Figure 2)**. Check inside the ports for cracking as well. Use a dye penetrant on an aluminum head to check any suspicious areas. If the head has been overheated many shops will automatically pressure test, magnaflux, or use dye penetrant on the cylinder head either in-house or at a machine shop. It is better to err on the side of safety and have a head tested rather than risk a customer's coming back for an undetected crack.

Clean the cooling passages in the head using brushes and shop air. If there are deposits in the larger visible passages you should assume there are also deposits in the smaller passages surrounding the valve seats. Ideally the head should be thoroughly cleaned professionally at the machine shop or in an in-house pressure washer. If a valve has burned for no apparent reason cleaning is crucial.

CHECKING FOR VALVE GUIDE WEAR

During cylinder head disassembly you should have cleaned the valve guides with a brush. If not do so now. Clean the valves on a wire wheel, and fit them into their respective guides. Pull the valve off the seat one inch, and check for play. You should just be able to detect movement. If the valve wobbles back and forth more than about five-thousandths of an inch (.005 in.) you will need to measure the valve stem with an outside micrometer and the valve guide with an inside micrometer or ball gauge to determine which, if not both, is worn beyond an allowable limit. Once you become experienced at checking valve guide wear you will be able feel what is acceptable and what is not. Often guides are replaced as a matter of course during a valve job. They have a tendency to wear in a bell-mouth fashion that is not always detected through guide measurement **(Figure 3)**. This can cause premature failure of the valve seals. Typical stem-to-guide clearance is .001 in. to .003 in., though some manufacturers specify a greater clearance.

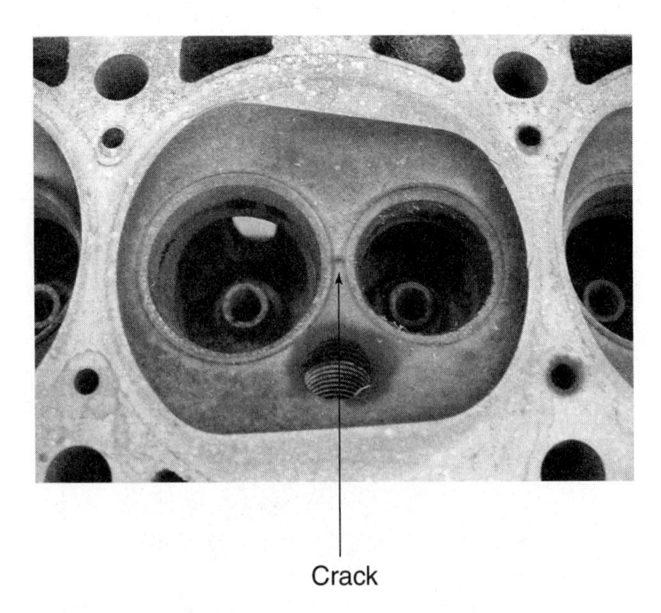

Figure 2. Look closely for cracks; this is a typical one between the seats.

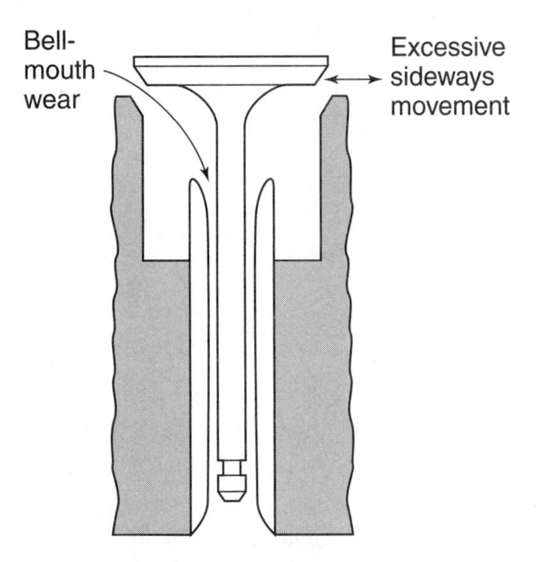

Figure 3. Rock the valve back and forth in the guide to check for excessive movement. This guide is worn at the top and bottom in a bell-mouth fashion.

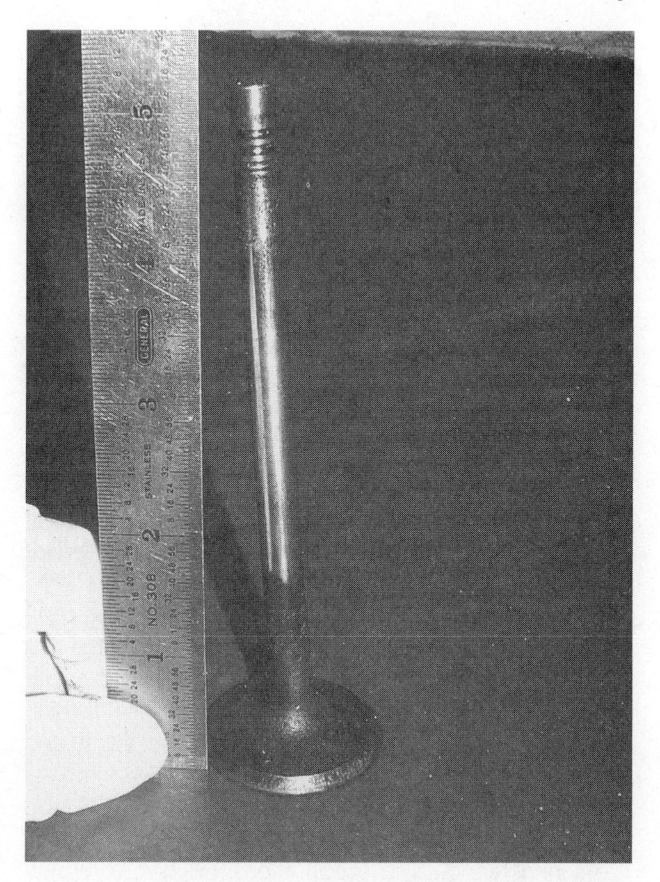

Figure 4. This valve is bent, which is clearly seen when placed next to a straightedge.

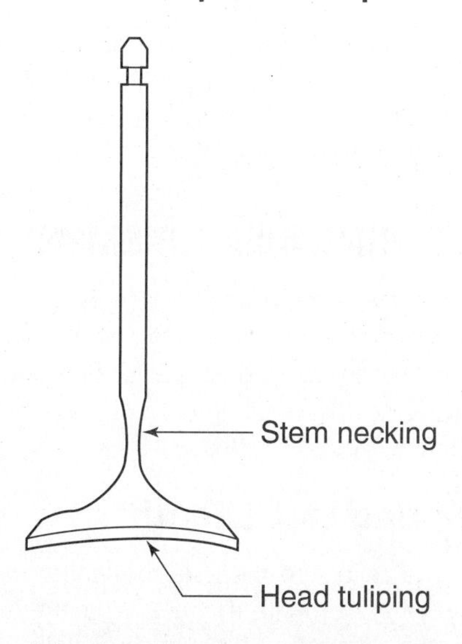

Figure 5. A valve that has a stretched neck or a tuliped head must be replaced.

Figure 6. This valve burned from a tight valve adjustment.

VALVE INSPECTION

Before investing the labor in reconditioning valves they should be scrutinized for irreparable damage. A bent valve, for example, cannot be reconditioned. Lay a machinist's rule parallel to the valve stem, and check the stem straightness visually **(Figure 4)**. You may also have access to a vee block designed to check valve straightness. Lay the valve in the vee block and place a dial indicator on the valve margin to check for a bent valve.

Examine the stem for scoring and measure the stem thickness for wear. Often you can see a wear ridge on the stem at the top and bottom of guide travel. Inspect the valve head for tuliping and the stem for necking. Valve **tuliping** is when the valve face and head curl up into a tulip shape. This reduces the diameter of the valve face and prevents proper sealing. Valve **necking** is when the valve stem stretches from the weight of the head. Both these conditions require valve replacement **(Figure 5)**. The valve margin must be greater than the allowable limit because as the valve is refinished some of the margin will be cut away. Inspect the keepers for any visible wear. Fit them around the valve and check to be sure they have a groove between the faces. Many technicians simply replace the valve keepers; they are inexpensive and a failure can completely destroy an engine. If the valve face is deeply grooved or has burned, the valve will not be salvageable.

When a valve is burned study the possible causes **(Figure 6)**. Was the valve adjustment too tight, keeping the valve open too long? Are there heavy deposits on the stem that could cause the valve to stick in the guide? Is the valve guide wear severe enough to cause the valve not to seat properly? Is there a crack in the head near the seat that was allowing combustion gases into the coolant and pushing the coolant away from the seat? Are the springs too weak or broken, allowing the valves to hang open? Try to determine the cause of the failure so you can be certain that your repair will be successful.

VALVE SPRING MEASUREMENT

There are several methods of determining the condition of the valve springs. Manufacturers may specify one or more methods to be used; follow their specific directions. A spring can be measured for free length, tension, and squareness when the springs are off the head.

Valve Spring Free Length

Use a dial caliper to measure the length of the valve spring **(Figure 7)**. Compare the measurement with specifications. Lay the springs on a flat surface, and place a straightedge on top of them to check for other springs that may be out of specification. If this procedure is not possible, or if one or more springs are suspect, measure each spring and compare the length with specifications. Replace a spring that is either too long or too short.

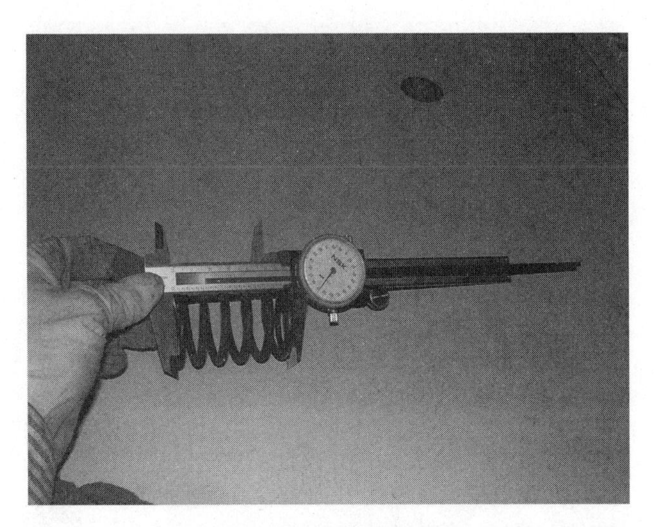

Figure 7. Measure the spring free length using a dial caliper and compare to specifications.

Spring Squareness

Place the valve spring on a flat surface, and lay a straightedge next to it. Turn the spring around against the square, and measure the maximum variation off square **(Figure 8)**. Check the manufacturer's specifications. In general a spring that is more than 1/16 in. unsquare should be replaced.

Valve Spring Tension

Valve spring tension is measured at a particular height; this may be at the valve-closed height (installed height) or at the valve-open height. Check the manufacturer's service information to determine which test you should perform. Two different types of valve spring tension gauges are commonly used. One has a tension gauge mounted on the tool **(Figure 9)**. You compress the valve to the specified height and observe the pressure reading. Another type of tension gauge uses a platform that you set to the proper height, and then you install a torque-o-meter on the test lever. Compress the spring until a tone sounds **(Figure 10)**. Observe the foot-pounds on the torque wrench, and multiply by two to find the spring tension. The specifications will give a tension range that the springs must fall within to be acceptable.

VALVE LIFTER INSPECTION

Many technicians will simply replace the lifters during a valve job or engine overhaul. That can be quite an expensive decision, especially if they are roller lifters **(Figure 11)**. Examine the lifters closely if you are considering reusing them. Lifters should be slightly convex on the end that rides on the camshaft. Over time the lifters wear and become concave; this condition mandates replacement **(Figure 12)**. The body

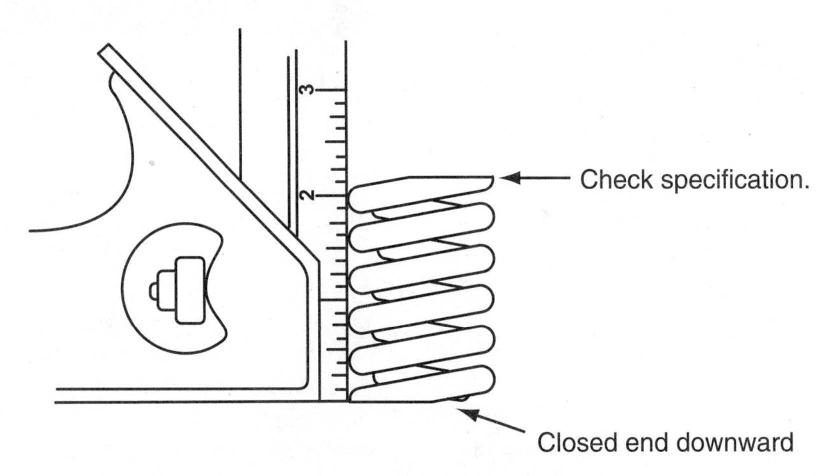

Figure 8. Place the spring on a flat surface and stand it next to a square to check its straightness.

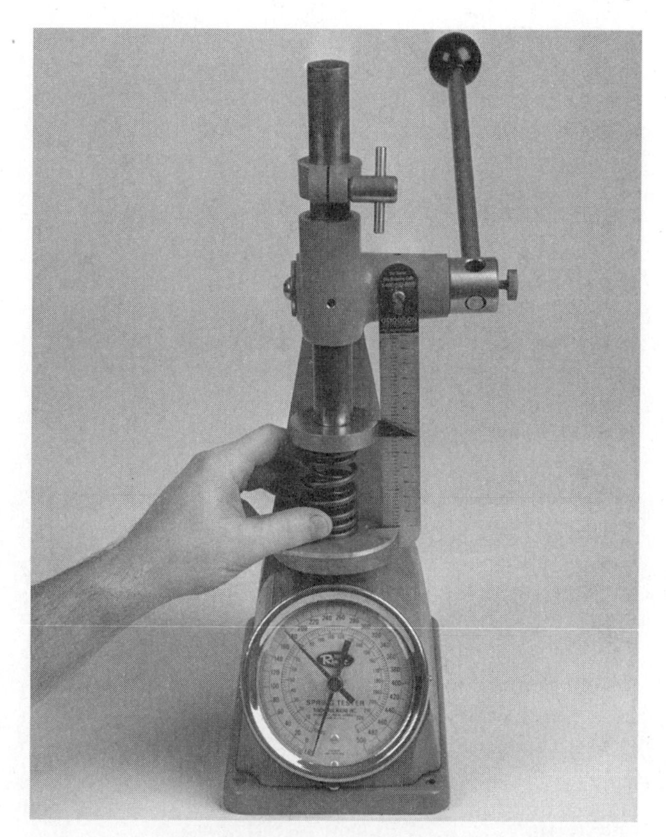

Figure 9. Compress the spring to the specified height and read the tension.

Figure 10. Set the height of the platform and compress the spring until the tone sounds. Read the torque-o-meter and multiply the force by two.

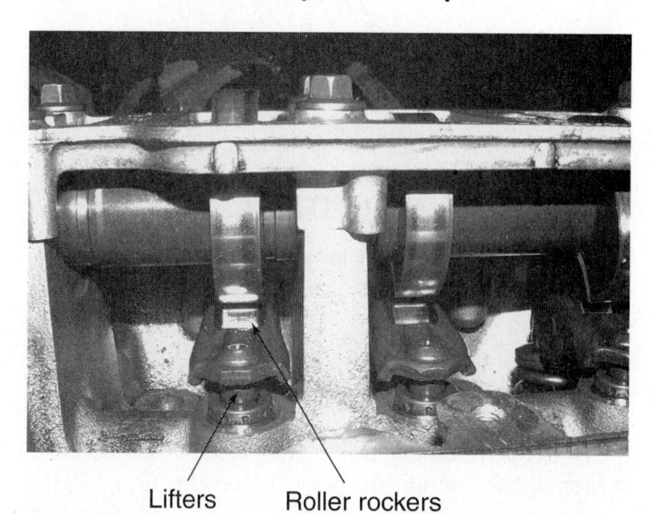

Figure 11. A cautious student replaced this set of lifters and roller rockers for quite a hit in the wallet.

of the lifter should also be checked for scoring. If the body is not smooth it should be replaced. When damage is seen on the surface of the lifter examine the lifter bore for scoring as well. When inspecting a roller lifter check the roller for smooth rotation and the surface for any flat spots. Check the oil passages for any blockage. Today lifters are generally replaced, not cleaned. Some shops still have a tester that checks the lifter leakdown rate. This tester ensures that the internal seals in the lifter are effective in preventing the lifter from collapsing. This testing is quite common on heavy-duty diesel engines where components are very expensive and built to endure many miles. When replacing lifters on a modern passenger car or light-duty truck it is customary to

Figure 12. This badly damaged lifter was similar to the rest. The engine required a new set.

replace the entire set of lifters, not just a suspect one or two. You must also replace the set of lifters whenever you replace the camshaft. Wear on an old lifter can prematurely damage the lobes of the new camshaft.

CAMSHAFT INSPECTION

The camshaft should be inspected for wear on the journals and on the lobes. It should also be checked for straightness, particularly if the head was overheated. A visual inspection is often all that is needed to condemn a camshaft.

Sometimes you can easily see that one or more lobes are worn down compared with other lobes. If in doubt use a micrometer to measure the height of the lobe. Compare the height with specifications, and replace a damaged camshaft. If the lobes are no longer at the specified height, valve lift will decrease, affecting volumetric efficiency and engine power.

Inspect the camshaft journals for wear and scoring. Measure the diameter of the journal in two spots 90° apart. The measurements should be equal and match the specification. If damage is found on the journals look over the cam bores closely.

To check the camshaft straightness place the camshaft in vee blocks, and set a dial indicator on the center journal. Rotate the camshaft, and make sure total indicator movement is not greater than .002 in. Camshafts are very brittle and cannot be straightened; replace a bent cam.

CAMSHAFT BORE AND BEARING SURFACE INSPECTION

You need to check the camshaft bearing bores and the bearings or bearing surfaces for cracks, scoring, out of round, and misalignment. If bearings are used inspect them for unusual wear that could indicate a bent camshaft. You will replace the bearings during your service. When working with an in-the-block camshaft you should replace the bearings as explained during the discussion of block service. Many heads use the camshaft saddle and mating caps as the bearing surface. These also need to be carefully inspected. If the cam bores or caps are worn considerably the head will need to be repaired if possible or replaced.

Torque the caps on in their proper spots, and check for bore stretch or out of round. Use a telescoping gauge or an inside micrometer to measure the bore side to side and then top to bottom. The measurements should be within .001 in. of each other. Some manufacturers will not tolerate any out of round, so check the specifications.

Sometimes damaged bores can be repaired in a fashion similar to rod big end bores. Some material is taken off the mating surfaces of the caps, and then the holes are rebored to the correct size. Manufacturers may reject this procedure and require head replacement instead.

ROCKER ARM AND PUSHROD INSPECTION

The rocker arms and pushrods, if equipped, are essential to proper valve opening. Check the rocker arm pivot points and the ends where they contact the pushrod and valve tip for wear. Inspect each rocker arm for cracks, particularly near the pivot point. If a valve tip is severely worn it is very likely that the rocker arm is also worn and should be replaced **(Figure 13)**.

Roll the pushrods on a flat surface, and check for any bends. Do not try to straighten pushrods; replace them. Clean the pushrod oil passage with solvent or a fine brush. The passages should be perfectly clean if they are to be reused. Even partial blockage can restrict oil flow enough to cause premature valvetrain wear.

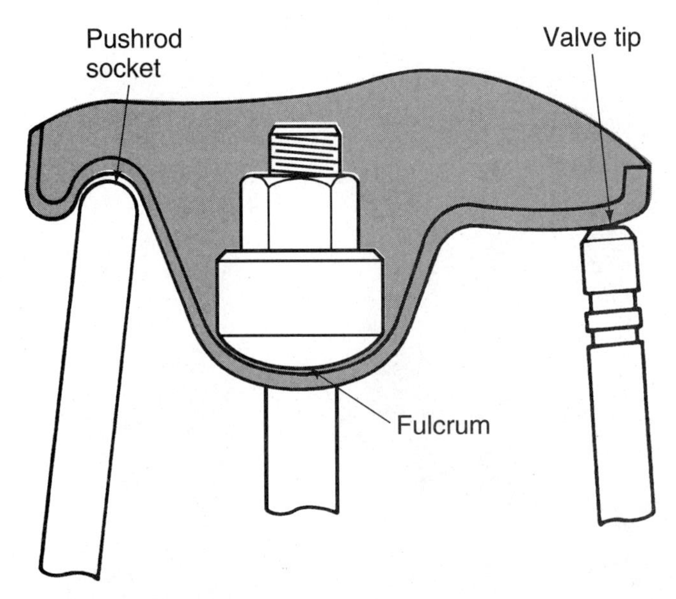

Figure 13. Check the rocker arm carefully for wear and cracks. Look at the pushrod and valve tip as well.

Summary

- Careful and thorough examination of cylinder head components can help ensure a successful repair.
- Check the cylinder head for flatness, cam bore alignment, and cracks.
- Clean the head cooling passages to ensure adequate cooling of all parts of the cylinder head.
- Valve guides have a tendency to wear in a bell-mouth fashion. This condition can cause premature failure of new valve seals.
- Replace guides that have clearance beyond the specified amount.
- Check the valves for bends or burning; these conditions warrant replacement.
- When a valve has burned consider all the possible causes so that you can correct the cause of failure.
- Examine the valve stem for wear or scoring, the valve neck for stretch, and the valve head for tuliping.
- Carefully inspect or replace the valve keepers.
- Measure the valve spring free length, squareness, and tension.
- Inspect the lifters for a concave bottom or excessive scoring on the face; replace as indicated.
- Look over the camshaft carefully for signs of wear on the lobes or journals. Measure them with a micrometer if you are uncertain about their condition.
- Replace a camshaft that shows a bend while set in vee blocks.
- Make sure that pushrods are straight and oil passages clean.
- Study the wear on rocker arms to determine if they can be reused. Check for cracks, particularly near the pivot point.

Review Questions

1. What inspections should you make on valve lifters to determine whether they should be replaced?
2. Describe three ways to evaluate valve springs.
3. Describe two methods of checking for excessive valve guide wear.
4. If the camshaft bores are warped beyond specification what are the repair options?
5. Pushrods are hollow and pretty light. Is it true or false that if they are bent you should be able to straighten them for reuse?
6. Cylinder heads typically crack in each of the following locations *except:*
   A. Between the valve seats
   B. Near the exhaust valve seat
   C. In the exhaust ports
   D. Near the thermostat housing
7. Technician A says that a valve that has stem deposits should be replaced. Technician B says that if a valve is tuliped it should be replaced. Who is correct?
   A. Technician A only
   B. Technician B only
   C. Both Technician A and Technician B
   D. Neither Technician A nor Technician B
8. Technician A says that deposits in the cooling passages can cause a burned valve. Technician B says that excessively worn guides can cause valve burning. Who is correct?
   A. Technician A only
   B. Technician B only
   C. Both Technician A and Technician B
   D. Neither Technician A nor Technician B
9. Technician A says that a worn cam lobe will cause reduced engine power. Technician B says that a bent camshaft should be straightened. Who is correct?
   A. Technician A only
   B. Technician B only
   C. Both Technician A and Technician B
   D. Neither Technician A nor Technician B
10. Technician A says that if the valve tip is badly damaged the rocker arm will usually require replacement. Technician B says that rocker arms should be closely inspected for cracks in the pivot area. Who is correct?
   A. Technician A only
   B. Technician B only
   C. Both Technician A and Technician B
   D. Neither Technician A nor Technician B

Chapter 27 Cylinder Head Service and Assembly

Introduction

Now that we have covered the cylinder head and valvetrain functions, disassembly, and inspection we are ready to discuss cylinder head service. You must be thorough in your repairs to the cylinder head in order to achieve high-quality, lasting work. Certain things must be undertaken in the correct order. Grinding of the valves and seats is precision work that requires patience and attention to detail. This chapter will cover cylinder head repairs, valve and seat reconditioning, valve and seat fitting, valve and spring measurements and assembly, and head assembly and installation. You must start with a solid cylinder head ready to provide thousands of miles of good service. You have already determined what repairs need to be made to the head; let us discuss those repair options first.

Interesting Fact

At one dealership I worked at we had old valve and seat grinding equipment. During the busy summer months we sent heads out to a machine shop for valve jobs. In the winter I particularly enjoyed the precision machining and adjustments of a valve job. A few general repair shops still do some or all of their head work in-house.

CYLINDER HEAD REPAIRS

The cylinder head repairs discussed here are usually performed at a machine shop or engine specialty shop. Very few general repair shops have the equipment needed to machine or straighten cylinder heads.

A cylinder head that is warped on the sealing surface can simply be resurfaced true if the cam bores are not warped. Make sure that the manufacturer allows the head to be shaved; some specify replacement if the warpage is beyond specifications. Others have a machining limit, so check the overall height of the head and compare it with specifications to be sure it can be machined again. If the cam bore is warped the cylinder head should be straightened before it is refinished. **Figure 1** shows a milling machine used to resurface heads. Consult the manufacturers' specifications for the required surface finish; these differ between head materials and change frequently with various head

Figure 1. This milling machine can be used to resurface cylinder heads, block decks, or flywheels.

designs. The surface finish is important for proper head gasket sealing.

To straighten an aluminum cylinder head you place shims with a thickness equal to half of the warpage under the ends of the head. The head is then pressed and heated for several hours at between 400°F and 500°F. The head should be cured in the oven overnight. You can repeat the process if necessary until the head is straight. Then you can refinish the head. If the cam bores are still misaligned bore or hone them, if recommended.

A cracked cylinder head can sometimes be repaired. Not all machine shops perform these repairs, especially on aluminum heads. Cast-iron heads are more easily repaired by drilling a hole at each end of the crack to limit expansion of the crack. Then holes are drilled so that overlapping tapered threaded plugs or pins can be installed for the length of the crack. Aluminum cylinder heads are tungsten inert gas (TIG) welded to repair cracks. This is a specialized process.

VALVE GUIDE REPAIR OR REPLACEMENT

When valve guides are worn they must be repaired or replaced *before* any work is done on the valve seats **(Figure 2)**. Typically, integral guides are knurled or drilled out and inserts are fitted. The **knurling** process is much like tapping a hole. A knurling tool raises and depresses metal in the bore like a thread **(Figure 3)**. Then a resizing tool is used to smooth off the edges of the "thread" **(Figure 4)**. Knurled valve guides provide good lubrication to the valve stem. Some technicians and machinists prefer knurling as a repair method because this added lubrication may extend the life of the fresh guide. If the guides are badly bell mouthed or worn, or if the machinist does not favor knurling, the guide can be drilled oversize and a new insert pressed in. In some cases the machinist may resize the existing guide and install a thin wall liner.

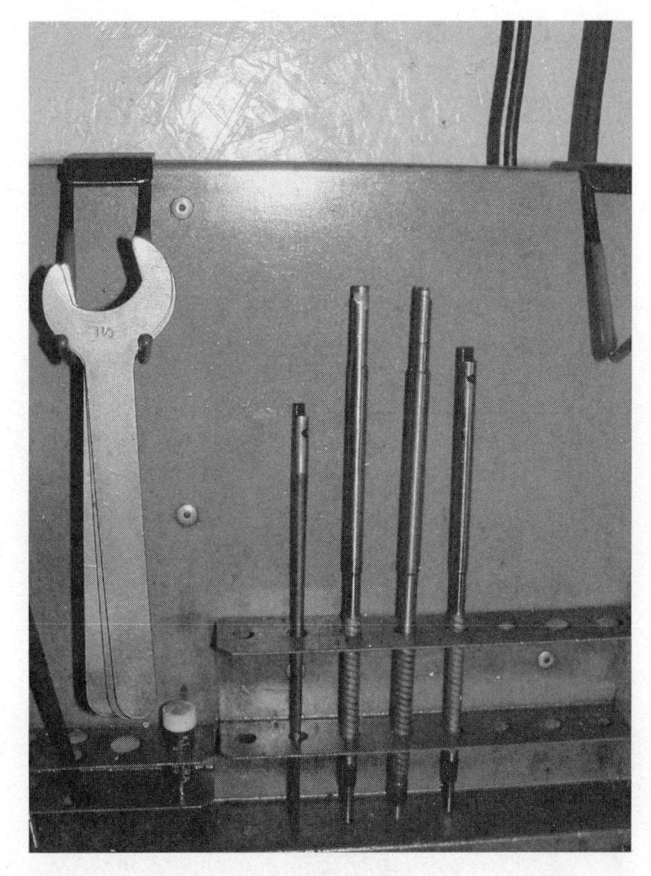

Figure 3. Knurling bits are used to raise metal within the guide.

Pressed-in guide inserts are merely pressed out and new ones pressed in. Often the head is heated slightly (150°F to 200°F), and a guide punch can be used to easily

Figure 2. This valve guide (half) was badly worn and came out of the head in two pieces.

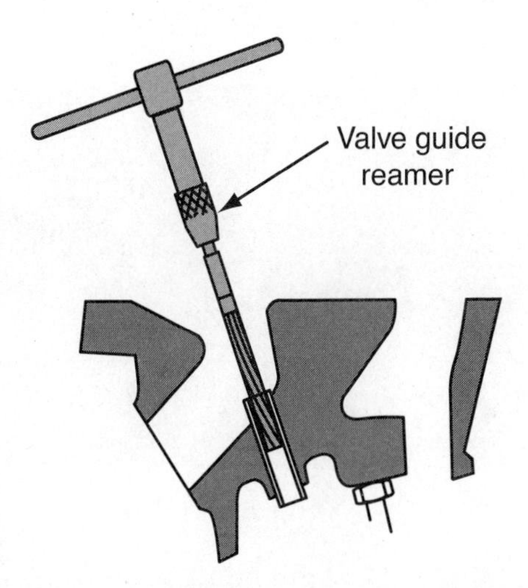

Figure 4. After knurling, a reamer is passed through the guide to smooth the sharp edges and size the guide.

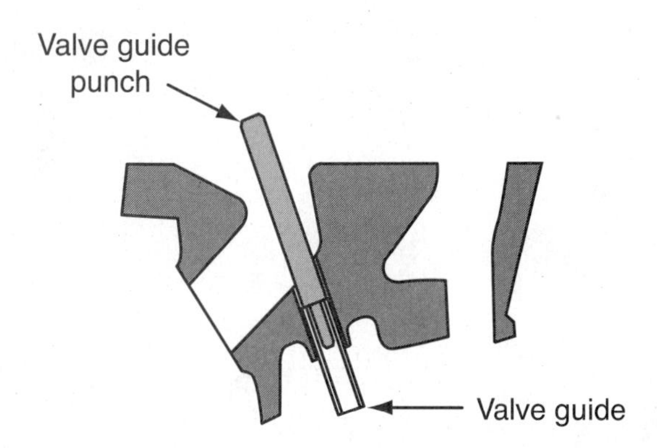

Figure 5. Use the correct valve guide driver to safely punch the old guide out and install a new one.

drive the guides out and punch the new ones in **(Figure 5)**. In many cases using the press is recommended because it applies more even pressure to the guide. Be sure to measure the depth of the guide before removal so you can install it to the same distance. Some guides have a chamfered edge on one end. Do not drive the guide from this end; the guide punch will not make adequate contact with the guide.

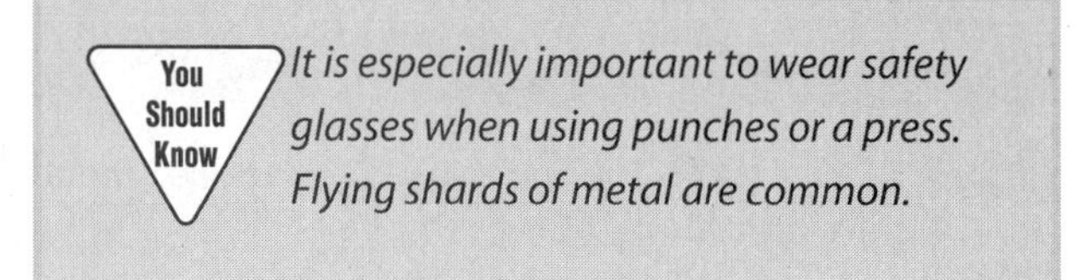

VALVE RECONDITIONING

A valve grinding machine is required to reface the valves **(Figure 6)**. You want to cut as little metal off the valve face as possible while still removing all **cupping** and pitting. The center of the valve face gets rounded out and looks like a C from repeated pounding on the seat; this is called cupping. When you have machined enough material off the valve face it will be flat again and provide a tight seal against the seat. Make sure the valve face and stem are clean. This ensures that the cutting stone will not get loaded with carbon and that the valve will sit squarely in the mounting chuck. Make sure you have enough margin left to cut the valve face. Several different types of valve grinders are available; you will have to become familiar with the one used in your shop. These guidelines apply to most valve grinding equipment.

1. Refer to the manufacturer's specification to determine at what angle the valve face should be cut.
2. Adjust the valve holding bench to the specified angle.
3. Use the diamond-tip stone dresser to put a good clean finish on the stone. Use lubrication while cutting. Take off as little material as necessary. Make several fast passes across the diamond while you are removing material; then finish the stone by moving it slowly across the diamond several times to smooth the surface finish **(Figure 7)**.
4. Place the valve in the holding chuck as close to the neck as possible. Be sure the balls of the chuck are contacting the machined part of the stem.
5. If possible, set the mechanical stops on the bench **(Figure 8)**. Move the valve in close to the stone. You want the valve to move all the way across the stone so as not to develop a groove in your freshly dressed stone. Adjust the stops so the valve does not travel beyond the end of the stone; this can nick the valve as it runs back up onto the stone. You also want to set the stops so that you prevent the stone from contacting the neck of the valve. If you nick the neck of the valve the valve must be replaced.
6. Turn the chuck and the stone on, and adjust the lubrication so it hits the valve face.

Figure 6. A typical shop valve-grinding machine.

Figure 7. Pass the stone across the diamond cutting tip to clean the face of the stone.

Figure 8. Adjust the stops on the bench so that you cannot nick the valve neck and so you use the whole stone.

7. Slowly move the valve into contact with the stone while simultaneously moving the stone back and forth across the valve **(Figure 9)**. If there is a depth indicator set it to zero. Move the stone quickly, and adjust the valve into the stone slowly until you can see no more pitting. Then put a nice finish on the face by making slow passes across the valve. Move the valve away from the stone, and turn off the machine. Look at the depth indicator to see how much material you removed from the valve.
8. Carefully inspect the valve, and continue grinding if pits or cupping are still seen.
9. Measure the margin, and be sure you have enough left to reuse the valve. If not discard the valve; if the valve is all right proceed to the next step.

Figure 9. Draw the valve into the stone while both are rotating. Set the oil so that it is running over the valve face.

Figure 10. Run the valve tip back and forth across the stone to dress the tip.

10. Next you need to dress the valve tip to make a clean, smooth surface to contact the rocker. You also need to remove the same amount of material from the tip that you did from the face so that you do not disturb the rocker arm geometry. If you have depth indicators simply match the amount of material you take off the tip with what you removed from the face. On machines without this measuring equipment you will have to gauge this by feel; a small difference will not impact the geometry adversely.
11. Dress the tip-cutting stone, usually on the end of the valve grinder **(Figure 10)**.
12. Place the valve in the holder, and adjust the lubrication to hit the tip. Move the stone into the valve tip slowly while rocking the valve back and forth across the stone. Proceed until the tip is fully freshened.
13. Finally, place the valve in the chamfering holder. Very gently push the valve against the stone to produce a slight chamfer. All you are trying to do is eliminate the sharp edge of the tip; a deep chamfer is not desired **(Figure 11)**.
14. Repeat the process for each of the valves. Many technicians will refinish the face of each valve and then refinish the tips to avoid resetting the stops on the stone bench.

The end result should be a perfectly cleaned-up valve face and tip with a margin at least as thick as specified.

VALVE SEAT REPLACEMENT

Sometimes the valve seats are so badly damaged or recessed into the head that they need to be replaced. Valve seat **retrusion** (recession) means that the seat actually gets pounded down into the head below the surface

Figure 11. Rotate the valve in the holder and put a light chamfer on the edge of the valve tip.

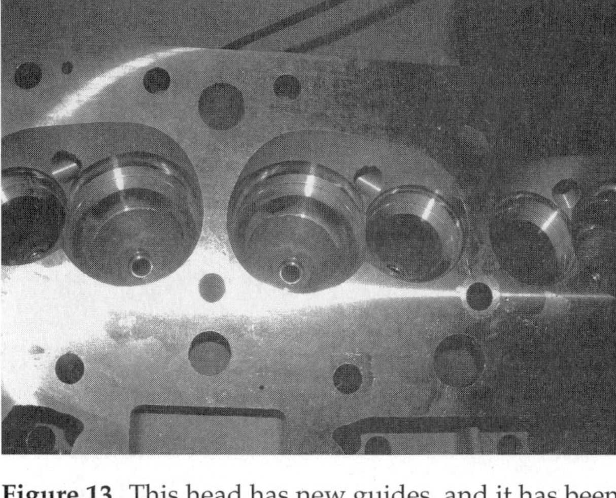

Figure 13. This head has new guides, and it has been drilled to accept new seat inserts.

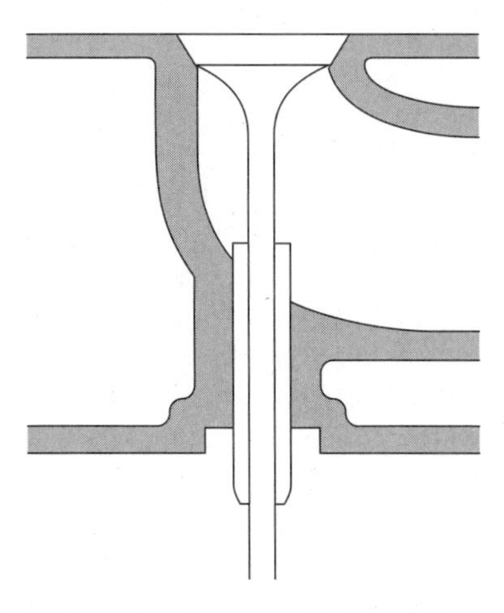

Figure 12. This valve has receded too deeply into the head. Unless the seat is replaced, the engine will not be able to breathe properly.

Figure 14. These diamond seat cutters are long lasting and do an excellent job of refinishing the seats.

of the combustion chamber **(Figure 12)**. This has a negative impact on airflow because even when the valve is open the path is obstructed by the edges of the combustion chamber. Valve seats are replaced by machining or cutting them out and pressing in new seats **(Figure 13)**.

VALVE SEAT REFINISHING

The valve seats must be fully refinished to provide a good sealing surface for the fresh valves. You will have to achieve a good finish on the seat. It is also essential to achieve the proper seat width. Remember, typical specifications are 1/16 in. on the intakes and 3/32 in. on the exhausts, but check your particular specifications. It is best to clean the valve seats and port area with a light wire brush if the head has not been professionally cleaned. This saves the stones and cutters from getting loaded up with carbon. There are several types of valve seat refinishing equipment. A diamond cutter type is shown in **Figure 14**. This bench includes valve guide pilots, knurling tools, diamond seat cutters, and an overhead drill for cutting the seats **(Figure 15)**. Another type of seat cutter is an adjustable carbide cutter that can cut three angles on the seat in one shot. Other seat cutting equipment uses stones of different sizes and angles. This has been the traditional method for years, but equipment is often being replaced by the other styles. Any type of equipment is designed to achieve the same effects on the valve seats.

Figure 15. The machine shop uses this bench for much of the seat work.

Carbide Seat Cutters

To use the carbide cutter adjust the cutters to fit the seat **(Figure 16)**. The cutter shown has a guide pilot in it.

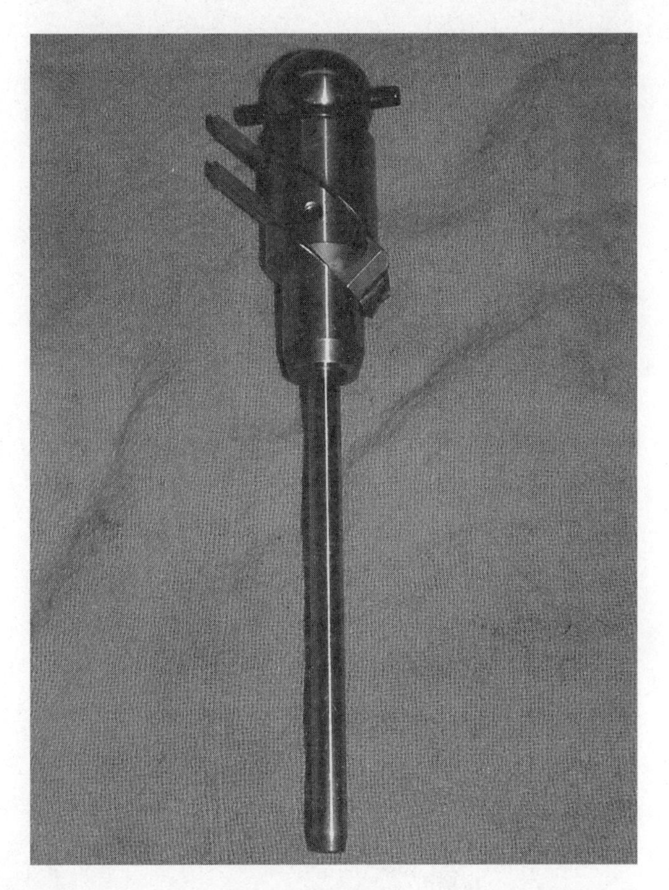

Figure 16. This adjustable carbide cutter is an efficient seat-cutting method; it cuts all three angles at once and provides the correct seat width.

With any cutting procedure you must select a proper fitting pilot that fits securely in the guide to ensure that you cut the seat squarely in relation to the guide. Some cutters are designed to be used with a drill; others you can turn by hand. Rotate the tool clockwise. Follow the equipment instructions. Remember, these cutters cut the seat and finish it to the correct width. The cutter performs a three-angle valve job. The seat is typically cut at 45°, and there is a 15° or 30° angle on the top of the seat and a 60° or 75° angle cut on the lower part of the seat, in the throat.

Stone Seat Cutters

The stone-type refinishing equipment, whether diamond or traditional, uses very similar processes. We will discuss the stone seat-cutting procedure here. The stones must be dressed like the stones on the valve grinding machine. Diamond cutters do not require refinishing; they should theoretically last forever unless the diamonds break free from the tool.

To refinish a seat using a stone cutter:

1. Select a 45°-angle stone (or a 30°-angle stone if that is the seat angle) that fits the seat without scraping on any portion of the combustion chamber. It must be large enough to cut the whole width of the seat **(Figure 17)**.
2. Fit the stone onto the drilling tool **(Figure 18)**.
3. Place a drop of oil on the stone dressing shaft, and place the stone and tool onto the shaft. Select the proper angle at which the diamond cutter will cut the stone.
4. Apply the power head (special drill) on top of the stone holding fixture, and spin the stone. Support the weight of the power head, and hold it squarely on the stone. Slowly move the diamond cutter into the stone while moving it

Figure 17. Stones are still a fine way to refinish seats; you just need some patience.

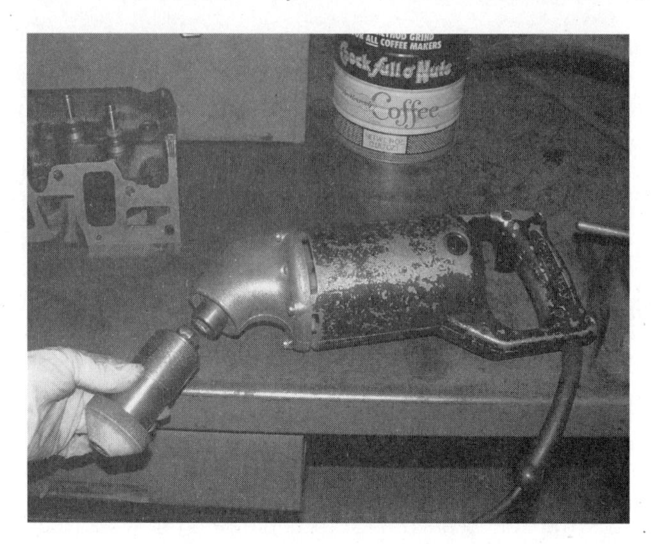

Figure 18. Fit the stone on its holder to get it ready for use with the power head.

quickly up and down across the stone **(Figure 19)**. Finish the stone by taking several slow passes across the stone.

5. Make sure the guide is clean, and select the proper pilot **(Figure 20)**. The pilots are tapered. They should fit

Figure 19. Pass the diamond across the stone to freshen up the stone.

Figure 20. Use this tool to secure the pilot into the guide. It is also very helpful when removing the pilot.

snugly and not quite bottom out on the guide. They are sized as 9/32 (as an example), 9/32 +1 (.001 in.), +2, +3, −1, and so on for each different size guide.

6. Place a drop of oil on the pilot, and place the stone and its holder over the pilot.
7. Spin the stone on the seat while supporting the weight of the power head and holding it squarely above the stone. Spin the stone for just a few seconds at a time; check the seat between cuts. You want to take little material off but remove all pitting from the seat **(Figure 21)**.

FITTING THE VALVE AND SEAT

Once the valve and seat are clean you need to check the seat width and gauge where the valve face is contacting the seat. You will usually need to narrow the seat after machining it. Check the manufacturer's specifications, and compare the seat width with the spec **(Figure 22)**. You want the area of contact to be near the center of the face, with some area above and below the contact patch still left

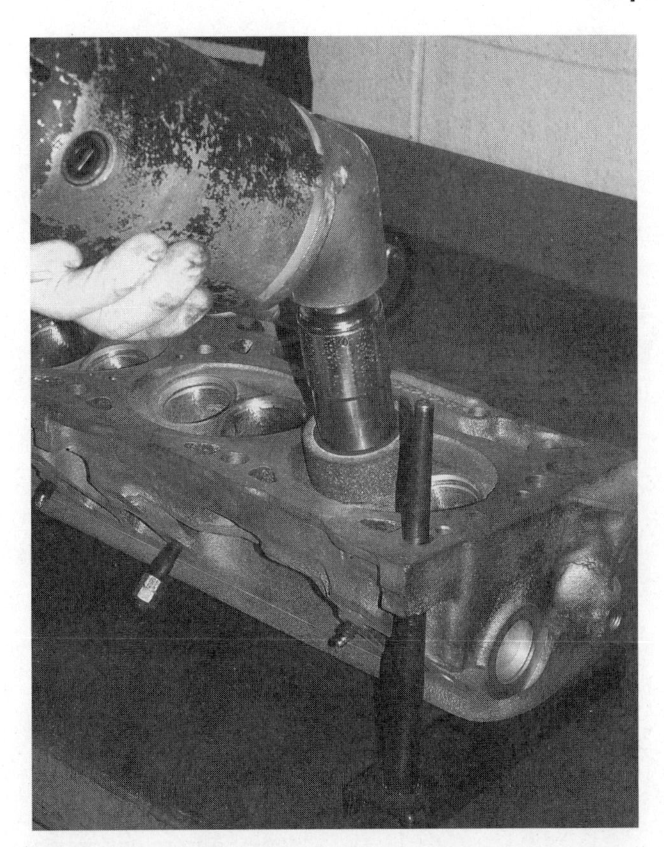

Figure 21. Make short cuts on the seat with the stone. Check the seat often; you want to remove as little material as possible.

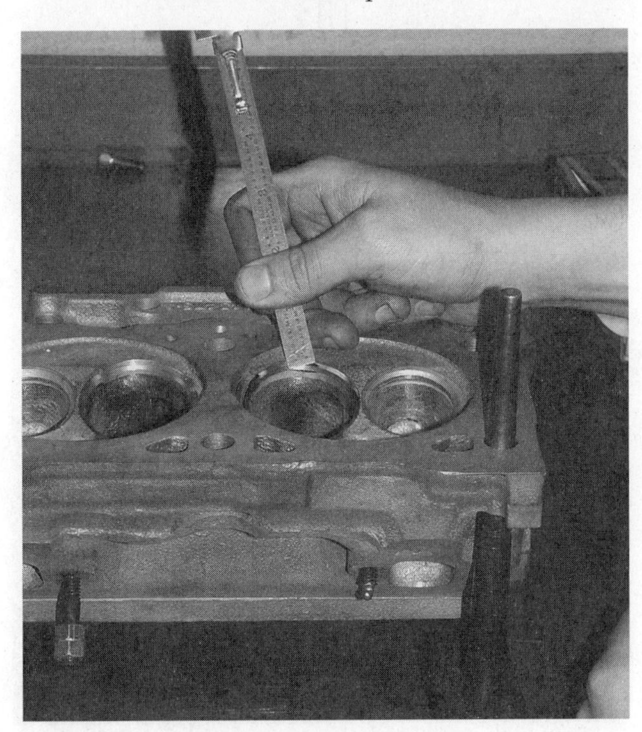

Figure 22. Measure the width of the seat using a machinist's rule. Intake seats are typically 1/16 and exhausts are often 3/32.

on the face. To check where the seat is contacting dab a little **Prussian blue** on the valve face, and use a lapping stick to spin the valve on the seat **(Figure 23)**.

The Prussian blue will be rubbed away from the contact patch. If the contact is high toward the head you will cut a 15° or 30° angle on the top of the seat to narrow the seat and adjust the valve contact down **(Figure 24)**. This is the most common requirement when using refinished valves because they tend to sink down in the head as they wear. If the seat is wide and the area of contact is low on the valve face, toward the tip, cut the seat with a 60°–angle or 75°–angle stone in the throat. This will narrow the seat and move the area of contact up toward the head. Continue adjusting until you have the correct seat width and contact area on the valve face. Always finish the seat with a short burst with the 45°–angle stone. This removes any burrs that may have been left on the seating surface by the other cuts. Make a final check of the contact area using Prussian blue. Next you should check the concentricity, or runout, of the seat. If it is concentric it means that it is centered to the valve guide. Use a dial indicator seated in the guide to check that the seat runout is less than .001 in.

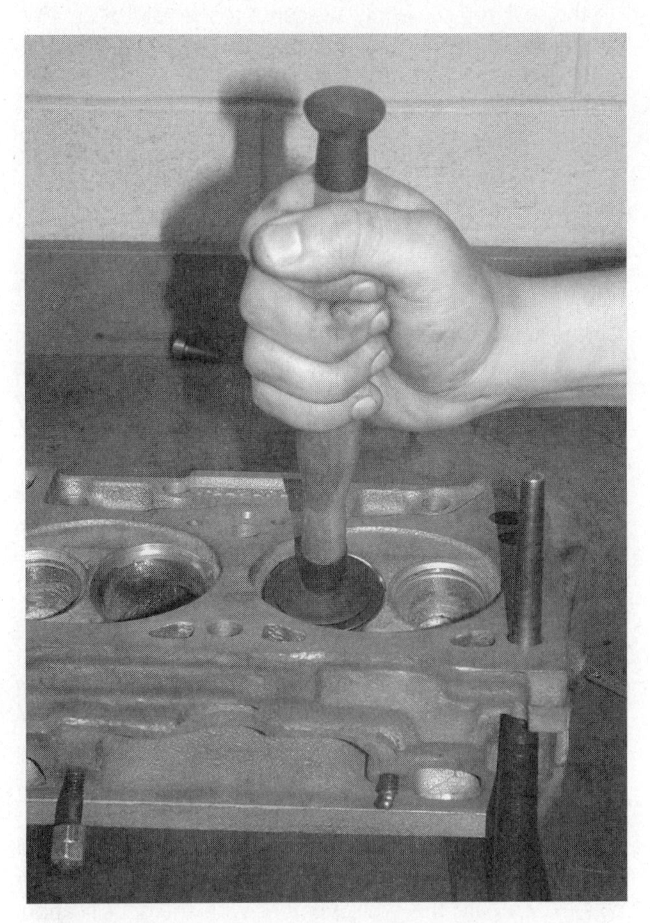

Figure 23. Rotate the valve in the head with a lapping stick to see where the seat is contacting the valve.

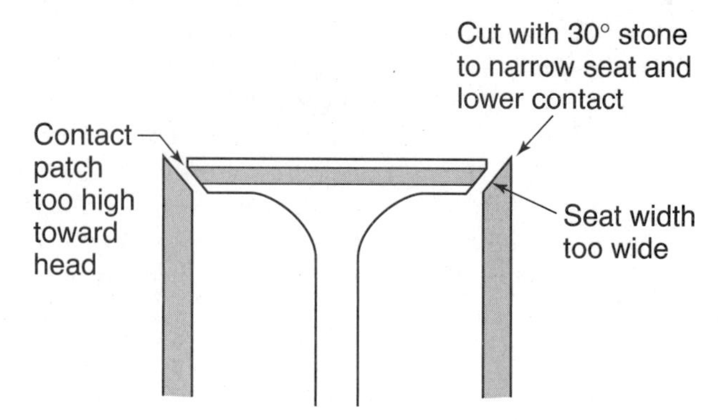

Figure 24. Cut a 30° angle on the top of the seat to narrow the 45° seat and move the contact down toward the tip.

Some technicians will perform a three-angle valve job on every job. This means they will cut the 45° seating angle, a 15°–angle or 30°–angle topping cut, and a 60°– or 75°–angle cut in the throat. In production work on a stock engine this is not required.

Finally, you need to check the integrity of the valve face to the seat. You can apply a vacuum to the valve seal stand or the top of the guide and make sure each valve can hold a vacuum. Alternatively, you can pour clean parts cleaner or mineral spirits on top of the valves and make sure none leaks down within one minute.

VALVE MEASUREMENT

Valve stem height should be measured to determine whether more material needs to be ground off the valve stem tip. After the seat and face refinishing, the valve stem will sit higher above the head. This measurement is particularly important on engines with nonadjustable hydraulic lifters. If you install the lifter and the tip is extending too far the lifter may not allow the valve to close. Use a machinist's rule to measure stem height, if specified, from the base of the head to the top of the tip **(Figure 25)**. Compare the measurement with specifications. If it is too tall grind the required amount off the tip.

Measure valve installed height to make sure that the spring will have the proper tension. As the valve moves up in the head the length of the spring when installed may be too long to provide adequate tension. You can measure the valve-installed height from the spring base to the bottom of the retainer where the spring rides. Alternatively, you may be asked to install the valve spring and measure its installed height from the bottom to the top of the spring **(Figure 26)**. Either measurement should yield the same value. Compare the measurement with the specification. You can adjust the spring-installed height by placing a valve shim on the spring base on the head **(Figure 27)**. Shims come in .010 in. increments up to .060 in.

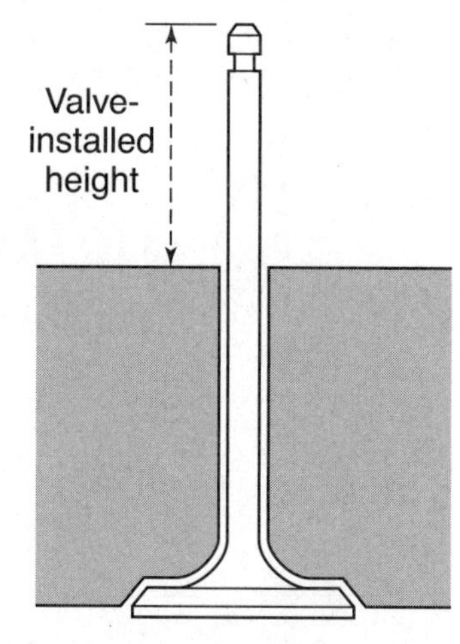

Figure 25. Measure the height of the valve stem from the cylinder head to the valve tip.

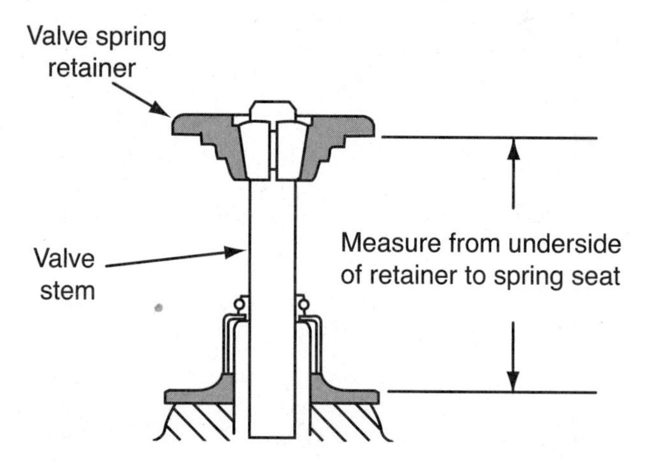

Figure 26. You can check valve spring installed height without the spring in place. Measure from the head to the bottom of the spring retainer.

Figure 27. The machine shop has a board of spring shims with varying thicknesses for different applications.

VALVE SEAL INSTALLATION

In preparation for assembling the cylinder head make sure it is fully cleaned of solvent and any seat grinding dust. Clean the head thoroughly in the parts cleaning tank, and blow the head and guides dry with shop air. Lubricate the guides to prevent corrosion. New valve seals should always be fitted when the head is disassembled. Each type of seal should be lightly lubricated with oil on its inner diameter, where the valve will ride. Umbrella-type seals are simply pressed down onto the guide. Positive-lock seals require more work. A seal installation tool is very helpful. It is just a long tube that fits onto the hard edge of the seal. If one is not available fit a deep socket of the proper diameter to achieve the same effect. Tap the seal into place using a plastic hammer. O-ring seals sit in a groove toward the top of the stem. They must be installed with the valve in the head before the spring is in place. Simply roll them into place, and make sure there are no twists in the seal.

VALVE INSTALLATION

Fit the proper valve into its seat, and pull it up through the seal. Lay the spring and retainer over the valve. Use the valve spring compressor to compress the spring below the keeper grooves. Apply a little petroleum jelly or grease to the keepers to help retain them in their grooves **(Figure 28)**. Fit them in place, and release the spring compressor squarely. Tap on the valve retainer with a plastic hammer or a socket over the retainer to be sure the valve keepers are fully seated in their groove **(Figure 29)**. If the valve keepers are installed so they can pop out you want to correct that before the engine is in service. Wipe away the excess grease. Repeat for all the valves.

CAMSHAFT, ROCKER ARM, PUSHROD, AND LIFTER INSTALLATION

The assembly procedure and sequence will obviously vary depending on the cam and head configuration. We will discuss a few of the basic guidelines. When installing a cam in the block use a cam-holding tool or the cam sprocket to give you leverage to carefully slide the camshaft into place. Lubricate the camshaft journals and lobes with engine assembly lube or oil. Install the thrust plate if equipped. On an overhead cam engine install lifters or followers as needed before installing the cam. To install the camshaft lubricate it and then fit the caps or housing over the cam. It is imperative that you follow the camshaft tightening sequence. Many camshafts and housings have been broken as a result of improper installation.

Figure 28. Install the keepers with a little grease to hold them in place.

Figure 29. Knock the valve retainer to be sure the keepers are fully seated.

Lubricate hydraulic lifters well with oil. It is best to place them into a can of oil and compress and release them a few times to get some oil into the body. Make sure the lifter bore is clean, and lubricate it with oil. When reusing lifters be sure they are returned to their original location. They may have been select fit to their bores at the assembly plant. The lifter should slide in smoothly. Install the pushrods and rocker arms as applicable. Leave the rocker arms loose for later adjustment.

HEAD GASKET AND HEAD INSTALLATION

When the head is fully assembled you are ready to install it. Check to be sure that the block deck and the head's gasket surfaces are perfectly clean. Most high-quality head gaskets are now installed dry, but check the service information and follow the manufacturer's recommendations. Sometimes the gasket manufacturer will specify that a sealant be used. Look at the gasket carefully, and look for a top marking; that will go toward the head **(Figure 30)**. Some gaskets are not marked, but you must observe the holes and passages carefully to make sure they line up appropriately. Other gaskets do not have a top or a bottom and can be fit either way. Lay the gasket on the dowels.

Make sure the head bolts are clean and have good threads. Always use new torque-to-yield bolts; a stretched bolt could destroy your fine work. Look up the appropriate torquing sequence and procedures. Thread sealant should be used on bolts that go into water passages. The service information indicates which bolts require sealant.

Place the head carefully onto the dowels, and finger tighten a couple of bolts to secure the head. Torque the head in the proper sequence **(Figure 31)**. The bolts are usually tightened in stages. It may be that you tighten each bolt in sequence to 35 ft.-lbs. and then repeat the sequence, tightening them to their final torque of 62 ft.-lbs. It is a good idea to run each bolt to final torque again to be sure you did not miss any in the process.

Torque-to-yield bolts differ. The sequence to tighten them would be something like this:

1. Tighten each bolt in sequence to 25 ft.-lbs.
2. Tighten each bolt in sequence to 40 ft.-lbs.
3. Turn each bolt in sequence an additional 65°.

You will need a torque angle meter or gauge to turn a bolt 65° **(Figure 32)**. When the turn is 45° or 90° it is easy enough to achieve the correct angle by sight.

Leave the valve cover off so that you can adjust the valves as discussed in Chapter 31. After the valves are adjusted install the intake and exhaust manifolds using the proper torque sequence and procedure. Some intake manifold bolts are torque to yield and will require replacement. Next, carefully reassemble all components, belts, hoses, vacuum lines, and electrical connectors in their marked positions.

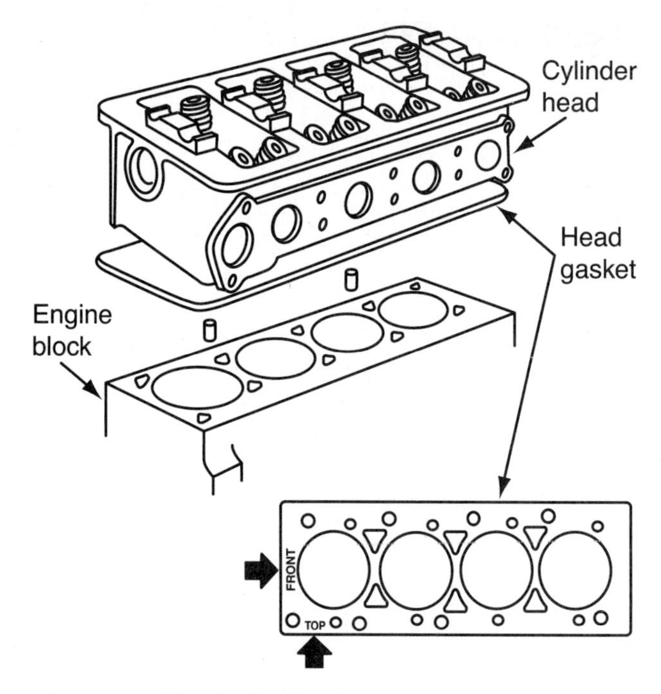

Figure 30. Check the head gasket for top or front markings.

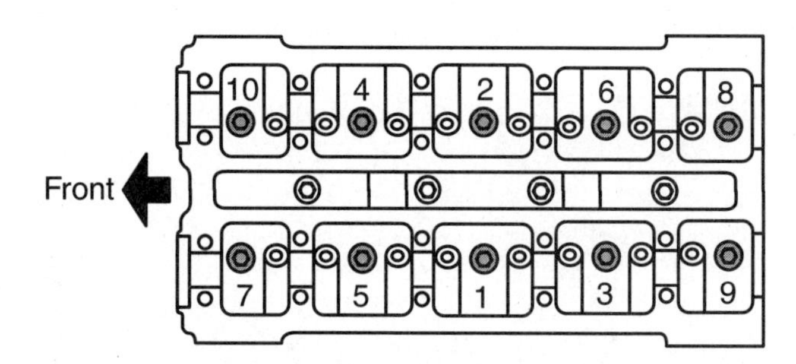

Figure 31. Use the correct tightening sequence on the head to avoid sealing problems later.

Figure 32. A torque angle gauge allows you to accurately measure the degrees of turning.

Summary

- Aluminum cylinder heads should be straightened if possible when the cam bore is misaligned.
- Cracks may be pinned or welded, though sometimes head replacement is the only option.
- Resurface the cylinder head to provide a good sealing surface for the new head gasket.
- Valve guides can be knurled or replaced when worn; this should be done before any valve or seat fitting.
- Resurface the valves to remove all pitting and cupping. Remove as little material as possible.
- Check to be sure the margin meets or exceeds specifications before returning the valve to service.
- Clean up the valve stem tip, and put a light chamfer on it.
- When valve seats are badly damaged or deeply recessed they should be replaced with new inserts.
- Select a snug-fitting guide pilot, and install it into a clean guide.
- Use a carbide, diamond, or stone cutter to refinish the valve seat angle.
- Measure the seat width after refinishing to determine whether other angles must be cut on the seat to narrow it.
- The valve seat should contact the valve on the center of the face.
- Top the seat with a 15° or 30° angle to narrow it and move the area of contact down toward the tip.
- Throat the seat with a 60°– or 75°–angle cutter to narrow it and move the area of contact up toward the head.
- Check the valves for leakage using vacuum or solvent.
- Measure valve stem height and valve-installed height to ensure proper lifter action and spring tension, respectively.
- Always fit the cylinder head with new valve seals.
- Install the head onto the dowels, and torque it into place using the proper sequence and procedure.
- Always replace head bolts when recommended.

Review Questions

1. What inspection should be made on the valve after refinishing before it is placed back into service?
2. What methods can be used to refinish the valve seat angle?
3. If the valve seat is too wide and the area of contact on the valve face is high toward the head use a __________ or __________ stone to narrow and position the contact patch.
4. Always replace the __________ when performing a valve job.
5. Technician A says that a head should be straightened before resurfacing. Technician B says that cracked aluminum cylinder heads are usually pinned to repair the crack. Who is correct?
   A. Technician A only
   B. Technician B only
   C. Both Technician A and Technician B
   D. Neither Technician A nor Technician B
6. Technician A says that valve guides can be pearlized to restore them to their original condition. Technician B says that worn valve guide inserts can be pressed or driven out and new ones installed. Who is correct?
   A. Technician A only
   B. Technician B only
   C. Both Technician A and Technician B
   D. Neither Technician A nor Technician B
7. Technician A says to try to take as much material off the tip as you did off the face. Technician B says you should put a light chamfer on the tip. Who is correct?
   A. Technician A only
   B. Technician B only
   C. Both Technician A and Technician B
   D. Neither Technician A nor Technician B
8. Technician A says that before repairing valve seats the valve guides must be cleaned and restored if they are worn excessively. Technician B says to pick a guide pilot that slides down to the bottom of the guide easily. Who is correct?
   A. Technician A only
   B. Technician B only
   C. Both Technician A and Technician B
   D. Neither Technician A nor Technician B
9. Technician A says that improper valve stem height can throw off the rocker geometry. Technician B says that improper valve or spring-installed height can reduce the spring tension. Who is correct?
   A. Technician A only
   B. Technician B only
   C. Both Technician A and Technician B
   D. Neither Technician A nor Technician B
10. Technician A says that some valve seals can be reused. Technician B says that the overhead camshaft must be torqued in sequence, or the camshaft can break. Who is correct?
    A. Technician A only
    B. Technician B only
    C. Both Technician A and Technician B
    D. Neither Technician A nor Technician B

Section 6

Timing Mechanism Construction, Inspection, and Repair

SECTION OBJECTIVES

After you have read, studied, and practiced the contents of this section you should be able to:

- Know the components of the timing mechanism.
- Understand the importance of valve timing to proper engine performance.
- Identify and understand the advantages and disadvantages of valve timing systems using a belt, chain, or gears to link the camshaft to the crankshaft.
- Understand the purpose of variable valve timing and/or lift systems.
- Know that dual timing chains may be used to drive camshafts individually on variable valve timing engines or on engines with balance shafts.
- Perform a thorough inspection of the timing mechanism to be sure you replace all the parts necessary to perform high-quality, long-lasting work.
- Recognize the symptoms of, and test for, a worn, jumped, or broken timing mechanism.
- Discuss the possible extent of damage on an interference engine with the customer before beginning the job or offering an estimate.
- Replace worn or broken timing mechanisms using all the new components needed to provide reliable service.
- Achieve proper valve timing using the manufacturer's service information.
- Correctly tension a timing belt.
- Verify proper idle quality and engine performance after replacing the timing mechanism.

Variable valve timing systems can reduce emissions while still improving power at higher rpms.

Chapter 28 Timing Mechanism Construction

Introduction

The front end of the engine consists of the valve timing mechanism. This is either gears or one or more belts or chains used to time the rotation of the camshaft to the crankshaft. Timing gears are used more on older engines; modern engines typically use either belts or chains to drive the camshaft(s) **(Figure 1)**. Perfect valve timing is essential for proper engine performance. The timing mechanism may also drive other auxiliary shafts, components, or balance shafts. The mechanism is covered by a timing cover. Many timing belts require replacement at maintenance intervals. Timing gears and chains are generally replaced during an engine overhaul or when they begin to show signs of wear. When a timing mechanism breaks engine damage can be severe. It is important that you understand how to prevent catastrophic failures. This chapter will expose you to the various design configurations commonly used. We will also discuss the purposes, advantages, and disadvantages of the traditional and variable valve timing and lift designs. All of this should help you as you inspect, diagnose, and repair front-end components.

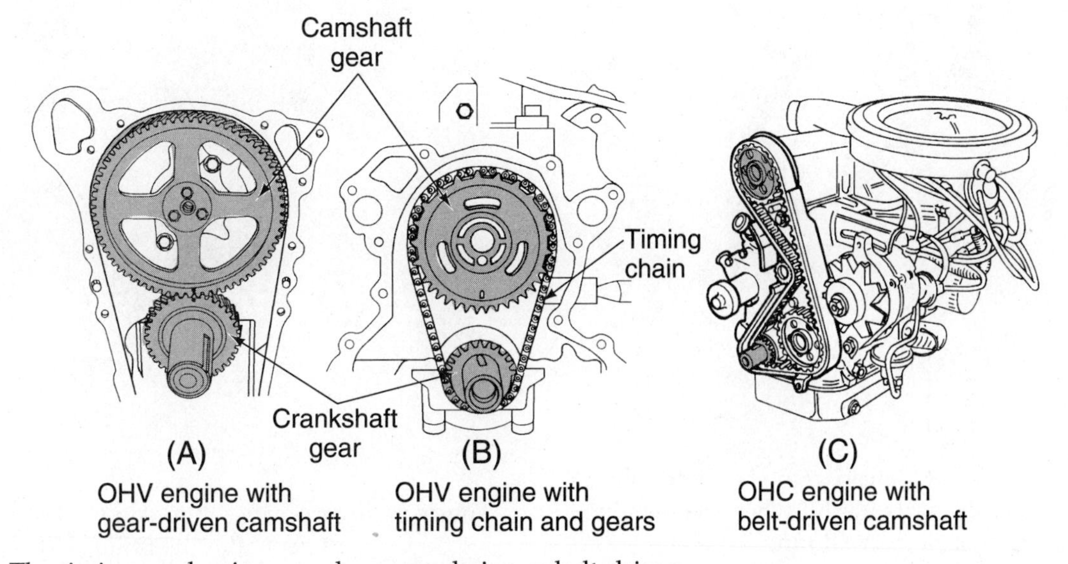

Figure 1. The timing mechanism may be gear-, chain-, or belt-driven.

Interesting Fact

As a technician it is important that you read technical service bulletins when performing service work. One manufacturer decided to require belt replacement every 15,000 miles because the belts were snapping so often! Other manufacturers have lengthened the service interval to over 100,000 miles due to improvements in timing belt materials.

VALVE TIMING SYSTEM

The camshaft rotates at half the speed of the crankshaft to allow the valves to open only for the desired length of time. The camshaft sprockets are twice the size of the crankshaft sprockets unless an intermediate shaft is used. The valve timing system must provide perfect correlation between the camshaft and crankshaft. If the cam timing is off by one tooth the engine will likely run very poorly; emissions will increase while power and fuel economy decrease. We will also discuss variable valve timing and lift systems. These too must start with the correct alignment between the crank and the cam to allow proper variations on valve timing. Some of these systems use a secondary timing chain that can allow for changes in a camshaft's timing **(Figure 2)**.

The engine may be a freewheeling or interference type. In a **freewheeling** or noninterference engine when the timing mechanism fails the valves will usually not hit the pistons. Even some freewheeling engines can allow the valves to contact the pistons if the engine rpm is high enough. On an **interference engine** it is very likely that when the timing mechanism breaks, the valves will contact the pistons **(Figure 3)**. At best valves will need to be replaced. At worst valves will be driven into the pistons, and the whole engine will need to be overhauled or replaced. You can help your customers by explaining the importance of proper timing system maintenance, including routine timing belt replacement.

Every valve timing system will have alignment marks on the sprockets or gears to make sure you can properly time the engine **(Figure 4)**. Camshaft or valve timing is usually established with cylinder #1 at TDC on the compression stroke.

CHAIN-DRIVEN SYSTEMS

Chain-driven systems may be used on either cam-in-the-block or overhead-cam engines. The timing chain is a strong, long-lasting way of linking the cam to the crank. A flat-link silent chain makes much less noise than a roller-style chain. The chain system is still noisy, however, which is

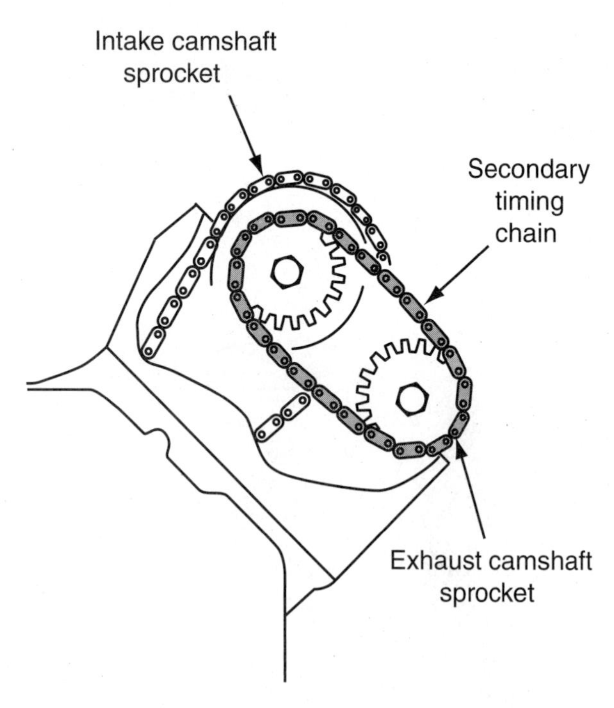

Figure 2. A secondary chain may be used for a camshaft particularly when the engine uses variable-valve timing.

Figure 3. Every valve that came out of this engine was bent after the timing belt snapped.

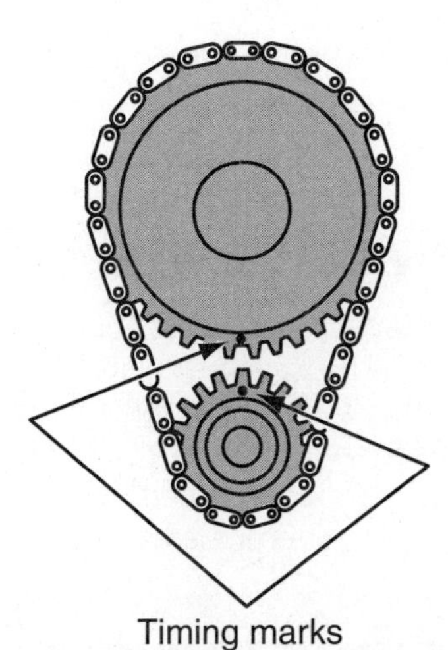

Figure 4. Timing marks on the sprockets allow proper alignment during service.

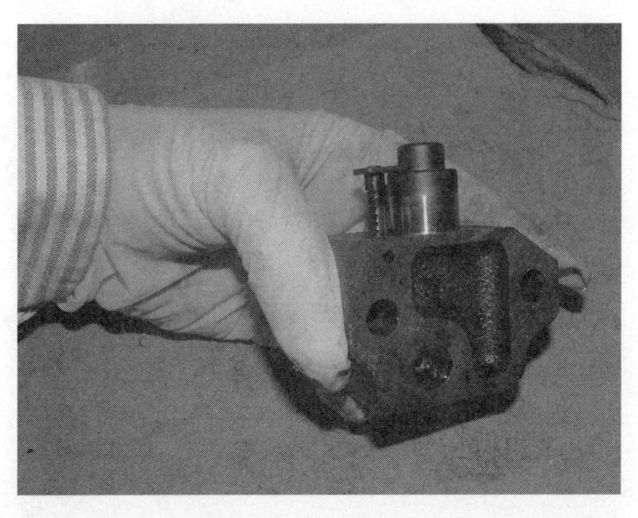

Figure 5. The tensioner uses spring pressure and oil pressure to maintain the correct tension on the chain.

objectionable to some consumers. It is also more expensive for the manufacturers to fit a chain than a belt.

When a chain is used steel, fiber-composite, or composite-plastic sprockets may be used to rotate with the chain. To drive one or more overhead camshafts one or more guides and a **tensioner** must be used. The guides hold the chain against a synthetic rubber or Teflon face. The timing chain tensioner holds proper tension on the guide to take up slack as the chain and guides wear. The tensioner is usually a spring-loaded plunger that may be fed with oil pressure to take up chain slack. The tensioner also has a spring-loaded ratcheting system. When the chain or guide is worn enough the spring will push the plunger out one more step on the ratchet. This also prevents excessive noise at startup before oil pressure is held behind the plunger **(Figure 5)**. Timing chains receive splash oiling often with a slinger placed at an oil passage.

Engines that use balance shafts may use two chains: one will drive the camshafts and the other will drive the balance shafts. **Figure 6** shows this design. Note the guides and tensioners as well. This engine has dual balance shafts. You can see that they will spin at twice the speed of the crankshaft, which is typical of balance shafts. You can see this by comparing the size of the balance shaft sprockets with that of the crankshaft sprocket.

Chains are generally replaced during an engine overhaul or if they begin to make excessive noise. As they wear you can often hear the chain slap during a quick deceleration. We will discuss diagnosis and inspection further in the next chapter.

Figure 6. This DOHC engine shows the tensioner, guides, and primary and balance shaft chains.

BELT-DRIVEN SYSTEMS

A timing belt is frequently used on overhead-camshaft engines. The timing belt is strong and very quiet. It is much more cost-effective for the manufacturer than gears or chains. As belt materials have improved belt replacement intervals have increased. This is good for the consumer because timing belt replacement can be a costly maintenance procedure. Some manufacturers do not even specify a replacement interval; they simply require the technician to inspect it periodically.

The timing belt runs from the crank sprocket to the cam sprocket(s) without a guide. Instead an idler pulley is used to guide the belt and provide a tension adjustment. Some timing belt systems use a spring tensioner that self-adjusts; others require periodic adjustment. The timing belt may also drive other components. On older engines they frequently drove the auxiliary shaft that turned the **distributor (Figure 7)**. Other manufacturers use the timing belt to drive the water pump. When servicing these belts replace the water pump along with the belt. Timing belts may also be used on engines with one or more balance shafts. Timing belts do not need lubrication; in fact one reason for premature failure is oil or coolant leaking onto the belt.

Camshaft sprocket
Belt tensioner pulley
Crankshaft
Distributor drive shaft

Figure 7. A timing belt is also used to drive the distributor drive shaft.

GEAR-DRIVEN SYSTEMS

Gear-driven systems are not as popular as they used to be, in part because they are used only on camshaft-in-the-block engines. They are a very reliable and precise system, if a little noisy. Because there is no chain or belt slack, valve timing is always very close to perfect. The gears are very durable as long as they receive adequate lubrication.

VARIABLE VALVE TIMING AND LIFT

Eager to reduce the limitations of the compromises made in camshaft design many manufacturers are turning to variable valve timing (VVT) or valve lift systems. One camshaft cannot produce fabulous low-end torque and also manage excellent high-rpm breathing. Variable valve lift and timing systems reduce those inherent compromises. Valve lift should be increased at higher rpms, when volumetric efficiency falls. The valve timing should be retarded at higher rpms to keep the valves open later in the stroke because the stroke is occurring so quickly. Many more manufacturers are adding variable valve timing and/or valve lift systems to their vehicles every year. Most manufacturers equip at least some of their models with a VVT system. GM says that 2.5 million of their engines will be equipped with variable valve timing systems by 2007. Honda has been using a variable timing and lift system since 1989.

There are many variations in the design of the systems. Systems either change the valve timing or alter the timing and the valve lift. Some manufacturers are using camshafts with three different lobes for engine operation at various rpms. Both the valve timing and valve lift are changed. The PCM determines when cam timing and lift changes should occur and activates an oil-control valve that moves a pin to lock in the different lobes. Other manufacturers may change timing only. In one example one cam sprocket may be driven by the crankshaft while the other, usually the intake, is driven by a variable chain or gear system that can affect valve timing in small steps by as much as 60°.

Honda took the lead in production variable valve timing with its VTEC (valve timing electronic control) system, designed to improve performance at higher rpms. Each pair of valves has three cam lobes and three rocker arms. During low-speed operation the low-speed rockers push directly on the two regular-speed lobes. At high rpms, when the PCM determines that the third lobe will improve performance, it signals an oil-control valve to engage the third rocker to the high-speed lobe. A pin slides out of the low-speed rockers and into the high-speed rocker to engage it to use the camshaft lobe with more aggressive timing, lift, and duration. **Figure 8** shows a simplified diagram of the pin actuating system.

GM's VVT system improves engine performance, fuel economy, and emissions. It also eliminates the need for an exhaust gas recirculation (EGR) system. On some applications it allows 90 percent of the engine's torque to be available between the wide range of 1900 and 5600 rpm. The VVT system uses a camshaft position actuator for each of the intake and exhaust camshafts that bolt on the end of the camshafts. The actuators lock the exhaust cam in its fully advanced position and the intake cam in its fully retarded position for startup and idle. This minimizes overlap to create a very smooth idle. Off idle the PCM controls the camshaft actuator solenoid valves to allow the position actuators to alter intake and exhaust timing by 25 camshaft degrees. The PCM looks at information about engine coolant temperature, engine oil temperature, intake airflow, throttle position, vehicle speed, and volumetric efficiency to determine the optimum positioning of the camshafts for any given set of conditions. The solenoid valve controls oil pressure to the position actuator to effect changes in camshaft timing. The camshaft position actuator has an outer housing that is driven by a timing chain. A rotor with fixed vanes lies inside the assembly and is attached to the camshaft. The solenoid valve, as commanded by the PCM, controls the direction and pressure of oil flow against the vanes. Increasing pressure against one side of the vanes will advance the camshaft variably until the desired position is achieved. Changing the oil pressure to the opposite side of the vanes will retard the camshaft. Whenever the engine is off idle the PCM is continually providing electrical pulses to the solenoid valves to modify camshaft timing or to hold it in the desired position. Under light engine loads the PCM retards the valve timing to reduce overlap and maintain stable engine operation. When the engine is under a medium load, overlap is increased to improve fuel economy and emissions. To improve mid-range torque under a heavy load and at moderate or low rpm, intake valve timing is advanced. Under a heavy load at high rpm intake valve closing is retarded to improve engine output.

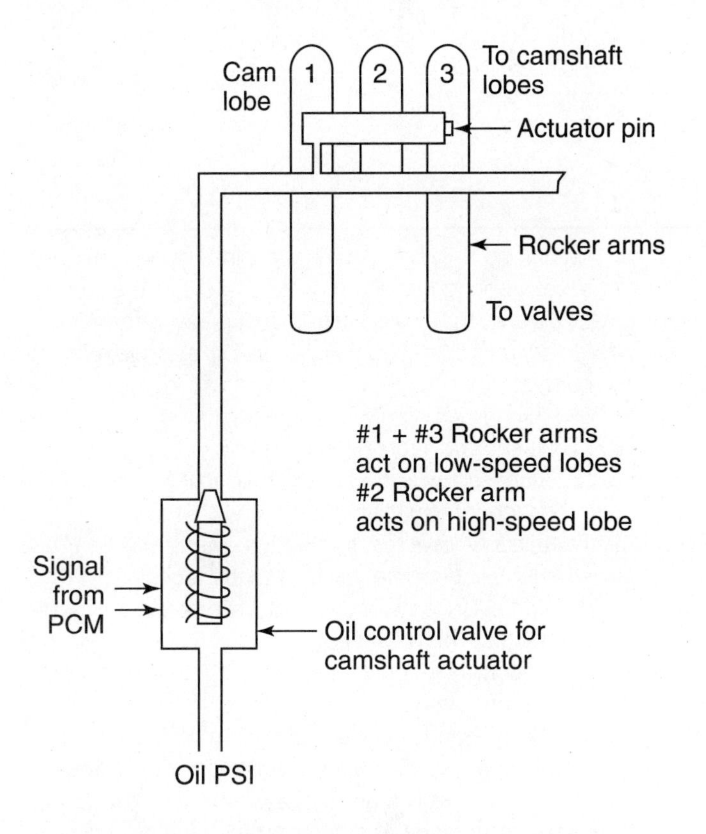

Figure 8. When a predetermined rpm and other conditions are met, the PCM signals the cam actuating oil-control valve. This valve allows oil pressure to push a pin through the rocker arms, activating the high-speed rocker.

Toyota is using a VVT system that changes the phase angle of the camshaft relative to the crankshaft to advance or retard the camshaft. A hydraulic actuator controlled by the PCM sits on the end of the camshaft to effect timing changes. The camshaft timing can be varied continuously up to 60°. The system also employs a variable lift and duration system by using two different camshaft lobes for each set of intake and exhaust valves. One rocker arm opens both intake valves, and another opens both exhaust valves. During low- to moderate-speed operation the "high-speed" cam lobes do not contact the rocker arms. At higher speeds a sliding pin is pushed by hydraulic pressure to take up the space between the higher-duration cam lobes. The increased rocker arm movement also increases valve lift.

The number and types of systems in use increase each year; the idea has proven to be effective in improving power and in lowering emissions. Designs in the works include valves opened by electrically operated solenoids. This system would offer infinitely variable valve lift, duration, and timing.

Summary

- The front end consists of the valve timing mechanism.
- Proper valve timing is essential for proper engine performance.
- Valve timing systems may use a belt, chain, or gears to link the camshaft to the crankshaft.
- Marks are provided on the sprockets, gears, covers, block, or head to properly time the engine, usually at TDC firing cylinder #1.
- A timing chain is a long-lasting drive mechanism that can be used on cam-in-the-block or overhead-cam engines.
- Timing chains are not generally replaced as a regular maintenance service.
- Timing chain disadvantages are noise and cost.
- Dual timing chains may be used to drive camshafts individually on variable valve timing engines or on engines with balance shafts.
- Timing belts are a very popular and cost-effective way to drive overhead camshafts.
- Timing belts usually require periodic replacement as a maintenance service.
- Timing gears may be used to time in-the-block camshafts to the crankshaft.
- Timing gears are very durable and provide precise control of timing.
- Manufacturers are increasingly using variable valve timing and/or lift systems to overcome the limitations of the compromise of one camshaft for all engine speeds and loads.

Review Questions

1. Is it true or false that timing marks are usually placed on the timing chain or belt?
2. A __________ engine usually prevents valves from contacting the pistons if the timing mechanism fails.
3. List three symptoms of improper valve timing.
4. On a timing chain mechanism __________ and __________ are used to provide proper chain operation.
5. Technician A says that the valves must be timed to provide proper spark timing. Technician B says that valve timing is generally set up with the cylinder #1 at TDC firing. Who is correct?
   A. Technician A only
   B. Technician B only
   C. Both Technician A and Technician B
   D. Neither Technician A nor Technician B
6. Each of the following is a typical valve timing mechanism except:
   A. Chain
   B. Belt
   C. Gear
   D. Band
7. Technician A says that timing belts are quieter than timing chains. Technician B says that timing chains generally last longer than timing belts. Who is correct?
   A. Technician A only
   B. Technician B only
   C. Both Technician A and Technician B
   D. Neither Technician A nor Technician B
8. Technician A says that timing chain tensioners usually have a spring-loaded ratchet to provide preliminary tension of the chain. Technician B says the tensioner is also tensioned hydraulically by oil pressure. Who is correct?
   A. Technician A only
   B. Technician B only
   C. Both Technician A and Technician B
   D. Neither Technician A nor Technician B
9. Technician A says that two timing chains may be used when the engine uses balance shafts. Technician B says that the balance shafts rotate at the same speed as the crankshaft. Who is correct?
   A. Technician A only
   B. Technician B only
   C. Both Technician A and Technician B
   D. Neither Technician A nor Technician B
10. Technician A says that a VVT system may change the camshaft duration at different rpms. Technician B says that a VVT system may retard the intake timing under heavy-load, high-rpm conditions to improve power. Who is correct?
    A. Technician A only
    B. Technician B only
    C. Both Technician A and Technician B
    D. Neither Technician A nor Technician B

Timing Mechanism Inspection

Introduction

In order to prepare for valve timing mechanism service you must carefully inspect the components and decide which need replacement. Even when the timing mechanism is being replaced during a major engine overhaul you should carefully inspect for component damage to determine whether there was a particular cause of the wear. Belts, chains, and gears must be inspected for wear. Tensioners, sprockets, and guides must also be thoroughly evaluated to make a good decision about which components should be replaced. You want to provide repairs in a cost-effective manner but without the risk of premature failure of your work. In many cases you will replace multiple components in the timing mechanism. You should use the manufacturer's specific procedures when that information is available or if you have any doubt about your diagnosis. There are also industry-standard procedures for determining the condition of belts, chains, and gears that we will discuss here. This chapter will give you an overview of the inspection process.

> **Interesting Fact**
>
> *"Crrraaaaaaaaank, crank, crank, crrraaaaaaaaank, crank, crank, pop, pop." That is an engine cranking over at irregular speeds and backfiring; it is a telltale sign of a valve timing mechanism that has jumped out of time. If the engine is old enough to have adjustable ignition timing check that out too; it can cause the same symptoms.*

SYMPTOMS OF A WORN TIMING MECHANISM

A worn timing chain may slap against the cover or against itself on a quick deceleration and make a rattling noise. Use a stethoscope in the area of the chain to confirm your preliminary diagnosis, and then proceed with the checks described in the next section to confirm the extent of the damage **(Figure 1)**.

> **You Should Know**
>
> *A noisy timing chain that is slapping against the timing cover should be replaced right away to avoid failure and the need for more costly repairs.*

A customer may also report that the engine seems to have poor acceleration from startup but that it runs well, perhaps even better than usual, at higher rpms. When there is slack in the chain the camshaft timing is behind the crank. This is called retarded valve timing. This improves high-end performance while sacrificing low-end responsiveness.

Timing gears may clatter on acceleration and deceleration. The engine does not have to be under a load; just snap the throttle open and closed while listening under the cover for gear clatter. The customer is unlikely to notice the reduced low-end performance before the gears break from too much **backlash**.

Customers with worn or cracked timing belts will notice nothing until their engine stops running or begins running very poorly. What you and they should be paying close attention to is the mileage on the engine and the recommended belt service interval.

Figure 1. Listen for a slapping chain on deceleration.

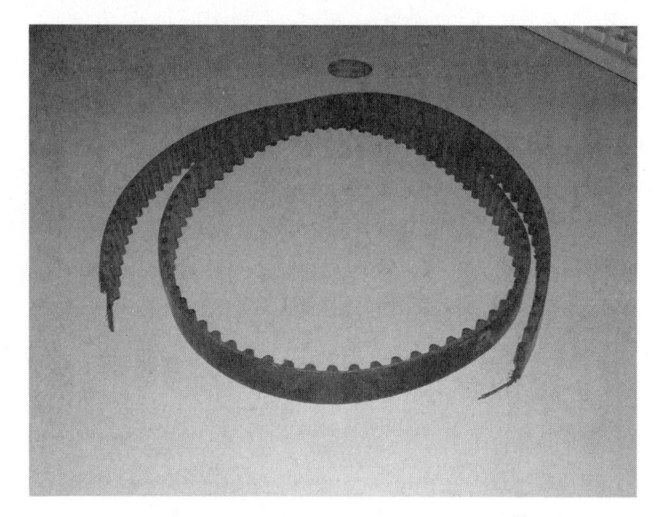

Figure 2. This snapped timing belt left the customer stranded on the highway.

SYMPTOMS OF A JUMPED OR BROKEN TIMING MECHANISM

When a timing chain, belt, or gear breaks and no longer drives the camshaft(s) you can often tell by how the engine sounds when it turns over. It cranks over *very* quickly, and no sounds even resembling firing occur **(Figure 2)**. The engine has no compression. Hopefully it has been towed to your shop; that is the only way it will run again.

When a timing chain, belt, or gear skips one or more teeth, or jumps time, it sets the valve timing off significantly. Some engines will run when the timing is off one or even two teeth, but not well. Usually you can hear the problem as the engine cranks over. It turns over unevenly; the engine speeds up and then slows down as you crank. Idle quality, acceleration, and emissions will be affected. If the timing is off significantly it will pop out of the intake or exhaust or backfire under acceleration. Often the engine will not even start.

Diagnosing a Jumped or Broken Timing Mechanism

If the engine just spins over fast and does not run at all you may be able to crank the engine over and look for valvetrain movement through the oil filler cap. If there is no movement the mechanism is broken. When you cannot see the valvetrain components through the oil fill cap, it may be easiest to pull the upper timing cover off. If a customer says his car was running fine yesterday and today it just stopped running a failed timing mechanism is a common cause. One good way to verify a jumped or broken timing mechanism is to perform a cranking compression test. If the engine is running poorly because of a jumped timing mechanism your diagnosis may require a compression test. Crank the engine over, and check each cylinder. When the timing has jumped the compression will be low on all the cylinders. Frequently it will be as low as 75 or 50 psi on every cylinder **(Figure 3)**.

Another quick check for a broken timing mechanism is to check cranking vacuum. Instead of the normal 3–6 in. Hg vacuum, you will see no vacuum if the mechanism is broken. The vacuum will be low if the belt or chain has jumped

Figure 3. Low compression on most or all of the cylinders can point to jumped valve timing.

Figure 4. Line the arrow on the cam sprocket up with the dot on the head to place the camshaft at TDC #1 firing.

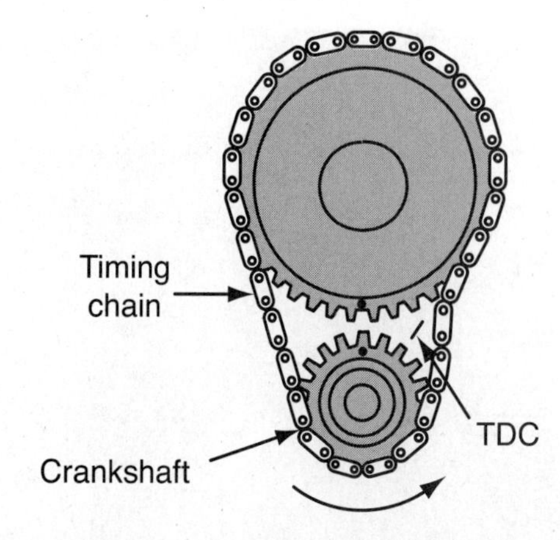

Figure 5. Rotate the crank backwards to get all the slack on one side. Next, measure the rotation before the camshaft moves as you rotate the slack out of the chain.

time. If the engine runs engine vacuum will be lower than the usual 18–22 in. Hg vacuum found at idle.

Sometimes when the engine runs fairly well but not correctly the compression may not be low enough to confirm jumped timing. If you still suspect that the valve timing is off rotate the engine to TDC #1 compression using the crankshaft or camshaft indicator. If you can locate TDC on the cam sprocket pretty easily you may be able to detect TDC of the #1 piston by a mark on the crankshaft pulley. If there is no mark on the crankshaft pulley you may still be able to pick out jumped time without removing the crankshaft pulley and accessories. Remove the upper timing cover. Rotate the engine by hand until the TDC #1 marking is lined up on the camshaft **(Figure 4)**. Remove cylinder #1's spark plug. Place a straw or a suitable plastic piece on top of the piston. Rotate the engine to ensure that the straw starts declining within a couple of degrees of crankshaft rotation. Rotate the engine two full revolutions, and carefully observe the straw as you approach TDC #1 on the camshaft; if the timing is on the piston must reach the top of its travel and stop briefly right when the camshaft timing mark lines up.

If you are not positive that the timing is off you have wasted very little time. Remove the crankshaft pulley and timing cover to confirm proper timing of the crankshaft, camshaft, and, if equipped, balance shafts or auxiliary shafts using the specific indicators found in the manufacturer's service information. We will discuss this in Chapter 30, "Timing Mechanism Repair and Assembly."

TIMING CHAIN INSPECTION

To inspect a timing chain for wear you can measure the amount of movement that the crankshaft makes before the camshaft or camshafts begin to move. Generally this should be less than 8°. Use the crankshaft timing marks when available to watch the degrees of crankshaft movement. Many newer engines do not have timing marks on the crank pulley and front cover because ignition timing is rarely adjustable now. If that is the case use a torque angle gauge on a ratchet as you turn the crankshaft. On an older engine that has a distributor remove the distributor cap, and watch for rotor movement as a sign that the chain is rotating the cam or an auxiliary sprocket. When an engine does not use a distributor it may be possible to detect movement of the camshaft through the oil fill cap. Remove the cap and see if you can clearly view the camshaft or a rocker. If you can, rotate the crankshaft counterclockwise then clockwise to measure the amount of slack before the cam or valvetrain moves. Sometimes it will be necessary to remove a valve cover to get a good view of the movement of the cam or rockers. If the crankshaft rotates more than 8° before the camshaft moves, the timing chain and/or tensioner is worn beyond usable limits **(Figure 5)**.

If you have taken the timing cover off a cam-in-the-block engine you can try to deflect the chain. If you can deflect it less than one-quarter of an inch the chain is satisfactory for continued service as long as the chain itself and the sprockets are in good shape. Deflection of one-half inch requires immediate attention.

You can use this same procedure on an overhead-cam timing chain by removing the valve cover. Replacement is recommended when there is more than one-half inch of deflection.

Inspect the chain itself for shiny spots on the edges of the links; sometimes they slap against the metal cover. Look for worn or flat rollers on a roller-type chain. On a flat-link silent chain look for wear at the attachment pins and joints; shiny spots indicate friction points.

TIMING BELT INSPECTION

To inspect a timing belt you really need to pull the top timing cover off the engine. This is usually a plastic

Figure 6. Remove the upper timing cover to get a look at the belt. This cover has clips holding it down.

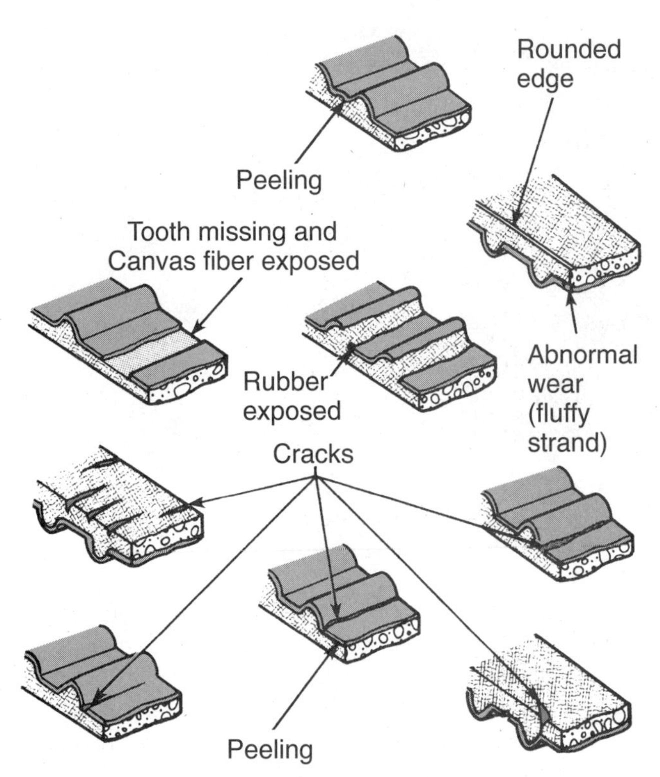

Figure 7. Any fault with the belt warrants replacement.

cover with an upper and lower section. For inspection purposes it will only be necessary to remove the upper cover **(Figure 6)**. You will be looking for cracks similar to those in serpentine belts. Rotate the engine around to be able to view the whole belt. Look on the underside of the belt at the cogs or teeth that hold the belt in the sprocket. Any broken or missing teeth dictate immediate replacement. **Figure 7** shows some typical timing belt failures. If the belt shows signs of oil or coolant contamination the belt should be replaced, and the leak must also be fixed as part of the repair. Be sure to diagnose the cause of the leak and include that procedure and any necessary parts in your estimate to the customer. As we have discussed there is no reason to try to eke out another few months on a damaged belt; it could result in a very expensive repair.

TIMING GEAR INSPECTION

To check timing gears for excessive wear remove the timing cover. You should listen before you take the cover off so you will begin to learn what good and bad gears sound like. Put a ratchet on the crankshaft bolt, and rock it lightly back and forth between the backlash of the gears. You do not want to turn it so much that the camshaft actually turns. Just listen for the clacking of the gears. It may be so bad you will know it the very first time, but take the cover off to confirm your diagnosis.

Worn teeth will be sharp, shiny, and pointed **(Figure 8)**. If you are not sure by visual inspection you can measure the gear backlash. This is the small clearance left between any gear set that allows room for expansion as the gears heat up. You can measure the clearance between any two teeth. Typical gear backlash is .004–.008 in.; you may be able to determine visually that this set of gears has more than that. If you are unsure take feeler gauges and rock the crank so the teeth are touching, and then backward as far as you can without rotating the camshaft. Measure the clearance between the gears, and compare your reading with the actual specification. Excessive backlash or visually detectable damage to the gears warrants replacement.

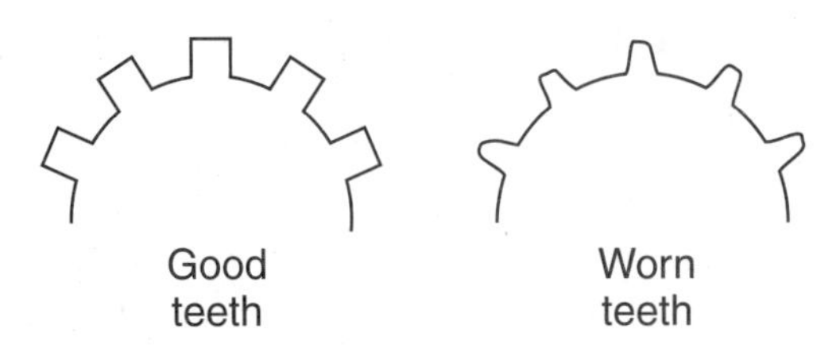

Figure 8. Worn teeth develop a sharp pointed edge.

Figure 9. This timing chain sprocket was replaced with the chain. Although the wear was not excessive, the timing mechanism had a lot of miles on it.

Figure 10. Replace the belt tensioner pulley if it has excessive mileage or any roughness or wobbling in the bearing.

SPROCKET INSPECTION

Recommended practice is to replace the sprockets used when a timing chain is replaced. The job is not usually a simple one, and the sprockets have typically been in service for over 100,000 miles. The teeth of the sprockets wear with the chain. If you had a low-mileage failure you would carefully inspect the sprocket teeth. Worn teeth will be shiny and sharper; good sprockets will have an unworn flat spot at the top of the tooth **(Figure 9)**. The teeth may actually be rounded on the face from wear with the rollers.

Sprockets are not normally replaced when the timing mechanism uses a belt. The belt is very soft compared with the steel sprockets and they do not wear rapidly. You should always carefully inspect them, however, because they can fail. A sharp edge on a timing belt sprocket will cause premature failure of a new belt. Compare each of the cogs with the others, and make sure they are in fine condition. If you have any doubt replace them.

TENSIONER INSPECTION

The timing chain hydraulic-type tensioner is commonly replaced with a high-mileage timing chain job. Again, it makes good sense to take care of all the components in the system. Most tensioners have a minimum length that the plunger should stick out from the body. Refer to the manufacturer's specifications, and if the plunger length is not within specifications replace it. You should also check for oil leakage past the front seal from which the plunger protrudes. If you see wetness around the seal you should replace the tensioner.

The pulley-type tensioner used with belts is also commonly replaced when the service interval is long, every 90,000 miles, for example. If the timing belt is being replaced for the second time and the tensioner was not replaced the first time you should definitely recommend replacement **(Figure 10)**. An engine that requires belt replacement every 60,000 miles and is being serviced at 120,000 miles deserves a new tensioner to prevent premature failure. To inspect the tensioner for faults check for scoring on the pulley surface. Scratches, sharp edges, or grooves can damage a new belt rapidly. Also check the pulley's bearing. Rotate it by hand and feel for any roughness. If you can hear it scratching or feel coarseness replace it. Also check the pulley for side play; grasp it on its ends and check for looseness. If it wobbles at all replace it.

TIMING CHAIN GUIDE INSPECTION

You should replace the guides on a high-mileage timing chain replacement. They are synthetic rubber, Teflon, or some other material that is much softer than the chain. It is recommended practice to replace the guides. To inspect a guide for wear look at the surface where the chain rides. If the grooves are deeper or wider, or if you can see any signs of wear, be safe and replace it.

REUSE VERSUS REPLACEMENT DECISIONS

When in doubt replace a worn valve timing mechanism component. A premature failure can cause extensive damage, and you could lose a customer. When you choose to reuse components make sure you are confident that they will last an appropriate amount of time for the customer and the engine. Your repair work should result in reliable and long-term service.

Summary

- An important part of timing mechanism service is thorough inspection to be sure you replace all the parts necessary to perform high-quality, long-lasting work.
- When a timing chain is worn you will often hear the rattling of the chain on deceleration.
- Sometimes a worn timing chain will reduce low-end acceleration and improve top-end performance.
- You can only definitively detect a worn timing belt by visual inspection; there is no audible warning.
- Worn timing gears will clack against each other as you accelerate and decelerate.
- An engine that has a broken timing mechanism will not run. The engine will turn over unusually fast due to the low engine compression with the valves open at the wrong time. Cranking compression will be low on most or all cylinders.
- An engine that has jumped time will run poorly, if at all. Check for low cranking compression all around or low vacuum.
- Occasionally, if the timing is off one small tooth only the idle quality and emissions will be affected; sometimes you will notice the reduced acceleration. Inspect the timing marks closely to verify proper timing of all the driven shafts.
- Worn timing chains will have flat rollers or links and shiny spots on the pivot points. The sprockets are usually pointed and shiny as well.
- Worn timing gears will show wear on the teeth; they will be sharp on the top rather than flat as they originally were.
- Inspect timing chains by movement of the cam or by visual inspection.
- Inspect timing belts visually; a worn belt will have cracks, fraying, or missing cogs.
- Check timing gears audibly for excessive backlash. Pull the timing cover and measure backlash to verify your diagnosis.
- Timing chain sprockets are normally replaced with the timing chain. Worn sprockets will have sharp, shiny tooth points.
- Timing belt sprockets should be carefully inspected for chipped or sharp cogs.
- Timing chain tensioners are frequently replaced with the chain. Check the length of plunger extension and signs of oil leakage to condemn them.
- Timing belt tensioner pulleys should be carefully inspected for scoring and grooves on the belt running surface. The bearing should be checked for roughness or wobbling. Replace the pulley if it has 90,000 miles or more on it.
- Timing chain guides should be replaced with the chain; they will normally have wear in the grooves and rounded edges.
- Make cautious decisions about replacement; timing mechanism repairs should provide reliable long-term service.

Review Questions

1. A worn timing chain will show __________.
2. Describe the symptoms of an engine that has jumped time but still runs.
3. List three other components that are typically replaced during a timing chain replacement.
4. Describe the procedure to verify excessive timing gear wear.
5. List three signs of an excessively worn timing belt.
6. Technician A says that worn timing belts will make the engine run better until they break. Technician B says that low-end acceleration may deteriorate with a worn timing chain. Who is correct?
   A. Technician A only
   B. Technician B only
   C. Both Technician A and Technician B
   D. Neither Technician A nor Technician B
7. Technician A says that worn timing gears may clatter on acceleration or deceleration. Technician B says to verify worn timing gears by checking for excessive end play. Who is correct?
   A. Technician A only
   B. Technician B only
   C. Both Technician A and Technician B
   D. Neither Technician A nor Technician B
8. Technician A says that timing chain tensioners are typically replaced with the chain. Technician B says you should measure the spring tension to determine whether they can be reused. Who is correct?
   A. Technician A only
   B. Technician B only
   C. Both Technician A and Technician B
   D. Neither Technician A nor Technician B

9. Technician A says that timing chain guides should last the life of the engine. Technician B says that worn timing belt pulleys will show scoring and/or roughness in the bearing. Who is correct?
   A. Technician A only
   B. Technician B only
   C. Both Technician A and Technician B
   D. Neither Technician A nor Technician B
10. When deciding whether to reuse or replace a timing mechanism component:
    A. You must carefully inspect it and be sure it will provide long, reliable service.
    B. You should be able to offer the customer a few choices.
    C. You should work with the customer to determine his expectations for the engine's longevity.
    D. All of the above are true.

Timing Mechanism Repair and Assembly

Introduction

The actual repair procedures for service of timing chains, belts, and gears vary immensely. Some OHC timing chains can be replaced in a few hours with the engine in place; others require that the engine be removed from the vehicle. It is common to have to remove the cylinder head and/or timing cover. Timing belts can generally be replaced with the engine in the vehicle. They can be simple and straightforward or time-consuming and a little tricky. A timing belt on a transversely mounted V6 engine in a minivan, for example, may leave you little room to access the front end of the engine. Replacing the timing mechanism on cam-in-the-block engines using either a chain or a gear set is usually uncomplicated. In every case it is critical that you set all the shafts into perfect time with the crankshaft. As we have discussed, setting the valve timing off by one tooth will impact engine drivability and emissions. Similarly, an engine with balance shafts not timed properly will vibrate noticeably. Often it is too easy to miss the timing of a shaft by one tooth if you are not paying close attention and double-checking your work **(Figure 1)**. This is important and time-consuming service work; you want to do it only once! Many technicians will replace the camshaft and crankshaft seals while they have access to them during the job. If the water pump is driven by the timing belt you will usually replace that as well. Always check idle quality and engine performance after a replacement job to be sure the engine is operating as designed. In this chapter we will offer a couple of specific examples of the procedures for replacing different timing mechanisms. You will need manufacturers' specific information to achieve proper timing mark alignment and to perform the job in the most efficient manner.

Figure 1. This engine is getting its second chance. The first time, the balance shafts were not timed properly and a noticeable vibration resulted.

You may also need special tools to hold the camshaft sprockets in position while replacing the belt or chain(s).

When repairing an engine with a broken timing mechanism it is important to know whether the engine is an interference engine. If it is advise the customer of the potentially very high added cost of removing the cylinder head and replacing valves. Also make sure the customer understands that even more serious damage, such as a valve stuck into a piston, could have occurred. Even on freewheeling engines it is wise to inform the customer that valve damage is possible. Discuss these issues with the customer or service advisor before offering an estimate for the work.

Interesting Fact

A student was traveling back to college in Colorado from her home in Vermont. She had just had her older vehicle serviced and a new timing belt installed. It seemed to run fine, and she made it to Colorado without trouble. When she went for her emissions test, however, the engine failed for high CO and HC levels. It took a technician a few hours to figure out that the timing belt was off one tooth and a few more hours to repair it.

TIMING CHAIN REPLACEMENT ON OHC ENGINES

Replace the timing chain(s) on an engine during a thorough overhaul, if noise or symptoms indicate excessive wear, or if the chain has broken. Timing chain replacement procedures vary significantly in the specifics of what must be removed in order to access the chain and sprockets. You will usually have to remove the valve cover(s), the front engine or timing cover, the engine drive belt(s), any accessories blocking access to the front cover, and the harmonic balancer. Sometimes you will need to remove the front engine mount as well. You must follow the manufacturer's specific instructions to perform the job most efficiently.

You Should Know

When rotating an engine you should always use the crankshaft bolt rather than the camshaft(s) bolt. You could break a camshaft trying to rotate the engine with it. Also, turn the engine in its normal direction of rotation unless specified otherwise. Engines rotate clockwise when looking at them from the front.

Figure 2. Remove the serpentine belt and accessories as needed to access the timing cover.

Below is a general outline of procedures for replacing the timing chain on a popular DOHC engine:

1. Disconnect the negative battery cable.
2. Drain the coolant and remove the reservoir.
3. Remove the serpentine engine drive belt **(Figure 2)**.
4. Loosen the generator and swing it back out of the way.
5. Remove the upper timing cover fasteners.
6. Remove the engine mount assembly. *The bracket bolts holding it may be torque-to-yield bolts and must be replaced with new special bolts.
7. Raise the vehicle and remove the right wheel.
8. Remove the inner fender splash guard.
9. Use a suitable puller to remove the harmonic balancer.
10. Remove the lower front cover fasteners.
11. Lower the vehicle and remove the front cover.
12. Rotate the engine clockwise until the camshafts' keyways are at 12:00 **(Figure 3)**. The holes in the camshaft sprocket should line up with the holes in the timing housing.
13. The crankshaft sprocket keyway should be at 12:00 if the valve timing is correct.
14. Remove the timing chain tensioner and guides.
15. Make a front marking on the cam and crank sprockets using a paint stick or correction fluid in case they will be reused.

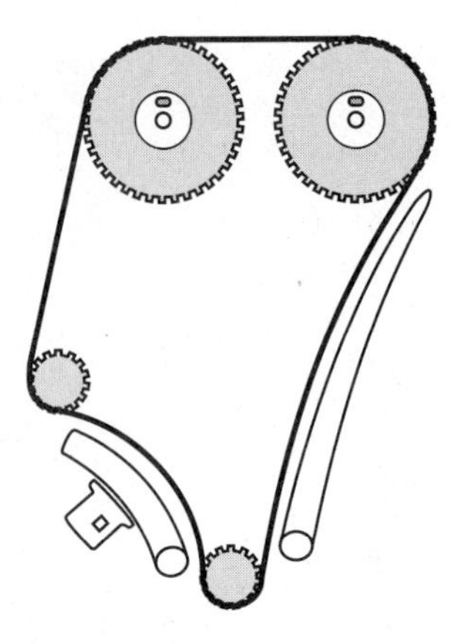

Figure 3. Line the timing marks up before disassembling the timing chain.

16. Remove the timing chain.
17. Carefully inspect all the components and determine which, if any, you will reuse **(Figure 4, Figure 5, Figure 6)**.

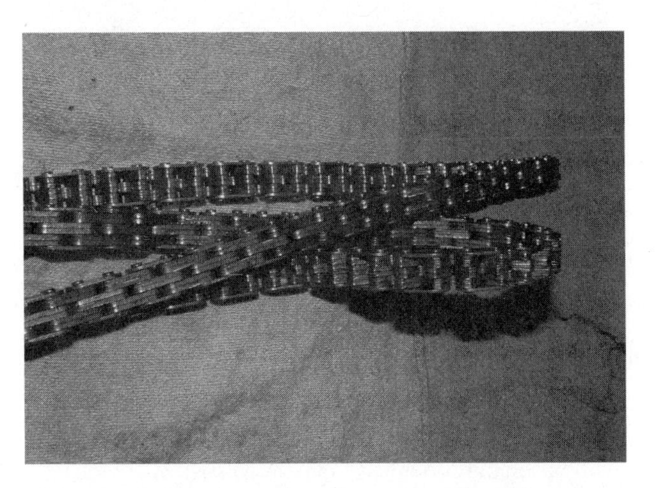

Figure 4. This chain shows wear on the shiny spots where the pins connect the links.

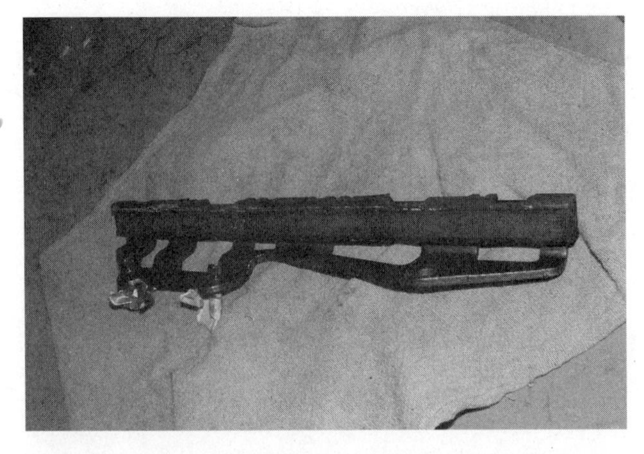

Figure 5. The timing chain guide has a groove worn in it from the chain.

Figure 6. The tensioner plunger did not protrude as specified; it was replaced.

18. Clean the front cover and mating surface thoroughly using a plastic scraper on aluminum surfaces.
19. Install the camshaft and crankshaft sprockets with the front markings out. Torque them to the specified value.
20. Make sure the camshaft sprocket holes are still lined up with the holes in the housing. If not rotate the crankshaft 90° off TDC to provide valve clearance, and turn the cams to line up the holes. Install a dowel through the sprocket into the hole in the timing housing to retain the sprockets **(Figure 7)**.
21. Turn the crankshaft backward (counterclockwise) to TDC until the keyway is in the 12:00 position and the marks on the sprocket and the engine block are aligned.
22. Install the chain around the crank sprocket, idler sprocket, and exhaust cam sprocket. Be sure all slack is on the tensioner side of the chain. Remove the intake cam dowel, and rotate the sprocket just a little to install the chain with no slack between the two cam sprockets. Relax the cam and refit the dowel pin. If it does not

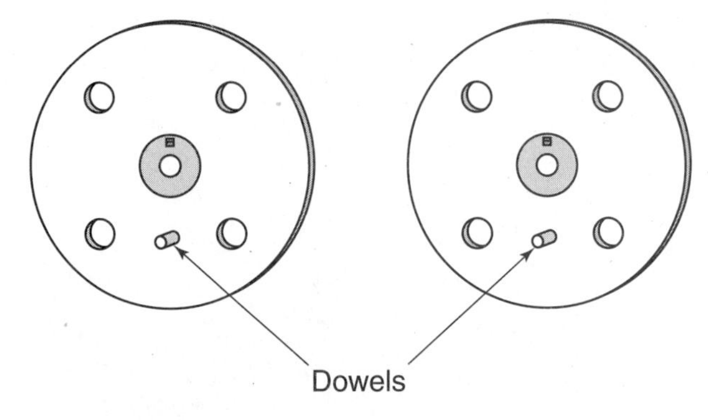

Figure 7. Insert the dowels through the sprockets and into the head. This holds the camshafts in perfect alignment while you install the chain.

slide in retime the sprocket and chain. All the chain slack must be on the tensioner side.

23. Raise the vehicle and recheck the crank sprocket alignment marks; adjust as needed.
24. Use a small screwdriver to release the ratcheting mechanism in the tensioner **(Figure 8)**. Fully depress the tensioner, and install a retaining pin through the hole in the plunger and housing **(Figure 9)**. Install the tensioner, and torque to specification.
25. Install the guides, and torque properly. Remove the retaining pin in the tensioner.
26. Remove the camshaft dowels, and rotate the engine two complete revolutions. Recheck the alignment marks to be sure they are *perfect*; adjust as needed **(Figure 10)**.

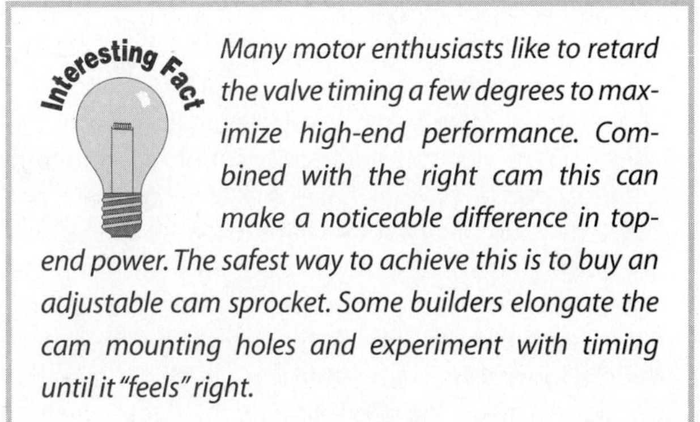

Many motor enthusiasts like to retard the valve timing a few degrees to maximize high-end performance. Combined with the right cam this can make a noticeable difference in top-end power. The safest way to achieve this is to buy an adjustable cam sprocket. Some builders elongate the cam mounting holes and experiment with timing until it "feels" right.

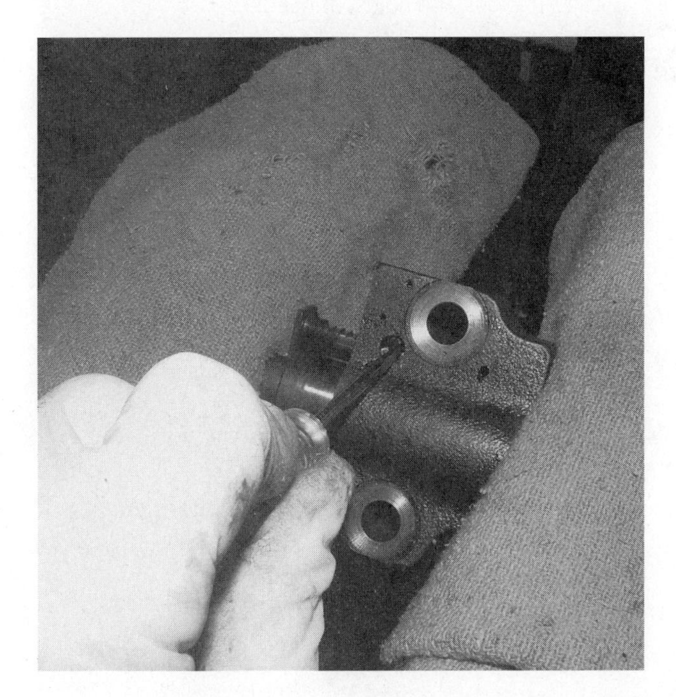

Figure 8. Use a tiny screwdriver to release the ratcheting mechanism while you retract the tensioner.

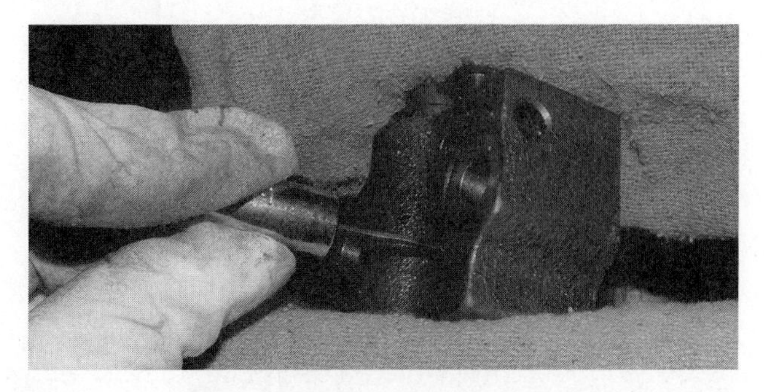

Figure 9. Slide a small Allen head or pin through the hole in the tensioner body to retain the tensioner in its retracted position.

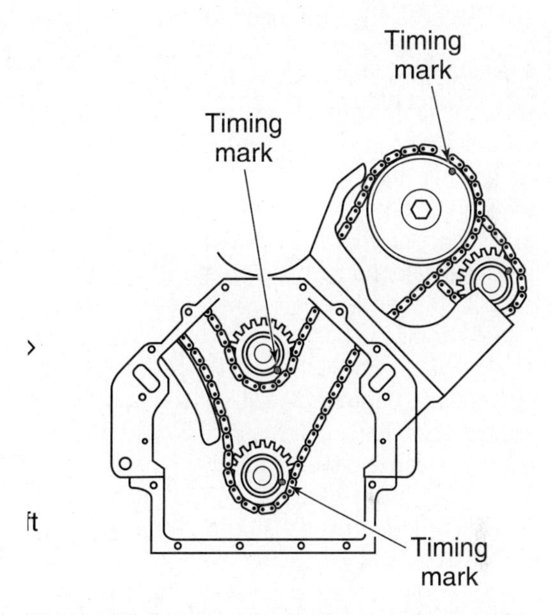

Figure 10. Line up the timing marks exactly as shown in the service information.

27. Replace the front seal and lubricate its lip with chassis grease or oil.
28. Reassemble in reverse order using appropriate torque specifications and procedures.
29. Idle the engine, and road test the vehicle to confirm proper performance.

As you can see from these instructions the work of replacing a timing chain requires time, concentration, and attention to the details of valve timing alignment. Follow the manufacturer's procedures for reinstallation to ensure all bolts are torqued properly and components are installed correctly.

TIMING BELT REPLACEMENT

Check the maintenance manual for the vehicle whenever a vehicle comes to your shop for service. Routine maintenance is often neglected by customers because many expect technicians to inform them when a procedure is due. This is critically important when it comes to timing belt replacement. Many consumers have no idea what a timing belt is, much less that it needs to be replaced periodically. Explain the importance of timing belt maintenance to your customers. The best way to determine whether a belt needs replacement is through a visual inspection.

The interval on timing belt replacement varies widely. On modern vehicles the interval typically ranges between 60,000 and 100,000 miles. There are some new timing belts that are not replaced on a schedule; inspect them at the requested intervals, and replace them when signs of wear are evident. Belts on some older vehicles may require replacement every 30,000 or 45,000 miles. These examples should make it clear that you will need to check the maintenance information to be sure. Timing belts are generally replaced during a thorough engine overhaul as well.

To replace the timing belt you will need to remove the upper and lower timing covers, the harmonic balancer, and any accessories that are in the way of those covers. The procedure will be similar to the timing chain replacement, but the plastic timing belt covers do not seal oil in. If the timing belt drives the water pump this is an excellent opportunity to replace it. Sometimes a timing belt replacement is a pretty quick and simple job. Read through the whole service procedure before beginning so that you will be familiar with all the steps required **(Figure 11)**. Let us look at a typical example of the procedure to replace a timing belt:

1. Remove the negative battery cable.
2. Remove the accessory drive belts.
3. Lift the vehicle and remove the right wheel and inner fender splash guard.
4. Remove the engine drive belt(s) and the harmonic balancer using a suitable puller.
5. Lower the vehicle and support the engine with a jack.
6. Remove the right engine mount and bracket from the engine.

Figure 11. Read through the service information first so you understand the procedure.

7. Remove the timing belt cover.
8. Rotate the engine to TDC #1 firing, and make sure that the marks on the cam and rear timing cover line up. Check the crankshaft sprocket to be sure its notch lines up with the notch on the rear timing cover. (If the engine is equipped with an auxiliary or balance shaft make sure that its marks are also properly aligned.) **(Figure 12)**
9. Rotate the tensioner pulley counterclockwise to loosen it until you can slide a small Allen wrench or pin through the locking hole in the pulley and into the hole in the tensioner bracket. This relieves the tension on the belt.
10. Remove the timing belt. Do not rotate the camshaft or crankshaft.
11. Scrutinize the sprockets, tensioner pulley, and water pump; replace components as needed.
12. Install the new components and new timing belt, ensuring that the belt fits tightly around all the components. The slack should be on the tensioner pulley side.
13. Rotate the tensioner pulley just enough to pull out the locking pin. Allow the automatic tensioner to properly tension the belt. Note that some tensioners are not automatic; on these you will have to loosen the locking nut, rotate the pulley on its eccentric bolt until the proper tension is achieved, and tighten the locking bolt. Use a belt tension gauge to confirm proper tension **(Figure 13)**. The belt tension gauge fits over the belt and deflects the needle farther the tighter the belt is. Read the tension on the gauge, compare it with specifications, and adjust as needed.

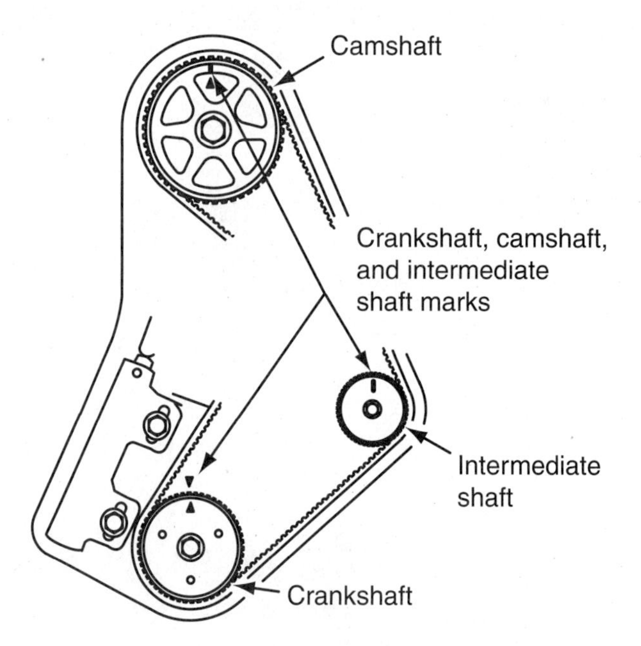

Figure 12. Check all the alignment marks before disassembly so that you will know how they should line up during reassembly.

Figure 13. Use a belt tension gauge when the tensioner is manual to be sure you make the correct adjustment.

Figure 14. The timing marks should line up perfectly after two revolutions of the engine.

You Should Know *It is critical that you replace the plastic timing covers securely on the engine. In addition to keeping debris off the belt they prevent water or ice from leaking in. If ice or even just a lot of water gets onto the belt it can skip teeth even when it is relatively new and properly tensioned.*

You Should Know *A timing belt that is adjusted too tight will make a whining or whirring noise when you start the engine up. Do not be tempted to let it "break in"; it could easily snap or damage engine or water pump bearings.*

14. Check the camshaft and crankshaft alignment marks. Rotate the engine two complete revolutions, and check the marks again. Adjust as needed **(Figure 14)**.
15. Reassemble in reverse order, following the recommended procedures.
16. Idle the engine and road test the vehicle to confirm proper performance.

TIMING CHAIN OR GEAR REPLACEMENT ON CAMSHAFT-IN-THE-BLOCK ENGINES

Replace an in-the-block timing chain or gear set when an engine is overhauled, when excessive wear is detected, or after the mechanism fails. Rotate the engine to TDC #1 using the mark on the harmonic balancer. To access the timing mechanism you will generally have to disconnect the negative battery clamp and remove the cooling fan if it is mounted on the water pump snout. You may have to drain the coolant and remove the water pump; if you do, look it over closely and replace it if it has defects or extended mileage on it. Remove the engine drive belts. Use a puller, in general, to remove the crankshaft pulley and any other accessories or brackets that may be blocking access to the timing cover bolts. Remove the timing cover bolts and timing cover, and note the markings, usually small dots, notches, or lines, on the gears or sprockets.

Inspect the chain or gears for damage so you are confident that your diagnosis will repair the original concern. Examine the sprockets for the timing chain; they are typically replaced along with the chain. Scrape the timing cover and mating surface clean. Replace the gears or chain and sprockets **(Figure 15)**. When replacing a timing chain place the chain around the sprockets, and adjust the sprocket until the timing marks line up; then reinstall the sprockets onto the crank- and camshafts. With gears mate them together so that the timing marks are aligned, and then slide them over the shafts **(Figure 16)**. In both cases the

Figure 15. Fit the chain over the sprockets so the alignment dots point toward each other.

Figure 16. Slide the sprockets over the shafts with the marks aligned. Recheck the marks after two engine revolutions.

shafts will have to be properly set up so the gears or sprockets will fit over the keys on the cam- and crankshafts. Rotate the engine two full revolutions, and be sure that the marks line up perfectly.

Reinstall the timing cover with a new gasket and any recommended sealant. Torque the bolts in the designated sequence, typically in a diagonal fashion around the cover. Reassemble the rest of the components, and verify proper engine performance.

TIMING CHAIN REPLACEMENT ON ENGINES WITH VVT SYSTEMS

In order to successfully replace the timing chain(s) on a VVT engine you will need to closely follow the manufacturer's procedure and have access to any special tools required. The Cadillac VVT system uses three timing chains to drive the four overhead camshafts. The primary chain drives each intermediate camshaft sprocket. A chain is used, one on each head, from the intermediate sprocket to drive the top camshaft sprockets with the camshaft position actuators.

The actual procedure is not dramatically different from other timing chain replacements, but there are more steps because of the multiple chains. Place the engine at TDC #1 on the compression stroke. After removing the timing cover you remove the secondary right-hand drive chain tensioner, guide, and chain. Next you move the tensioner and guide for the primary chain, and then you can remove the main chain. Finally, you remove the left-hand secondary tensioner guide and chain. Once the secondary chain is removed you can replace a faulty camshaft position actuator (they are bolted onto the end of the camshaft) or timing mechanism. Carefully review the manufacturer's directions and diagrams to properly time the chains. The illustrations will clearly show the locations of the sprocket markings when the timing is set up correctly.

Summary

- Perfect timing mechanism service is essential to proper engine performance.
- Replace timing chains or gears when they are worn excessively or broken, or during a major engine overhaul.
- Replace timing belts at the appropriate service interval, when they are broken or visibly worn or damaged, and during a major engine overhaul if they have not been recently replaced.
- Discuss with the customer the possible extent of damage on an interference engine before beginning the job or offering an estimate.
- Rotate the engine to TDC #1 firing, and study the alignment markings before removing the chain, belt, or gears.
- Replace components that show signs of wear; this is not a job that you or the customer would like to have to do again soon. If you work flat-rate you could do it for free!
- Compress a hydraulic tensioner by hand or in a soft jaw vise if necessary, and slide a pin or small Allen wrench through the tensioner body to release the tension for installation.
- Many tensioner pulleys are now automatic, but you may have to adjust the belt to a particular tension using a belt tension gauge.
- When installing a belt or chain be sure that all the slack is on the tensioner side of the sprockets.
- Carefully examine the alignment of the timing marks when reassembling the mechanism; they must line up precisely. Rotate the engine two complete revolutions to verify your accuracy.
- Always verify proper idle quality and engine performance after replacing the timing mechanism.

Review Questions

1. What damage is likely to occur when a timing mechanism fails on an interference engine?
2. Most timing mechanisms are timed with the engine at __________ on the __________ stroke.
3. After setting up the valve timing what step should you take to verify that it is perfect before reassembly?
4. Is it true or false that most timing chains use an automatic tensioner?
5. Technician A says that a timing chain is generally replaced every 60,000 or 90,000 miles. Technician B says the timing chain is typically replaced during a major engine overhaul. Who is correct?
   A. Technician A only
   B. Technician B only
   C. Both Technician A and Technician B
   D. Neither Technician A nor Technician B
6. The best way to properly determine whether a timing belt needs replacement is to:
   A. Ask the customer when it was last done
   B. Look at the maintenance information to determine the recommended interval
   C. Remove the top timing cover to inspect the condition of the belt
   D. Replace it at every tune-up to be safe
7. Technician A says the timing chain is removed before the tensioner. Technician B says that to install the tensioner you have to lock the plunger into the body. Who is correct?
   A. Technician A only
   B. Technician B only
   C. Both Technician A and Technician B
   D. Neither Technician A nor Technician B
8. Technician A says that some tensioner pulleys automatically adjust belt tension. Technician B says that on manually adjusted belts you set the belt with one inch of deflection per foot of open belt. Who is correct?
   A. Technician A only
   B. Technician B only
   C. Both Technician A and Technician B
   D. Neither Technician A nor Technician B
9. Technician A says that the timing chain cover seals oil in. Technician B says the plastic timing belt covers are not essential. Who is correct?
   A. Technician A only
   B. Technician B only
   C. Both Technician A and Technician B
   D. Neither Technician A nor Technician B
10. Technician A says you need to rotate the engine two complete revolutions to make a final check of the timing marks. Technician B says you should always road test a vehicle after timing mechanism replacement. Who is correct?
   A. Technician A only
   B. Technician B only
   C. Both Technician A and Technician B
   D. Neither Technician A nor Technician B

Section 7

Engine Assembly and Break-In

Technicians take a lot of pride in starting and driving a smooth and powerful engine after installation. It takes careful work right through the break-in process to ensure that the engine performs properly.

SECTION OBJECTIVES

After you have read, studied, and practiced the contents of this section you should be able to:

- Recognize the importance of careful and thorough reassembly.
- Torque components using the proper torque specifications and sequence.
- Properly fit a new, or occasionally a used, oil pump and oil pan using sealant only if specified.
- Install an engine in an RWD and FWD vehicle.
- Work underneath the engine to attach the exhaust, starter, hoses, and lower electrical connectors.
- Reinstall the radiator and cooling fan as needed, as well as any accessories to the engine such as the generator, power steering pump, air pump, and air compressor. Install a new drive belt and tension it properly.
- Fill the engine, transmission, cooling system, and accessories with the proper fluids.
- Connect all vacuum lines, hoses, and electrical connectors.
- Prime the oil pump or lubrication system before starting the engine whenever possible.
- Understand and perform the important break-in process during the first twenty minutes of a new or rebuilt engine's operation.
- While the engine is running for its first twenty minutes at 1500 to 2000 rpm, check for leaks, noises, proper thermostat and fan operation, and adequate charging voltage. Monitor the oil pressure and coolant temperature.
- Road test the vehicle properly to help seat the rings. Check for any unusual noises or performance concerns.
- Clean the interior and exterior before returning the vehicle to the customer. Reset the clock, the radio, and any other memory functions.
- Perform a five-hundred-mile service with oil change and appropriate inspections and tightening.
- Communicate effectively with the customer to explain your work and the driving procedures for the first 3000 miles.

Valve Adjustment

Introduction

Valve adjustment can be performed before the engine is installed after an overhaul or replacement. To perform a periodic adjustment you will remove any valve covers and follow the same procedures. The adjustment must be correct to provide long valve life and excellent engine performance. If there is too much clearance (lash) in the valvetrain it will clatter, and the valves will not open as much as they should. Too little valve lash can cause valve burning and poor engine performance. Some hydraulic lifters are adjustable and are set after replacement or reinstallation. Other hydraulic lifters are not adjustable. Mechanical lifters require periodic adjustment. One method of adjustment involves loosening or tightening an adjustment screw on the rocker arm **(Figure 1)**. Another process involves measuring the lash and adjusting the clearance by changing the size of a shim installed between the camshaft and the follower. We will look at the general methods of adjusting each type of valvetrain.

Many stock-car race technicians will adjust the valves after each race to get optimum performance out of the valvetrain.

ADJUSTMENT INTERVALS

Adjustable hydraulic lifters should be adjusted during installation of new or used lifters or whenever the valvetrain has been disassembled. An example would be after an on-the-car valve seal replacement. After the rockers are reinstalled the lifters must be adjusted. This initial adjustment should last as long as the valvetrain components or until they are disassembled again. If during normal service the valvetrain becomes noisy check the lifter adjustment before condemning the lifters if no visible damage is seen.

Mechanical or solid lifters require periodic adjustment. They should be adjusted at the specified interval. This may be as little as every 15,000 miles or as long as every 90,000 miles. Adjustment every 30,000 or 60,000 miles is a common recommendation. If the valves clatter they should be adjusted regardless of the specified mileage.

Figure 1. These rocker arms have adjusting screws held in place with a locking nut. Loosen the locknut and use a feeler gauge under the rocker arm. Tighten the screw to the correct setting. This setup may be used on an engine with solid lifters.

SYMPTOMS OF IMPROPER VALVE ADJUSTMENT

When the valvetrain has too much lash it cannot open the valve as far as it is designed to. You can identify the problem by listening under the rocker cover for the telltale valvetrain clatter. It is a light, fast tapping or ticking noise from under the valve cover(s). Loose valve adjustment will also affect engine performance. When the valves do not open fully or for as long as they should volumetric efficiency decreases. As you know, this will cause a loss of power, particularly at higher rpms.

If there is less than zero clearance it means the valves are tight. They are held open more than they should be and may not close fully when on the camshaft's base circle. This is a very dangerous situation; the valves can burn quickly. The valves need to seat fully in order to dissipate their heat. If the exhaust valves open too early in the combustion stroke they are subject to too much heat. In severe cases the engine will run poorly as the valves stay open too long, and overlap will be too great. The engine may backfire through the intake and exhaust. The engine might also turn over irregularly, as though the valve timing is off.

ADJUSTING HYDRAULIC LIFTERS

Some manufacturers use nonadjustable hydraulic lifters. In these cases the proper "adjustment" method is to torque the rockers on their stands to the proper specification. If serious changes, such as valve and seat grinding, have been made that could affect the rocker geometry, different-length pushrods or rocker stand shims can be fitted. Follow the manufacturer's procedures for determining when you must perform these "adjustments." It is not common on modern passenger-vehicle or light-duty truck engines to have to make any changes to the existing setup. The valves are smaller than they used to be, and you usually cannot take enough material off the valves to alter the valvetrain geometry significantly before the valve needs replacement.

Commonly, hydraulic lifters are adjustable. The goal in adjusting hydraulic lifters is to center the plunger in the lifter bore **(Figure 2)**. If the plunger is centered it can adjust the valvetrain tighter or looser as needed without bottoming out at one end. Centering it gives the plunger equal travel in both directions. If it were adjusted too tight the plunger would sit near the bottom of the bore and have little adjustment left as the valves wear.

Most vehicles have similar procedures for adjusting hydraulic lifters, but it is critical to look at the manufacturer's service information to determine how much to tighten the rocker. When reassembling an engine after a rebuild you should perform a preliminary adjustment during engine assembly and then make a final adjustment

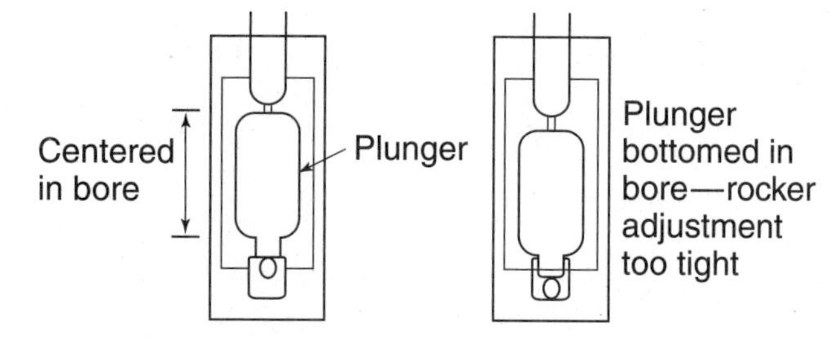

Figure 2. Follow the correct procedure to center the plunger in the bore. This will allow the lifter to adjust the valvetrain both looser and tighter.

after the engine has run. You can perform the adjustment procedure while the engine is off or when it is running. It is messier with the engine running because oil is splashed around. Oil deflectors fit on top of the pushrod end of the rockers to help minimize the oil splash **(Figure 3)**. In either case you will need to remove the valve covers. To adjust the valves with the engine running:

1. Loosen a rocker bolt until you can hear the valve clatter; this raises the plunger toward the top of its bore.

You Should Know *Do not use a generic specification when adjusting a lifter. Adjusting a rocker nut one and a half turns if the specification is one-half turn will significantly misadjust the lifter.*

Figure 3. These oil deflectors clip over the rocker arms and around the push rod, blocking the oil splash hole on top of the rocker.

2. Tighten the nut until the clatter goes away. Now the valve is at zero lash with the plunger still near to the top.
3. Tighten the rocker nut however many turns are specified by the manufacturer, typically one-half to one and a half turns. This centers the plunger in the bore.
4. Repeat for each valve.

To adjust the valves with the engine off:

1. Rotate the engine to TDC #1. One-half of the rockers should be loose with the cam on the base circle; these can be adjusted. You should retrieve the service information for this adjustment and have it with you. For example, one engine may require that you adjust intake and exhaust on #1, intake on #5 and #4, and exhaust on #2. Then you rotate the engine 360° and adjust intake and exhaust on #6, intake on #2, and exhaust on #4 and #5. As an alternative method you can rotate the engine until you reach TDC #1 compression and adjust those valves. Then rotate the engine and watch the rockers for the next cylinder in the firing order to come up to TDC on compression. You will watch the intake rocker come up and then become loose as the piston approaches TDC on the compression stroke. This puts the camshaft on the base circle for all the valves in that cylinder. Adjust them and continue to rotate the engine through the firing order, adjusting as you go.
2. To make the adjustment tighten the rocker, if necessary, just until the lash is removed between the rocker arm and the valve and the pushrod does not spin freely, or until there is no lash between the rocker and the valve **(Figure 4)**.
3. Tighten the rocker arm nut the specified number of turns.
4. Repeat for each valve.

When the valves are adjusted replace the valve cover gasket, and properly torque the cover. Wipe up all excess oil that may have spilled on the exhaust manifold or other components. Start the engine. Make sure that it cranks over and starts and runs normally. Listen for any valvetrain clatter. Readjust if necessary.

Figure 4. Tighten the rocker nut on its stud until the pushrod resists turning. Rotate the nut the specified number of turns beyond that point.

ADJUSTING MECHANICAL FOLLOWERS WITH SHIMS

Engines that use mechanical followers with the camshaft riding above them may use shims to adjust the valves **(Figure 5)**. When checking the clearances first locate the clearance specifications in the service information. In addition, the clearances are often listed in the engine information on the decal on the underside of the hood. The specifications will be given with the engine either hot or cold. An example could be .008–.012 in. intake, and .016–.020 in. exhaust when hot. Be sure to adjust the valves with the engine properly heated or cooled. First you will check the clearance, and then you will

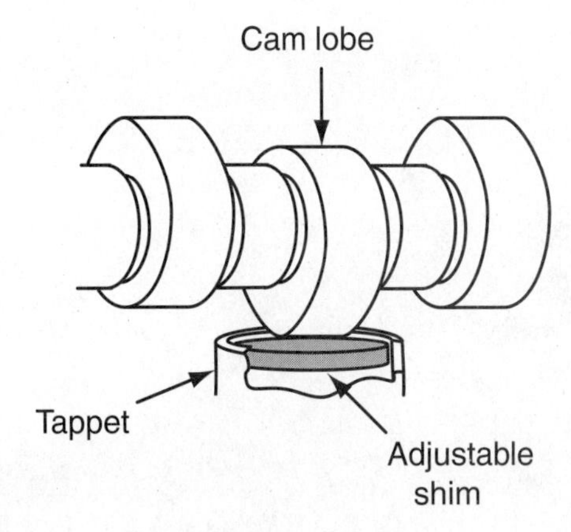

Figure 5. On this engine the shim sits on top of the tappet or follower.

tighten or loosen the clearance by fitting a thicker or thinner shim, respectively.

On some engines you can remove the shim by depressing the follower and using a magnet to pull the shim out.

There are several different tools available for this job, depending on the vehicle **(Figure 6)**. On other engines you may have to remove the camshaft to access the shims; they may be placed under the follower. On engines with removable shims you can adjust one valve at a time. When you have to remove the cam to replace shims you will need to carefully record all your clearances first. Then you can remove the cam and bring all the clearances within specification. On either type always double-check your adjustment after the new shim is installed.

To begin adjustment rotate the engine, using the crank bolt, to TDC #1. You will be able to check half of the valves. Run a feeler gauge under the camshaft on top of the follower. When you fit a gauge that has a little drag as you pass it under the cam you have found the clearance **(Figure 7)**. Record the clearance. You can see which clearances can be checked by looking for the ones where the base circle of the camshaft is on the follower. Rotate the engine another 360°, and you can check the remaining valves. You can find the specified valve sequence in the service information if you have any difficulty determining which should be adjusted. Other technicians will use a remote starter and "bump" the engine over until each cylinder in turn is at TDC, or until each camshaft lobe is facing straight up.

To adjust the clearance you will change the thickness of the shim. Use a micrometer to measure the thickness of the current shim. Calculate the needed change in clearance for the valve, and select a shim that much thicker or thinner. Let us look at an example:

1. Cylinder #1 exhaust clearance is .009 in. The specification is .016–.020 in.

Figure 6. This tool can be used on some engines to depress the follower. Then you can use a magnet to extract the shim.

Figure 7. Carefully check the clearance between the shim and the camshaft while it is on its base circle.

2. The valve clearance should be .007–.011 in. greater.
3. The existing shim size is 0.156 in.
4. The clearance needs to be .007–.011 in. greater, so the shim must be that much thinner.
5. Select a shim: 0.156–.010 in. = 146 in.
6. Install the new shim, and recheck the clearance.

After you have adjusted and checked each of the valves crank the engine over for a few seconds, and recheck the clearances. You need to be sure that all the shims are properly seated in the followers. Readjust as needed.

ADJUSTING VALVES USING ADJUSTABLE ROCKER ARMS

Another common method of adjusting mechanical valvetrains is through the rocker arm. One side of the rocker may ride on the camshaft or a solid lifter. The other side can work on the valve. One end of the rocker arm has a locknut and adjusting screw. To adjust the valve you can loosen the locking nut and turn the screw in or out to tighten or loosen the valve adjustment, respectively **(Figure 8)**. The clearance specifications will be similar to those with shim-adjusted valve lash. To adjust the valves:

1. Obtain the lash specifications, and be sure the engine is at the desired temperature.
2. Rotate the engine to TDC #1, and use the service information to note which valves can be adjusted.
3. Using feeler gauges determine the clearance on the valve you are adjusting. The feeler gauge should fit in the clearance under the adjustment screw with light drag **(Figure 9)**.

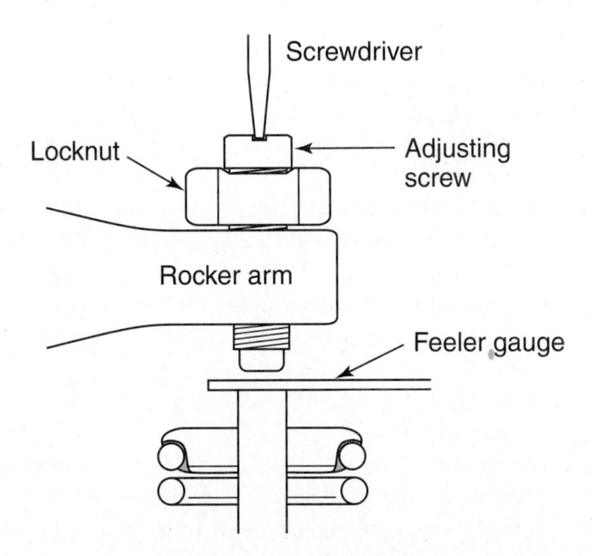

Figure 8. Loosen the locknut on the adjusting screw, fit the appropriate feeler gauge between the screw and the valve tip, and adjust the screw until some drag is felt on the feeler gauge.

4. If the valve needs adjustment loosen the locknut just enough that you can turn the screw. Insert the desired thickness (clearance) feeler gauge under the lash adjuster and alternate tightening the adjuster screw *lightly* with feeling the clearance. When there is just the right amount of drag tighten the locknut **(Figure 10)**.
5. Recheck your adjustment. This is essential as the adjustment can change as the locknut pulls the adjuster screw up. It may be necessary to adjust the valve a little snugly and then have it loosen up as you tighten the locknut. It is essential to recheck the adjustment after tightening the locknut.
6. Repeat for all the other valves adjustable at TDC #1. Rotate the crank 360°, and complete the adjustment of the valves.
7. Properly install the valve cover, and start the engine. It should turn over smoothly.
8. Allow the engine to warm up, and listen for any excessive valvetrain clatter and feel for rough running.

Figure 9. Check the clearance first to be sure adjustment is needed.

Figure 10. Three hands to hold the wrench, the screwdriver, and the feeler gauges at the same time would make this job simpler.

Summary

- Hydraulic lifters require adjustment, if possible, whenever the valvetrain has been disassembled.
- On a valvetrain with nonadjustable hydraulic lifters complete valvetrain assembly by torquing the rocker arm nuts to specification.
- Mechanical lifters will require valve lash adjustment after the valvetrain has been reassembled.
- Hydraulic lifters do not require periodic service after the initial adjustment.
- Mechanical lifters require periodic valve adjustments, typically every 30,000 or 60,000 miles.
- Valves that are too loose will cause valvetrain clatter and a loss of high-end power as the valves will not open as far or as long as they should.
- Valves that are too tight can cause the engine to crank over irregularly, to backfire through the intake and exhaust, and to burn valves.
- When adjusting hydraulic lifters you are trying to center the plunger in its bore to maximize adjustment in either direction.
- To adjust hydraulic lifters adjust the rocker until lash is at zero; then further tighten the rocker nut the specified number of turns.
- Mechanical valvetrains require valve adjustment to a specified clearance. This may be with the engine hot or cold.
- Some clearance adjustments are made with removable shims.
- Rocker arms with adjustment screws provide another common means of valve lash adjustment.
- Always recheck your adjustments; misadjustment could cause burned valves or noise and leave a customer dissatisfied.

Review Questions

1. When do hydraulic lifters require adjustment?
2. Is it true or false that mechanical lifters require periodic adjustment?
3. What causes valvetrain clatter?
4. What are you trying to achieve when you adjust hydraulic lifters?
5. Technician A says that too much valve lash is likely to cause the valves to burn. Technician B says that too much valve lash can cause backfiring. Who is correct?
   A. Technician A only
   B. Technician B only
   C. Both Technician A and Technician B
   D. Neither Technician A nor Technician B
6. Technician A says that too little valve lash can cause the engine to crank over irregularly. Technician B says that too little valve lash can cause burned valves. Who is correct?
   A. Technician A only
   B. Technician B only
   C. Both Technician A and Technician B
   D. Neither Technician A nor Technician B
7. Technician A says that nonadjustable hydraulic lifters may have to be shimmed to provide the correct clearances. Technician B says that nonadjustable lifters have a torque specification for the rocker arms. Who is correct?
   A. Technician A only
   B. Technician B only
   C. Both Technician A and Technician B
   D. Neither Technician A nor Technician B
8. Technician A says that adjustable hydraulic lifters should be adjusted during engine assembly. Technician B says you should readjust them after the engine has run. Who is correct?
   A. Technician A only
   B. Technician B only
   C. Both Technician A and Technician B
   D. Neither Technician A nor Technician B
9. Technician A says that on engines with mechanical valvetrains adjustments are made with the engine hot. Technician B says that adjustments will be inaccurate if made at the wrong temperature. Who is correct?
   A. Technician A only
   B. Technician B only
   C. Both Technician A and Technician B
   D. Neither Technician A nor Technician B
10. A shim-type valvetrain is being adjusted. The exhaust clearance is specified at.016–.020 in.; the existing clearance is .027 in. The shim measures 0.136 in. To correct the clearances install a shim __________ thick.
    A. 0.146 in.
    B. 0.142 in.
    C. 0.130 in.
    D. 0.126 in.

Chapter 32 Final Assembly

Introduction

After performing a major engine overhaul you will have quite a task of reassembling the rest of the engine and accessories. Much of this information will also apply if you have just completed a head gasket job, valve job, or timing chain replacement. In many cases you will install a "crate" engine rather than overhaul the one in the vehicle; when you take it out of the box it will have none of the accessories or manifolds and perhaps not even the oil pan on it. You will have to transfer all the parts from the old engine or get new parts to complete the assembly. Whatever the task you are completing, the final assembly is as important as the rest of the job. After any major engine work you should take the time to evaluate each component that you install. Do not risk ruining a new engine due to an overlooked part such as a cracked intake or a used water pump. This is also a good time to replace the spark plugs, wires, and air and fuel filters, and to perform a radiator service; no need for the customer to have to come back for engine maintenance in the near future. While you are performing the final assembly on the engine fully charge the battery so that it is ready when the engine is installed.

Look over all of the fasteners. Clean rusty ones on the wire wheel, and examine the threads. Replace any fasteners that have worn or damaged threads or heads. Clean parts that you reinstall on a new engine; do not install components with oil built up on them or those with significant corrosion **(Figure 1)**. Take the time to refurbish used parts by thoroughly cleaning them. When the customer opens the hood after major engine work she expects to see it looking fresh. Before you begin the reassembly process lay the parts out on a clean bench, and approach the reassembly process logically and methodically.

Figure 1. This filthy intake should be thoroughly cleaned, if not painted.

You Should Know *Take the time to clean all the fasteners, and line them up with the clean component they hold. This can actually save time and minimize risk during the assembly process. The greater your level of organization, the less likely you are to forget a critical step.*

OIL PUMP INSTALLATION

Whenever the engine is out of the vehicle the oil pump should be evaluated. A visual inspection will usually give you enough information to determine whether to reuse it or replace it. Always err on the side of caution; if there is any doubt replace it. When you perform an engine overhaul on a high-mileage engine, or if you are replacing the engine and the new one comes without an oil pump, install a new pump. In many cases on newer vehicles when you are performing a timing chain replacement you will have access to the oil pump driven by the crankshaft. Inspect and replace the pump as needed. Install the new oil filter now. Put some oil in it, and remember to lubricate the seal with oil.

Installing Cam-Driven Oil Pumps

Many oil pumps driven by the camshaft or distributor drive shaft will not come with a new screen and pickup tube. Replace these parts too; they may contain debris or be partially plugged **(Figure 2)**. Replace the pickup tube o-ring, and thoroughly clean the screen if you will reuse it. Securely attach the pickup tube and screen to the new pump. Rotate the pump in fresh oil before installing it. You will prime the pump again during engine startup and break-in, but this will help ensure that it never runs dry. Install the oil pump drive gear. Make sure the mating surfaces of the pump and block are perfectly clean. Do *not* use RTV to seal the oil pump. Small pieces of the sealant can be sucked into the pump and can restrict it or get pushed into passages and block them. Most oil pump gaskets, if used, are installed dry.

Carefully check the oil pump drive shaft and gear for damage. If the shaft is twisted or the gears are worn replace the shaft; do not risk engine failure. Fit the oil pump drive shaft into place in the block, and make sure it seats securely into the pump **(Figure 3)**. Be sure to properly torque the oil pump using the specifications. This is another case where carelessness would cause catastrophic engine failure. Fit any pump securing brackets into their proper position, and tighten them securely.

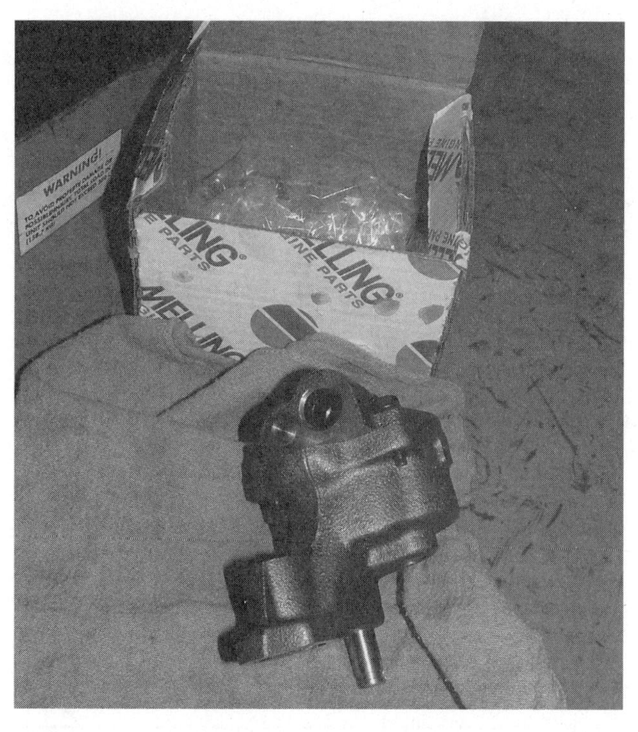

Figure 2. The new oil pump comes without a pickup tube or screen. Be sure that the old parts are clean and strong before reusing them.

Figure 3. Check the oil pump drive gear for cracks or twisting and for wear on the gear or drive end. Replace it if you see any damage.

Figure 4. This oil pump tightening specification was given in inch pounds! Torque to the correct value; the manufacturer has engineered the engines carefully.

On the popular old Chevy 350 small blocks many technicians will spot weld the pickup screen into place to ensure it does not fall off. Be careful that no metal gets into the pump.

Installing Crank-Driven Oil Pumps

On a crank-mounted oil pump check the key on the crankshaft to be sure it is in good condition and properly seated in the crankshaft. Check the keyway on the pump; if it is rounded or worn replace the pump. Pour some oil into the feed hole, and rotate the pump a few times to circulate oil into the new or used pump. The oil pump and its mating surface must be spotless. Most of these pumps seal with an o-ring; always fit a new o-ring. Again, do *not* use RTV sealant, even to hold the o-ring in place; a dab of assembly lube will hold it in place if necessary. Torque the pump into place using the proper torque sequence and specification; if a sequence is not given, tighten in a diagonal pattern **(Figure 4)**.

OIL PAN INSTALLATION

Clean the oil pan thoroughly inside and out. Dry it fully to remove any solvent. Inspect the pan for rust; it can rust from the outside through to the inside **(Figure 5)**. If it is a steel pan carefully inspect the flange for any irregularities. You can straighten the flange by placing it on a flat surface and rapping on the flange with a piece of flat steel stock or a block of wood. Replace a pan if it is not in good shape. On a cast-aluminum pan be sure there is no corrosion on the gasket surface or cracks anywhere on the pan. Clean the sealing surfaces of the oil pan and the block thoroughly. *Install the oil drain plug and tighten it securely.* You may want to repaint or replace a stamped-steel pan if it shows corrosion.

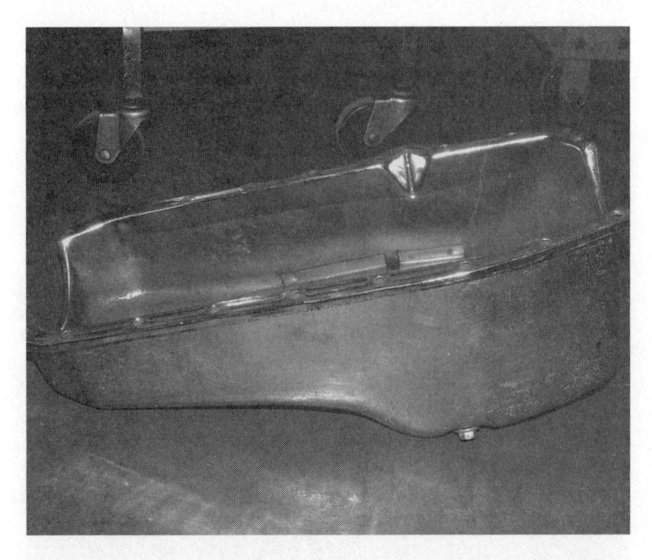

Figure 5. This chrome oil pan replaced a dangerously rusty old pan.

Refer to the manufacturer's service information to determine whether any sealant should be used on the gasket. On a multiple-piece oil pan gasket a small dab of RTV is often used where the gasket ends meet **(Figure 6)**. In other applications the oil pan is sealed with RTV silicone. In these cases it is critical that you apply a 1/8 in. to 3/32 in. bead of continuous sealer; more is *not* better. Using too much sealant can plug up the oil pump screen. Many cast-aluminum pans use a rubber gasket that fits into a groove in the pan; be sure it is not twisted during installation. Place the pan gasket into position, and start each of the bolts by hand. Be sure the gasket is fitted properly; look for any parts of the gasket that are out of place. Tighten the oil pan to the proper specification. Overtightening a cork gasket will deform it and cause a leak. The specification may seem low, but the manufacturer has done its research. Do not tighten it just a little more because you think it should be tighter. If a tightening sequence is given use it. If not follow a diagonal pattern to work your way across the pan.

VALVE COVER AND SPARK PLUG INSTALLATION

The guidance for installing the valve cover(s) is very similar to that for mounting the oil pan. Straighten any stamped steel flanges that are bent. Use sealant only when recommended by the manufacturer. Sometimes the valve cover will be two pieces with a rubber plug or seal near the front of the camshaft. It is typical to apply a spot of RTV silicone at the points where the gasket meets this seal. You know that using the proper torque specification and

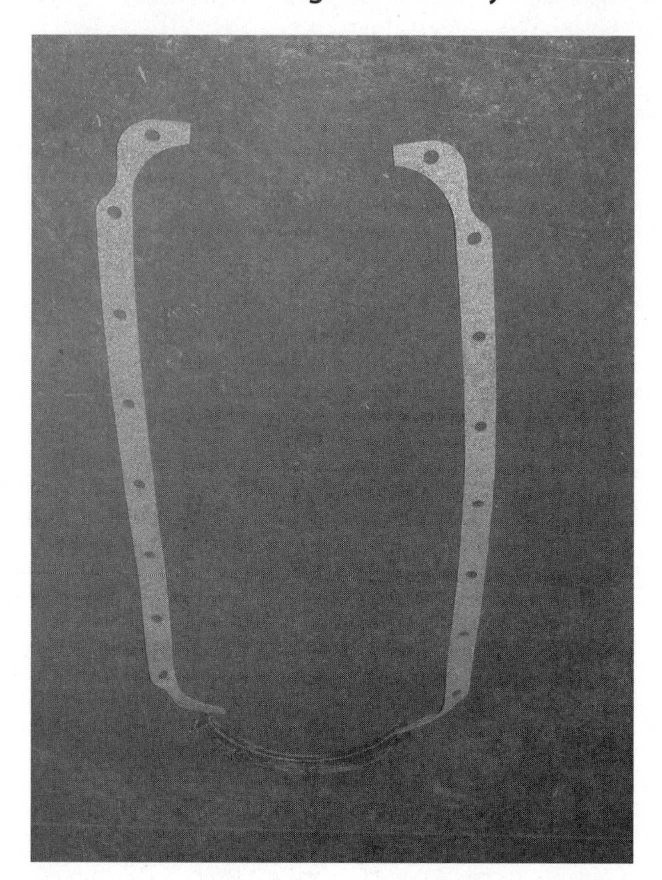

Figure 6. This cork gasket must not be overtightened. Apply a dab of silicone where the gaskets meet the rubber ring.

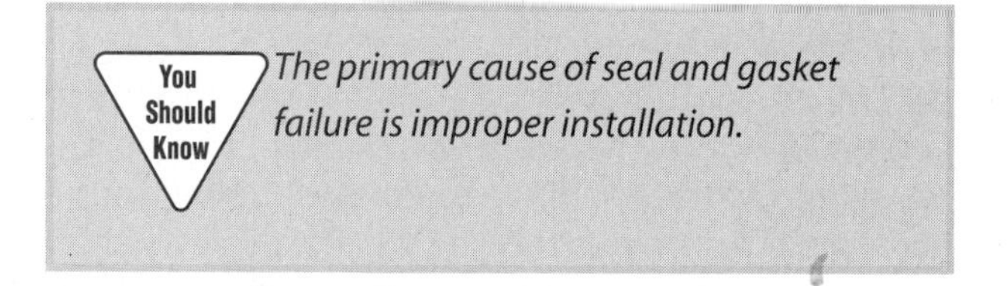

sequence will make the difference between a leaky gasket and one that provides a good, long-lasting seal.

Install new spark plugs. Use the proper plug for the engine. Many technicians are relying only on OEM spark plugs on newer vehicles. Aftermarket plugs have been causing many problems on late-model OBDII vehicles that have sensitive misfire diagnostic capabilities. The manufacturers spend a lot of time and money researching the best plug for the application. Unless you are sure that the aftermarket plug can meet those stringent specifications do not use them.

INTAKE MANIFOLD INSTALLATION

Clean the intake, block, and head manifold mating surfaces. Give a final inspection to a plastic manifold to be sure it has no cracks. Look at the gasket closely to be sure the ports and passages line up properly **(Figure 7)**. In some engine gasket sets you will be given two different intake gaskets. You will have to pick the correct one for your application by looking closely at the gasket and the head. On a vee-type engine you may need to use a little sealant to hold the gasket in place while you fit the manifold. Use an aviation sealant or high-tack adhesive rather than RTV. A dab of RTV sealant may be required at the ends of the gaskets; use it if specified **(Figure 8)**. Alignment tabs may be used on the head and intake gaskets; line them up to properly install the gaskets **(Figure 9)**. Install all the fasteners by hand, and then torque them to specification. Remember that a plastic intake can crack if torqued im-

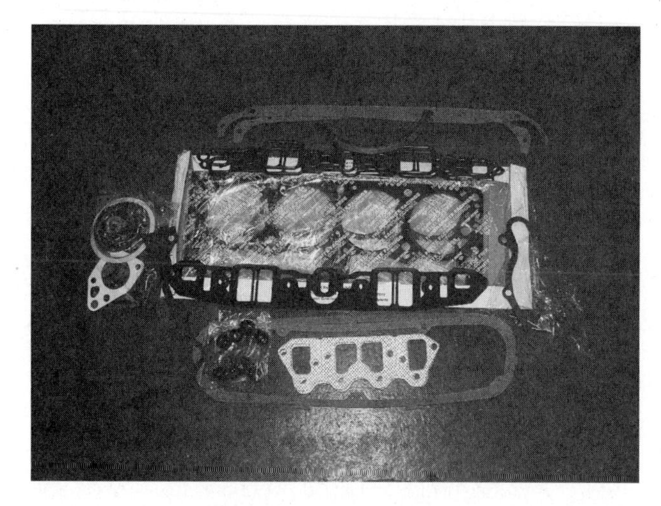

Figure 7. A complete engine gasket kit may come with two different sets of intake gaskets like this one.

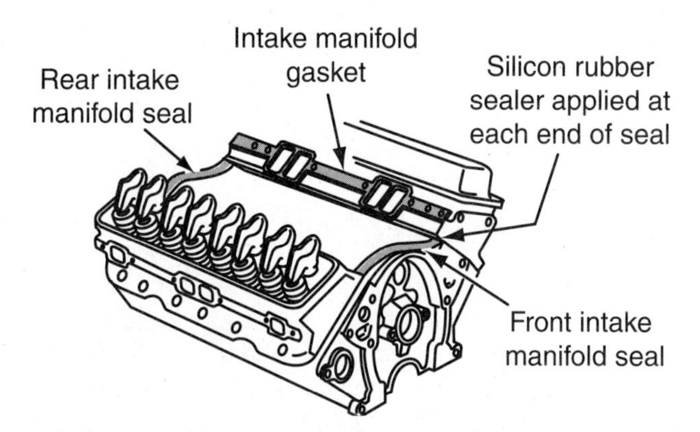

Figure 8. A dab of RTV silicone sealant at each of the corners of the manifold gaskets is necessary to provide a good seal on some engines.

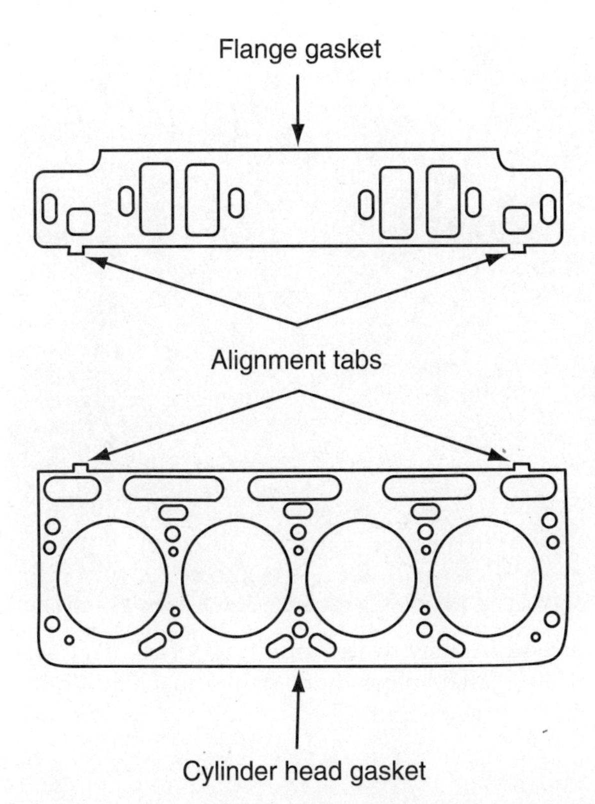

Figure 9. Line up the tabs on the intake manifold gasket with those on the head gasket.

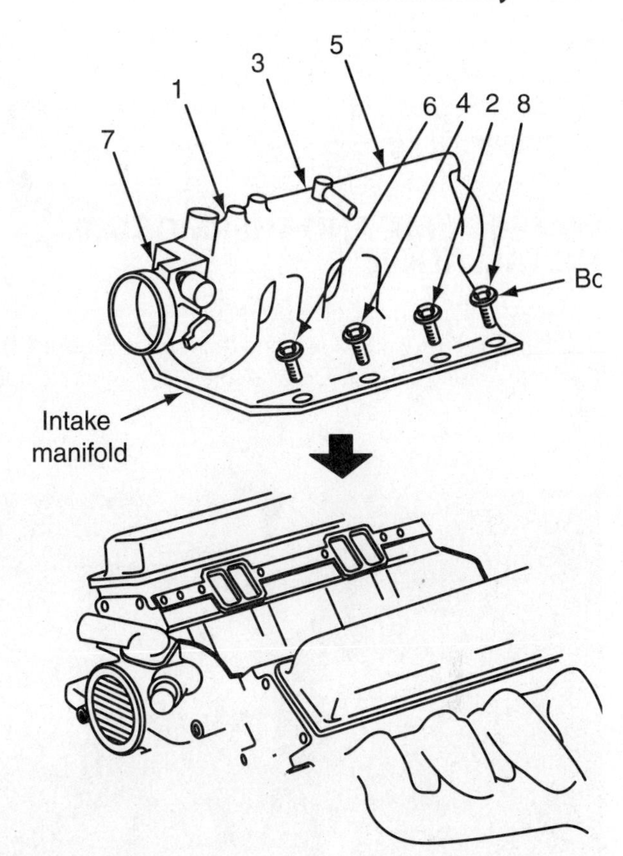

Figure 10. This shows a typical diagonal tightening sequence.

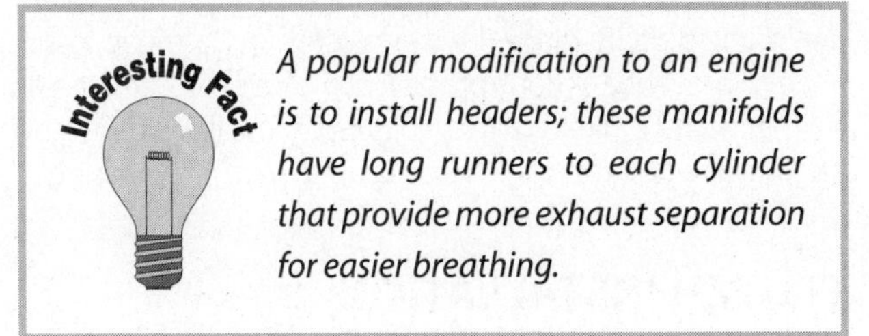

A popular modification to an engine is to install headers; these manifolds have long runners to each cylinder that provide more exhaust separation for easier breathing.

properly. If a sequence is stated use it; if not use another appropriate sequence **(Figure 10)**.

EXHAUST MANIFOLD INSTALLATION

When cleaning the mating surfaces of the exhaust manifold use a scraper or razor blade rather than an abrasive disc or wire wheel; the exhaust gasket could contain asbestos. You may be instructed to use only a plastic scraper on an aluminum cylinder head mating surface. Make sure you have replaced rusted studs and repaired any damaged threads. You may have to drill a broken stud and use an extractor to remove it if it broke off inside the head **(Figure 11)**. Be sure to tap the threads

Figure 11. You may need to use a drill and an extractor to remove exhaust studs broken off in the head.

after removing a damaged stud **(Figure 12)**. Inspect the manifold carefully to be sure it is not cracked. Install it with new fasteners, and torque the manifold to specification **(Figure 13)**.

WATER PUMP AND THERMOSTAT INSTALLATION

After a thorough engine overhaul it is common practice to replace the water pump and thermostat **(Figure 14)**. If you are reusing a water pump be sure that the fins are not

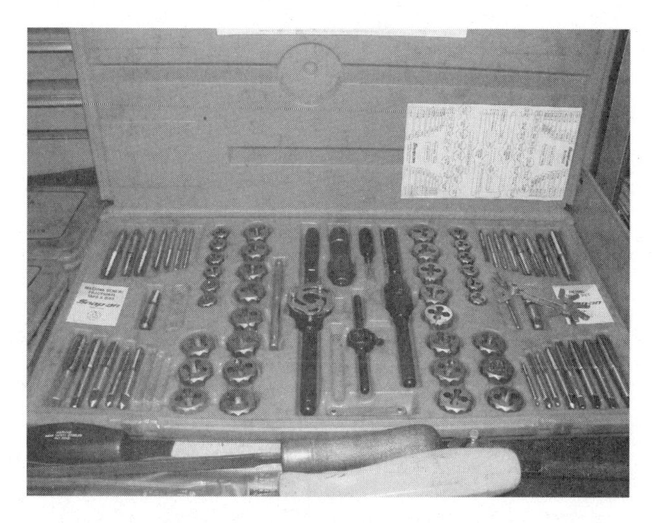

Figure 12. Select the correct tap to refresh the threads after removing a stud. Do not repair stud threads; replace the studs.

Figure 13. These headers combined with a "hot" cam will help the engine breathe, which makes it more powerful.

Figure 14. A new water pump is good insurance after major work or the installation of a new engine.

corroded or bent and that the bearing and seal are good. Look for evidence of leaking from the weep hole, and feel the bearings and shaft for looseness or roughness when rotating. When reusing a pump remove all the old gasket material from the pump and the block. Most pump gaskets are installed dry. Replace the thermostat; this is cheap insurance to prevent a serious problem. Be careful to fit the t-stat in the correct direction. The spring typically sits in the engine, whereas the jiggle valve is up toward the housing.

CLUTCH INSTALLATION

On an engine with a manual transmission evaluate the **clutch** assembly. Unless it has pretty low mileage on it you should replace it now to save the labor of doing it six months or a year from now. You will need a clutch aligning tool to properly install the clutch disc. If the disc is not centered you will never get the input shaft of the transmission to slide into the disc. This can be very frustrating when installing the engine **(Figure 15)**. You should also replace the pressure plate and **release bearing**. Torque the flywheel onto the crankshaft. Replace the small pilot bearing, if equipped, where the transmission input shaft rides. It may be in the crankshaft or in the flywheel. Fit the clutch disc first, being careful to install it in the correct direction; the flatter side without protruding springs faces the flywheel. Mount the new pressure plate over the clutch disc, and slide the alignment tool into the pilot bearing. You can warp the pressure plate if you do not torque it in a diagonal pattern. Torque the flywheel

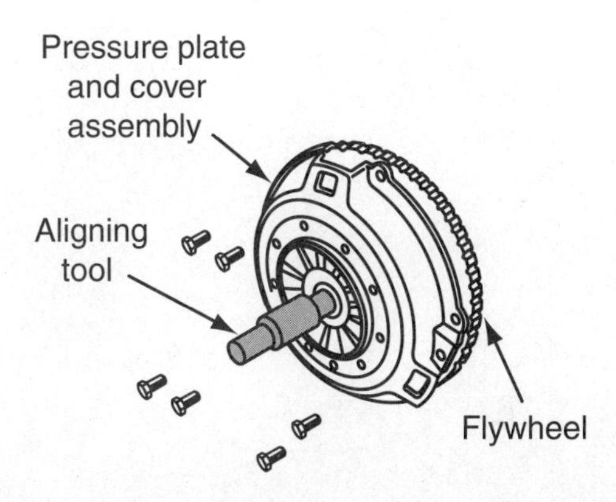

Figure 15. This clutch aligning tool slides through the clutch disc and into the pilot bushing. Torque the pressure plate on while the tool is holding the disc in place.

and pressure plate to specification; it is important. If the clutch **slave cylinder** mounts inside the bell housing install a new one now.

> **Interesting Fact**
>
> *I almost removed a freshly rebuilt engine once because a deep knocking noise started shortly after I started the engine. As it turned out it was loose flywheel bolts. And I got even luckier because I could access them by removing a cover. I would not want to relive that period of time when I thought I had mismeasured the bearing clearances.*

ACCESSORY INSTALLATION

Install any accessories and brackets that you were able to leave on the engine during removal. If you marked the brackets during disassembly this should be a simple task. You may be able to install the generator, power steering pump, and/or the AC compressor. The more you can assemble now the easier it will be, but think back to when you removed the engine to remember what had to come off the engine to get it out of the vehicle.

Mount the harmonic balancer. There are a few different ways to install a harmonic balancer. The best way is to use a special installation tool that fits into the crankshaft threads and around the inner ring of the balancer to pull the balancer on. If that is not available you may be able to use the main holding bolt for the harmonic balancer and carefully pull the balancer onto the crankshaft. Use a flywheel locking tool to keep the crankshaft from turning; you may need an assistant **(Figure 16)**. When the harmonic balancer is press fit onto the crankshaft without an attaching bolt you may have to use a hammer. Find a very large socket, a press tool, a special round tube installation tool, or a proper-size block of wood to distribute the force equally around the inner portion of the ring.

If possible, install any new accessory drive belts. Many newer vehicles use one serpentine belt to drive all the accessories. The newer the vehicle, the more likely it is to have an automatic tensioner. Usually there is a designed point on the tensioner where you can use a lever or a breaker bar to pull back against the tensioner spring and easily install the belt **(Figure 17)**. Check the gauge usually

Figure 16. This flywheel holding tool gives you the leverage to keep the crankshaft from spinning.

Figure 17. Pull back on the tensioner with the wrench. This will give you room to slip the belt over the pulleys. Notice the belt routing diagram on the belt cover.

found on the tensioner to be sure the new belt is running within the correct tension range. When the tensioner is not automatic use a belt tension gauge, when available, and tighten the belt to specification. You can also tighten it using some general guidelines. Check the area of the belt where there is the longest length of free belt between pulleys. For every foot of free belt it should deflect about one-half inch. Other technicians use the two-finger twist method **(Figure 18)**. You should be able to twist the belt one quarter turn with two fingers but not much more. Do not put every ounce of power into your two fingers; a belt that is too tight can damage the bearings in the accessories and even prematurely wear the main bearings.

Install the rest of the components in your box. You may have an EGR valve that should be installed. Clean the pintle of the EGR valve before reinstalling it onto the engine. A new gasket should have come with the engine gasket kit. You will probably have a coolant temperature sensor and other vacuum or electrical switches that should be put back on the engine. This is where your careful marking during disassembly will really help you. It can be very difficult to remember exactly how a small bracket for a solenoid mounts onto the valve covers or intake manifold.

Look over the engine carefully for anything that might be missing or loose. This is your last chance to correct many problems; next you will install the engine in the vehicle. Check the engine for leaks from gaskets, seals, or plugs; correct them as necessary. Run back through the whole procedure in your mind and remember your work. Concentrate hard, and sometimes you will think of that one little (or big) thing that you have not done or forgot to do at the time. When you are ready install the engine into the vehicle.

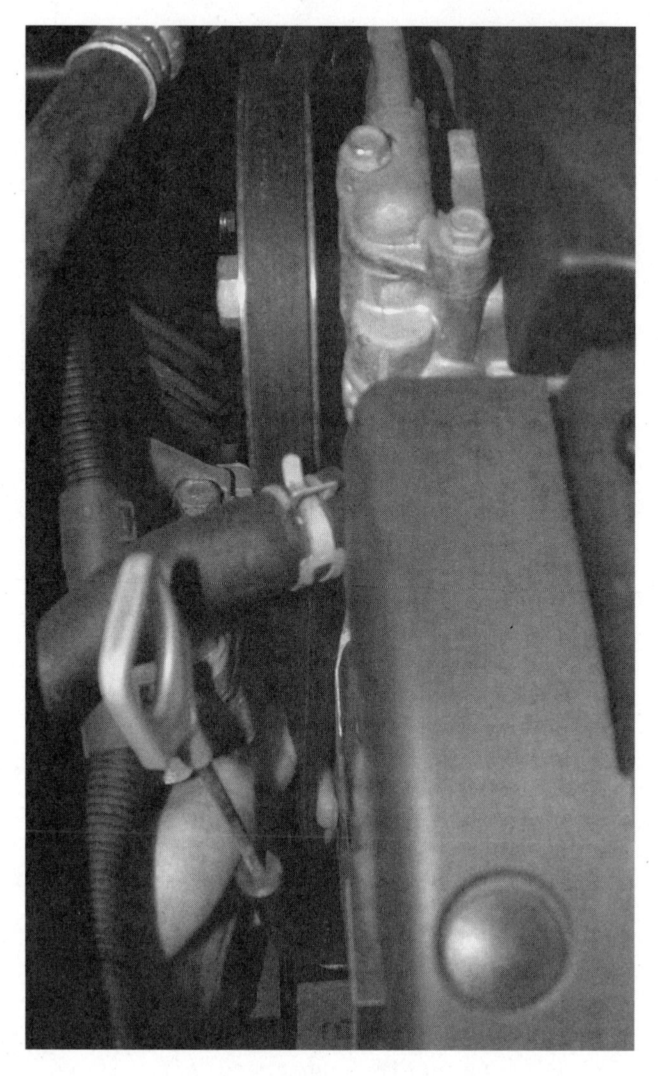

Figure 18. Use two fingers to turn the belt one quarter turn.

Summary

- After major engine work the final assembly process is critical to the success of your work.
- Clean all components and fasteners so they perform properly and look fresh.
- Torque pans and covers using the proper torque specifications and sequence. Leaks can cause a customer to lose confidence in your work and become irritated. Many leaks can be time-consuming to repair.
- Evaluate all components before you reinstall them. If they are worn or faulty now is the time to replace them.
- Be cautious after major work; replace components that are marginal so the customer can enjoy good engine performance for an extended period.
- Carefully fit a new, or in a few cases the used, oil pump. Lubricate it fully with oil.
- Install the oil pan using sealant only as specified. This is an important component to torque on in the proper sequence.
- Install and tighten the oil drain plug with the oil pan.
- The intake manifold gasket must be fitted properly using the specified sealant, if any. Use an appropriate torque sequence to tighten the manifold properly.
- Replace the exhaust manifold studs, and refresh the bores' threads before reinstalling the manifold.
- Replace the thermostat. On a vehicle with significant mileage you should also replace the water pump.

- If a manual transmission is used check the clutch and related components carefully; replace them if indicated.
- Mount the harmonic balancer before engine installation, using an appropriate technique.
- Fit as many accessories as possible before installing the engine into the vehicle.
- Review your work carefully before lifting the engine onto the crane; it will be easier to correct anything you may have missed while the engine is still out.

Review Questions

1. Is it true or false that in most cases you should replace the oil pump during an engine overhaul?
2. List four components you should carefully evaluate before reinstalling them on the engine.
3. What maintenance items should you replace during engine assembly after major engine repairs?
4. Is it true or false that you should seal the oil pump with a continuous bead of RTV?
5. Technician A says that a new oil pump should be packed with grease. Technician B says that a new oil pump may not come with a new screen. Who is correct?
   A. Technician A only
   B. Technician B only
   C. Both Technician A and Technician B
   D. Neither Technician A nor Technician B
6. Technician A says that many multipiece oil pan gaskets require a spot of RTV at the gasket ends. Technician B says that you can straighten a cast-aluminum pan using a punch and hammer. Who is correct?
   A. Technician A only
   B. Technician B only
   C. Both Technician A and Technician B
   D. Neither Technician A nor Technician B
7. Technician A says that a cork valve cover gasket should be tightened to specification and then one turn further. Technician B says that a synthetic rubber valve cover gasket should be lubed with RTV to hold it in place. Who is correct?
   A. Technician A only
   B. Technician B only
   C. Both Technician A and Technician B
   D. Neither Technician A nor Technician B
8. Technician A says that generally components should be torqued using a diagonal pattern from the inside outward. Technician B says that torquing components in a circle can crack the part. Who is correct?
   A. Technician A only
   B. Technician B only
   C. Both Technician A and Technician B
   D. Neither Technician A nor Technician B
9. Technician A says that you should replace exhaust manifold studs. Technician B says that the exhaust manifold gasket may contain asbestos. Who is correct?
   A. Technician A only
   B. Technician B only
   C. Both Technician A and Technician B
   D. Neither Technician A nor Technician B
10. Technician A says that replacing the clutch during engine reassembly may save the customer extensive labor charges later. Technician B says the flywheel really needs to be torqued to specification. Who is correct?
    A. Technician A only
    B. Technician B only
    C. Both Technician A and Technician B
    D. Neither Technician A nor Technician B

Chapter 33 Engine Installation

Introduction

If you have just rebuilt an engine you are getting close now to finishing the job; resist the temptation to rush. After paying the bill the customer is not going to be happy to come back again with other problems. If you are installing a rebuilt engine the customer will expect thousands of miles of trouble-free service. For whatever reason you have removed the engine, paying attention to detail during installation can make the difference between a successful job and a customer comeback. The customer will assume that you have taken care of any problems that could cause him a return trip to your shop in the near future; repair any potential problems as you come across them during the installation process. You may need to replace hardware, connectors, hoses, vacuum lines, and gaskets to ensure a trouble-free powertrain.

Reinstalling an engine into the vehicle is a job in and of itself. Take your time and concentrate on making sure everything is connected properly. Reinstall the fender covers to prevent any damage to the paint. If you were careful to mark wires, hoses, vacuum lines, and components during engine removal the installation process should be relatively straightforward. Read through the manufacturer's procedure for engine installation; this is often simply the reverse of the removal process, but it will remind you of all the steps involved.

PREPARATIONS FOR ENGINE INSTALLATION

Your engine should now have all its covers properly installed. It should be clean and fully reassembled

Some technicians use a checklist for the reinstallation process. This can help to ensure that no small task is left undone. You can use the manufacturer's procedure or your own list of steps you recorded during removal.

(Figure 1). Install a chain to the lifting hooks of the engine using suitable hardware, lift the engine onto the crane, and slide the mounting flange out of the engine stand **(Figure 2)**. With the engine hanging remove the engine stand mounting flange from the engine. Be careful not to place any part of yourself underneath the engine while it is hanging. The flywheel and clutch assembly or ring gear on an automatic transmission should be torqued into place. If you removed the engine with the transmission attached now is the time to reconnect the two pieces. On a car with a manual transmission line up the engine and transmission input shaft carefully. You must not hang the weight of the transmission on the input shaft or force the two units together. When the two pieces are properly aligned they will slide into place. When you are installing an automatic transmission onto the engine be certain that the torque converter is fully mounted on the pump. Then use the access cover to properly torque the flexplate to the torque converter. Never use the attaching bolts to try to draw the transmission onto the engine; you could break the bell housing. Use the appropriate torque specification when mounting the transmission to the engine. Overtightening the bell housing bolts can cause the rear cylinders to become out of round!

Figure 1. This fully rebuilt engine looks great and is ready for installation.

Figure 2. Use the engine lifting hooks to support the engine on the crane.

RWD ENGINE INSTALLATION

Prepare the engine cavity for installation; move all wires, lines, accessories, and hoses out of the way. Raise the engine above the hole, and slowly lower it into place. In many cases your first step will be to line up the engine and transmission. It may be helpful to place a jack under the transmission to raise it up and tilt it a little bit to mate with the engine. You will likely have to rock and twist the engine until it is ready to slide into place. Tighten the bell housing bolts by hand or lightly with a ratchet. Once the vehicle is up in the air you can access all of them and torque them to specification in a diagonal sequence.

Once the transmission is mounted to the engine lower the engine down onto its mounts. If the engine was already on a mount but not properly aligned use a prybar to raise the engine back into the proper position. You may have to support some of the weight of the engine with the crane as you work the engine into position. Lower the engine fully, and then rock the engine to settle it comfortably into place. Properly tighten the engine mounts **(Figure 3)**. If the engine is not settled into position you can twist the rubber in a mount accidentally and pretension it that way. This can cause vibration and premature failure of the mount.

FWD ENGINE INSTALLATION

Often you will install the FWD engine from the bottom. Lower the engine carefully onto the engine cradle. Rock the engine back and forth a bit to settle it comfortably into the mounts. You do not want to torque the mounts down when they are stressed or twisted in one direction. The same is true when you bolt the cradle back up to the vehicle. Make sure each mount is in place and the engine is settled before tightening the mounts down. A preloaded suspension cradle mount can cause the vehicle to pull right or left as you drive down the highway. Lower the vehicle down onto the

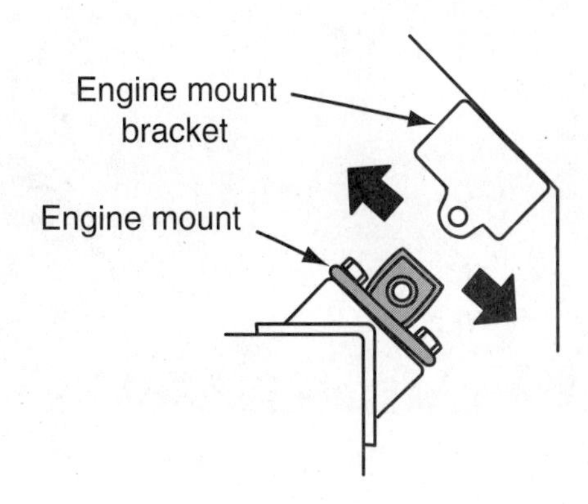

Figure 3. Line up the mounts and tighten them down when the engine has settled into place.

cradle, and loosely attach each of the suspension mounts. Rock the assembly to relax the bushings before tightening them into place.

TRANSMISSION INSTALLATION

With the engine mounted into place remove the crane and its chain. Raise the engine overhead. Properly torque the bell housing bolts. On an automatic transmission you must also tighten the torque converter to specification in a diagonal pattern. Replace any covers around the bell housing, ring gear, or flywheel. Reinstall shift linkage that had to be removed from the transmission during engine removal. If the shift linkage and/or clutch cable was disturbed locate the manufacturer's procedure to properly adjust it after installation. If you had to remove hydraulic components to the clutch system reinstall them now **(Figure 4)**. Follow the manufacturer's procedure for bleeding the system if the lines were opened. Fill the clutch or brake fluid reservoir with the proper brake fluid. Make sure the transmission is properly filled.

UNDER-VEHICLE OPERATIONS

On a front-wheel-drive vehicle install one or more axle shafts back into the transmission if they had to be removed. Reinstall the tie rod ends, lower ball joints, and axle nuts. Install any wheels. Lower the vehicle to properly torque the axle nut; the torque will be too high for you to just hold the wheel. Torque the wheels to specification. On a rear-wheel-drive vehicle reinstall the driveshaft if it had to be removed.

Reconnect the exhaust system to the manifold. Use a new front pipe gasket or donut and new hardware **(Figure 5)**. Make sure the exhaust is hanging properly all the way back to the tailpipe. It could have fallen off a hanger during these maneuvers.

Figure 4. Reinstall the clutch slave cylinder and bleed as needed.

Figure 5. Use new hardware and gasket to install the front pipe onto the manifold.

Recheck that the oil drain plug is snug. Many technicians will replace the coolant and heater hoses on a high-mileage overhaul or replacement. Connect the lower radiator hose to the engine. Replace any worn or bent hose clamps. Connect the transmission cooler lines if equipped. Reinstall the starter and its wiring and brackets. Inspect the wiring and connectors closely, and make any needed repairs. Locate any lower wiring connectors, and connect them while underneath the vehicle, as applicable. This may include connections to the oxygen sensors **(Figure 6)**, reverse light switch, vehicle speed sensor, and oil level indicator, among others. Make sure no wiring is pinched between the engine and the bell housing. Use any looms or wire holders to secure wires into place. Use

Figure 6. Be sure to connect the oxygen sensor connectors or the engine will not operate properly.

tie wraps to secure the wiring appropriately if the original holders had to be broken to be removed. Install the inner fender splash shields, if applicable, and then lower the vehicle to the ground.

Interesting Fact

A customer came back just one week after a thorough engine overhaul with the malfunction indicator lamp (MIL), or "check engine" light, on. The technician found a DTC (diagnostic trouble code) for both the front and rear oxygen sensors. Suspicious that both the sensors were unlikely to fail at the same time he used common sense and rechecked the wiring to the sensors. The harness insulation had been crimped in between the bell housing and engine, and a couple of wires had chafed through to ground.

TOP ENGINE OPERATIONS

Locate the oil pressure switch and remove it. Install an oil pressure gauge and route it to the top of the engine so you can see it when you crank the engine over. Be sure the connector for the switch is available for reinstallation after testing.

Reinstall the radiator and fan if they had to be removed. Connect all the coolant hoses to the radiator, thermostat, reservoir, and heater core. Use new clamps if the old ones are bent, corroded, or stripped. Fill the system with a 50/50 (or 60/40 in cold climates) mixture of antifreeze and water. Follow the manufacturer's procedure for filling the cooling system to minimize the risk of air being trapped in the system. Leave the cap off and let the coolant settle; check it again to see if more can be added. Pressure test the cooling system for leaks, and repair any leaks or seepage now **(Figure 7)**.

Figure 7. Fill the cooling system and pressure test it for leaks.

Reinstall the power steering pump, generator, air pump, or AC compressor as needed. If the air conditioning refrigerant was removed from the vehicle remember to properly charge the system, and verify proper operation before returning the vehicle to the customer. Refill the power steering system. Make sure that the brackets are in their proper positions and are tightened properly. Install one or more new drive belts, and tension appropriately using an automatic tensioner, a drive belt tension gauge, or the two-finger twist method.

Install all the solenoids, switches, actuators, and other small components and brackets back to their positions on the valve cover(s), intake manifold, or cylinder head(s). Locate all the vacuum lines, and reinstall them into their proper location. If you have any doubt about correct positioning refer to the vacuum routing diagram under the hood or in the service information **(Figure 8)**. A misrouted vacuum line can cause stalling and poor engine performance.

Interesting Fact

A customer returned his vehicle to the shop a week and a half after the engine had been replaced, complaining that it had been losing power. Careful diagnosis revealed a loose generator bracket. This rattling noise fooled the knock sensor, which sent a knock signal to the PCM. The computer responded by retarding the timing. The "knocking" continued, so the PCM continued to retard the timing until the customer could feel a significant loss of power.

Reattach the throttle cable and adjust it **(Figure 9)**. The cable should start to move the throttle with just a quarter inch or so of free play at the accelerator pedal. When the pedal is fully depressed the throttle should be wide open **(Figure 10)**. Clean the throttle bore and plate thoroughly with carburetor cleaner. Attach the automatic transmission **kick down cable** or throttle valve linkage, if

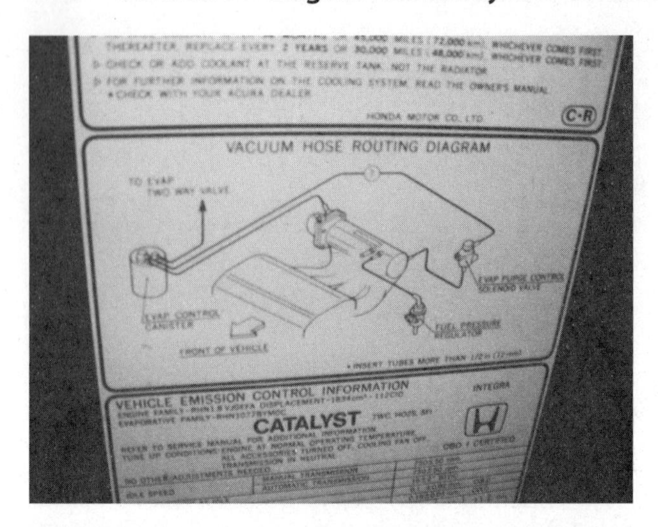

Figure 8. Use the vacuum routing diagram to be sure you connect the vacuum lines properly.

Figure 9. Secure the throttle cable by snapping it into place.

Figure 10. Be sure that the throttle opens fully when the accelerator pedal is pushed to the floor.

equipped. Generally the cable should begin moving as soon as the throttle moves. Move the throttle to wide open, and make sure the cable does not bind or stick. The cable should have just a little slack left after wide-open throttle. Refer to the manufacturer's service information for more specific recommendations.

Connect all the wiring. Pull the harness into position, and be sure that you find the proper home for each connector. Some connectors may not be used on your application. Look closely at the connector and terminals; you can usually tell if it had been connected prior to engine removal. Check over each electrical component to be sure it has the right connector properly attached. Be sure that the engine and chassis ground straps are back in place **(Figure 11)**.

Reconnect any fuel lines to the engine. If you had to remove the injectors use new injector o-rings. Very *lightly* lubricate the o-rings with petroleum jelly, and reinstall them carefully into their ports. Install a new air filter and fuel filter. Mark the mileage and date that this was done in the maintenance log, which should be found in the glove box. Although this may change the maintenance schedule it makes sense to start fresh with a new engine. Reinstall the intake ducting and throttle housing, as applicable **(Figure 12)**.

Figure 11. Look closely to be sure you secure all the connectors in their proper location.

Figure 12. Secure the intake ducting from the air cleaner housing to the throttle bore.

Interesting Fact

A technician was reinstalling a late-model engine into a pickup truck after an overhaul. The engine turned over well but would not start even though it had spark and fuel. He carefully rechecked the vacuum lines and electrical connections. He found that he had forgotten to reconnect the ground strap that is tucked away on the back side of one of the cylinder heads. He was smart enough to double-check his installation work and not panic about his engine overhaul.

Fill the engine with the specified type and quantity of oil. If the engine calls for 5W-30 engine oil, that is what the engine should be broken in with. Raise the engine, and be sure there are no fluid leaks.

FINAL STEPS

Take a step back and think about your work **(Figure 13)**. Try to think of anything you may have forgotten. Take another look at the bottom and top of the engine for anything loose or not connected. Double-check all the fluid levels. Be sure the battery clamps and posts are clean and in good condition **(Figure 14)**. When you are ready reinstall the freshly charged battery, and make the negative terminal connection last. Wash your hands, and install a seat cover on the driver's seat. Make sure there is a floor mat installed on the floor of the vehicle. Do not crank the engine over yet. There are several special steps that must be followed to properly start up and break in a new engine. Read on to Chapter 34 to become familiar with those procedures.

Figure 13. Look closely over the engine compartment to find any missing connectors, hoses, or lines.

Figure 14. Make every step count. Clean a dirty battery terminal to prevent a no start, and make sure the battery is fully charged.

Summary

- Take your time and pay attention to detail as you undertake the job of installing the engine.
- Repair any problems that could cause the customer a return trip to the shop in the near future.
- Replace any rusted hardware, and repair any damaged wiring or connectors.
- Hang the engine from its hooks onto the engine crane; then remove the engine stand mounting flange.
- Slowly lower the engine into place, and mate it with the transmission. Use patience and movement to align the engine and transmission; never force the transmission into place using the mounting bolts.
- Lower the engine onto the mounts, and settle the engine before tightening the bolts. Use the same technique to fasten the engine cradle onto the frame.
- Work underneath the engine to attach the exhaust, starter, hoses, and lower electrical connectors.
- Install an oil pressure gauge into the oil pressure switch port or another appropriate passage.
- If applicable, reassemble the front axles and suspension and steering components. Lower the vehicle and properly torque the axle nuts and wheels.
- Fill the engine, cooling system, and transmission with the proper fluids.
- With the vehicle on the ground reinstall the radiator and cooling fan as needed. Mount any accessories to the engine such as the generator, power steering pump, air pump, and air compressor. Install a new drive belt and tension it properly.
- Connect all vacuum lines, hoses, and electrical connectors. Look over the engine to be sure each component has its vacuum line or electrical connector attached.
- Step back from your work and think about your process. Take a final look around the engine compartment, looking for any unconnected components.

Review Questions

1. List four parts that should be replaced during a rebuilt or replacement engine installation.
2. Is it true or false that you should use the engine crane to keep tension on the engine while tightening the engine mounts?
3. On an automatic transmission torque the flexplate to the __________.
4. What can improper suspension cradle mounting cause?
5. Technician A says that when mounting the transmission onto the engine you will need to use the bell housing bolts to draw the two components together. Technician B says that improper torquing of the bell housing can cause cylinder out of round. Who is correct?
   A. Technician A only
   B. Technician B only
   C. Both Technician A and Technician B
   D. Neither Technician A nor Technician B
6. Each of the following is likely to be connected underneath the vehicle *except* the:
   A. Oxygen sensor
   B. Vehicle speed sensor
   C. Reverse light switch
   D. Accelerator cable
7. Technician A says to reconnect the oil pressure switch wiring before starting the engine. Technician B says to install an oil pressure gauge. Who is correct?
   A. Technician A only
   B. Technician B only
   C. Both Technician A and Technician B
   D. Neither Technician A nor Technician B
8. Technician A says that some electrical connectors on the harness may not be used. Technician B says that careful inspection should show you whether the terminals had been connected. Who is correct?
   A. Technician A only
   B. Technician B only
   C. Both Technician A and Technician B
   D. Neither Technician A nor Technician B
9. Technician A says to replace the fuel injectors. Technician B says to lubricate the new injector o-rings. Who is correct?
   A. Technician A only
   B. Technician B only
   C. Both Technician A and Technician B
   D. Neither Technician A nor Technician B
10. Each of the following is a step you should perform in engine installation *except:*
    A. Lower the engine slowly into position.
    B. Replace any rusted hardware.
    C. Fill the engine with oil, start the engine, and then drain and refill the engine.
    D. Connect the engine and chassis ground wires.

Chapter 34 Engine Startup and Break-In

Introduction

The first twenty minutes of an engine's life is a critical time. Lubrication must be flawless, and coolant must be circulated efficiently. Worn parts have already lost their high spots and sharp edges, but new components must wear in; they need proper lubrication to avoid excessive scuffing. The increased friction of new parts creates higher engine temperatures that only a fully functional cooling system can control. The rings begin to break into the cylinder walls and develop a seal **(Figure 1)**. The whole valvetrain must be properly adjusted and functional to allow the valves to wear into their seats. The bearings will be conforming to the shape of the journals **(Figure 2)**. During the first couple thousand miles of operation the bearings will have some fine metal particles embedded into them as the sharp rings scrape the cylinder walls to seat. The new lifters and perhaps new camshaft will also experience some critical mating during the first half hour of operation.

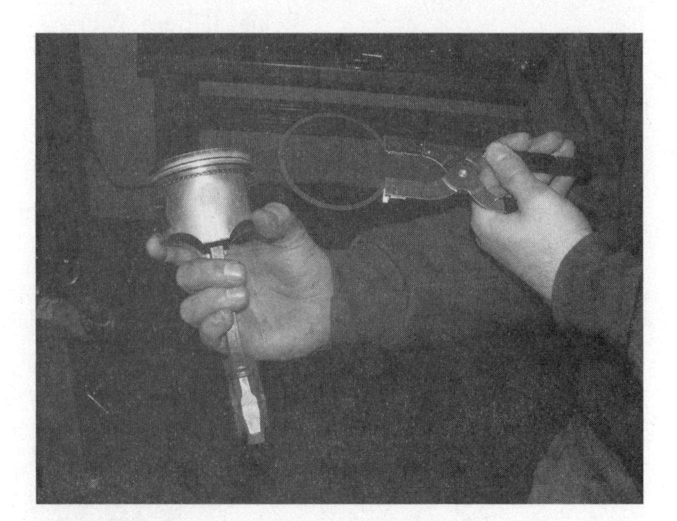

Figure 1. Sharp new rings will scrape against the freshly honed walls, sacrificing metal as they begin to wear in together.

If you are installing a crate engine you may receive specific instructions to follow. Manufacturers may also include guidelines about engine startup and break-in. Camshaft manufacturers may include additional instructions to follow for the break-in process. This chapter gives generalized advice that should apply to startup and break-in for most new or freshly rebuilt engines. We will also discuss the important conversation you should have with the customer before handing over the keys.

The process you use to break in a new engine can affect engine performance and oil consumption during the first few thousand miles.

STARTUP AND INITIAL BREAK-IN

The initial startup and the first twenty minutes of operation are critical to the health of the new or rebuilt engine. Before starting the engine double-check the fluids to be sure they are at the proper level; low coolant could cause engine overheating during the first twenty minutes of operation. A low oil level could be disastrous **(Figure 3)**. The lubrication system should be primed before the engine

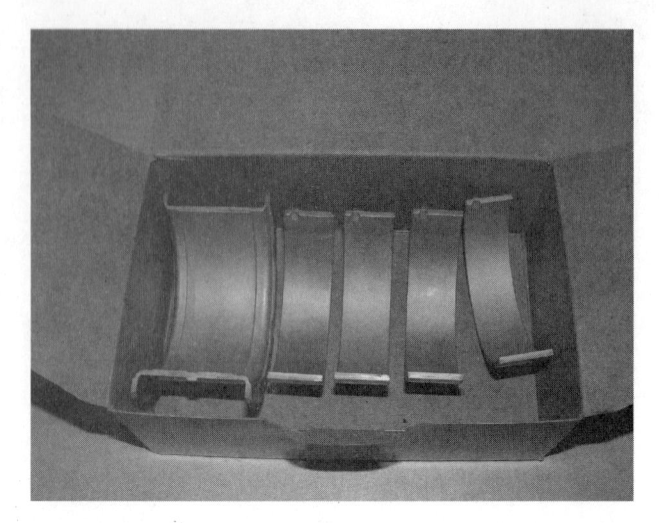

Figure 2. The new bearings will have to wear into the high and low spots on the crankshaft.

Figure 3. Be sure the fluids are properly filled. Even the washer solvent should be full; this one needs topping off.

actually runs, if possible, or the first minutes of operation could cause undue wear on new components. You should be prepared to evaluate the engine operation and sealing during its initial break-in period.

PRIMING THE LUBRICATION SYSTEM

If the oil pump is driven by the crankshaft you will not be able to rotate the pump physically in a way that is effective enough to prime the system. It is best to use an oil priming tool that will deliver oil under pressure to the engine through an adapter on the oil filter housing or in the oil pressure sending unit bore. Some parts stores carry pressurized aerosol priming oil cans that can spray oil throughout the engine via the oil sending unit port. If no priming tool is available start the engine and immediately verify that the engine is generating adequate oil pressure. If it is not turn it off quickly and locate the cause before damaging the engine.

On an older engine that drives the oil pump through the distributor shaft you can physically rotate the shaft to prime the lubrication system. You should already have an oil pressure gauge installed on the engine. If not remove the oil pressure switch and install a gauge now. What you want to do is rotate the oil pump until some oil pressure is seen on the gauge. You will not meet normal oil pressure specifications because you will not be able to spin the pump at a high-enough rpm. You will know that oil has reached all areas of the engine. To rotate the pump use a special drive tool to attach to the drive point of the pump. You may be able to use a one-quarter-inch drive extension, a special tool with the correct flat drive on the end, or an old distributor driveshaft from another engine. Use a drill or a speed handle to rotate the pump until you can feel significant resistance. Continue rotating the pump for a minute or so to be sure that oil has reached the top of the engine.

STARTING THE ENGINE

Set wheel chocks in front of and behind a wheel. If the shift linkage has been disturbed and is not properly adjusted the engine could start in gear and lurch forward. Set the parking brake as an added precaution. Install appropriate exhaust extraction equipment; the engine will smoke while it runs for the first several minutes. Recheck the oil and coolant levels; you should be positive that the engine will have adequate lubrication and cooling. Crank the engine over until it starts or for a maximum of thirty seconds. If everything in and on the engine is properly connected and adjusted it should start within two thirty-second cycles of the starter. If the engine does not start take another look at electrical connectors and vacuum lines **(Figure 4)**. Check the engine to be sure it is getting proper spark and fuel.

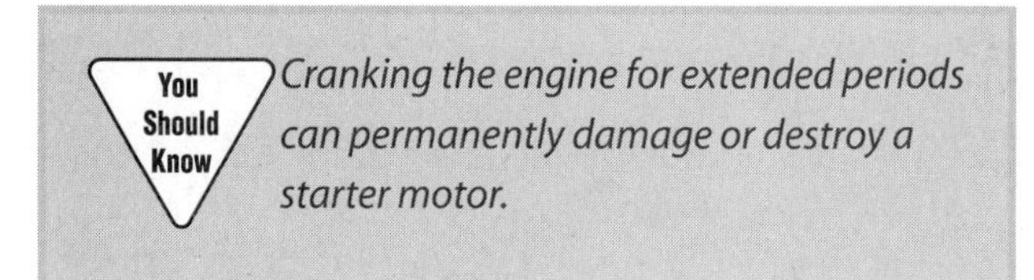

Cranking the engine for extended periods can permanently damage or destroy a starter motor.

Once the engine starts it is important to hold the rpm up to between 1500 and 2000 rpm for the first twenty minutes of operation. Block the throttle open, or use an adjusting rod on the accelerator pedal so that you can make other checks while the engine runs. The increased engine rpm ensures that good oil pressure will lubricate all areas of the engine adequately. Make sure that the oil pressure is within specifications.

The new rings will need good lubrication to prevent scuffing as they break in to the cylinders. A new camshaft

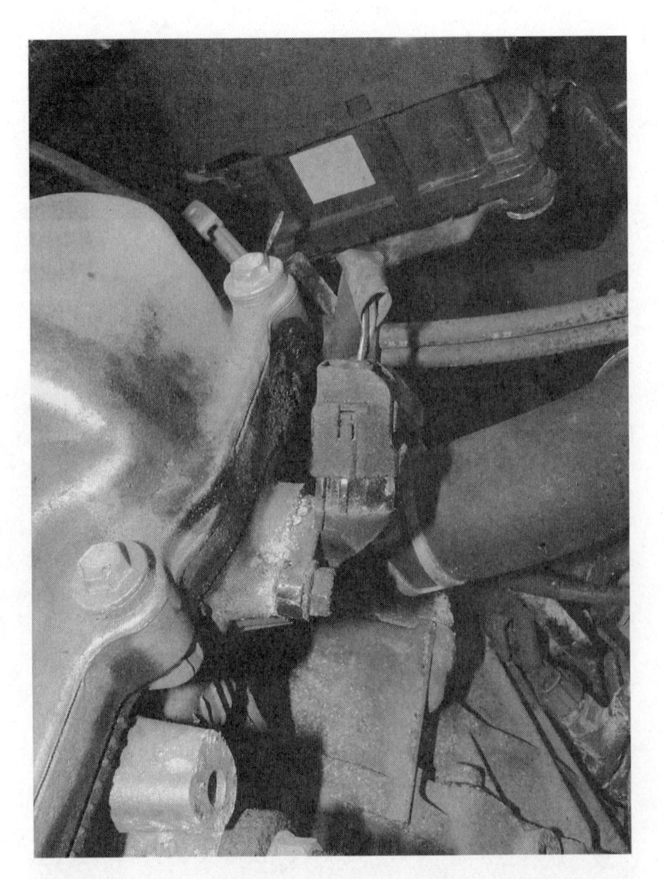

Figure 4. This loose camshaft position sensor connector prevented the engine from starting.

and lifters will endure a lot of friction as they wear in together. Engine bearings will conform to their journals during the first minutes of operation. These areas of increased friction, and others, require the higher oil pressure and flow delivered at the increased engine rpm. Do not let the engine idle.

Monitor the oil pressure, and be sure it stays within the specified psi range **(Figure 5)**. After a few minutes of operation top off the coolant level once it has circulated throughout the engine. Replace the reservoir or radiator cap before the thermostat opens. Check the radiator inlet hose as the engine heats up; it should get hot and full as the engine warms up to temperature. If problems occur with oil pressure or cooling, or if the engine makes unusual noises, shut the engine down and investigate. Prevent extensive engine damage that could occur if you continue to run the engine while obvious problems exist.

While the engine is running at 1500 to 2000 rpm check the engine for leaks. Raise the vehicle, and be sure that no seals, gaskets, or plugs are leaking. If you find leakage stop the engine, and repair the leaks now before damage occurs. It may be frustrating to go backward, but it is a lot better than having to redo major engine work. Listen for unusual noises all over the engine. Use a stethoscope if needed to isolate rattles or belt or bearing issues. Make

Figure 5. Monitor the oil pressure gauge closely throughout the first twenty minutes.

Figure 6. The engine should be running smoothly. Remove all traces of your work. The injector connector tags should be removed.

sure that the engine is running on all its cylinders. The engine should look steady and sound smooth under the hood **(Figure 6)**. The exhaust should be emitted in regular pulses. Listen for vacuum or compression leaks, and correct them if needed.

Continually check the engine oil pressure and cooling system. Verify that the cooling fan comes on when the engine reaches the prescribed temperature. Use a pyrometer **(Figure 7)** for the best accuracy and definitely if there is no coolant temperature gauge to monitor engine temperature. Check for proper charging system operation by placing the leads of your voltmeter across the positive and negative posts of the battery. If the generator, voltage regulator, and wiring are all good it should read between 13.5 and 15.0 volts.

Figure 7. Take readings on the engine near the thermostat housing to get an accurate engine coolant temperature measurement.

Figure 8. Be certain that the oil pressure warning system is functioning properly.

After twenty minutes of high-rpm operation return the engine to its normal idle speed. Set the ignition timing as specified, if applicable. Shut the engine down, and remove the oil pressure gauge. Install the oil pressure switch, and verify proper operation. The light must come on with the key in the RUN position; it should go out promptly after engine startup **(Figure 8)**. Make sure there is no oil leakage by the switch. Check all fluid levels before going on a road test.

ROAD TEST

Prepare yourself for a fifteen- to twenty-minute road test. Turn the radio off; you need to listen closely for rattles, knocks, or squealing. If any abnormal noises occur make a mental note to repair them upon return. On modern engines a knock sensor can pick up vibration from a rattle or knocking, and the PCM will retard the timing and reduce power. If a rattle or knock is left alone engine performance will suffer. If you can hear a noise the customer will also likely pick it up. Figure out the cause, and repair it when you return to the shop. Pull over and stop the engine if a serious knocking noise occurs or if the oil light comes on. Attempting to limp the vehicle back to the shop could destroy your work **(Figure 9)**.

During the road test you need to begin the process of breaking in the rings. Some ring manufacturers say this is no longer necessary, but it is still a generally recommended procedure. Run the vehicle on the highway or open road. Accelerate to 50 mph (as the speed limit allows), and allow engine braking to bring the speed back down to 30 mph. Accelerate aggressively but not to wide-open throttle. You should stay below 75 percent of maximum engine load while still pushing the engine close to that threshold. Repeat this cycle a dozen times. The acceleration will help the rings seal with good pressure around them. The high vacuum during deceleration helps keep the cylinder walls and rings well lubed as the rings break in. Do not over rev the engine or drive at excessive speeds. The smoking from

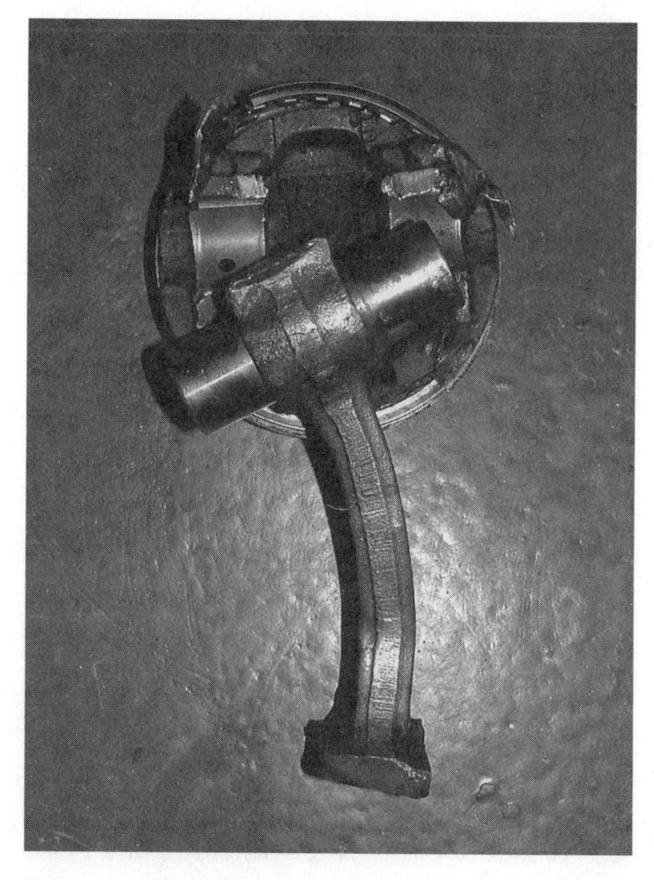

Figure 9. This would be difficult to explain to your boss or a customer eager to have the vehicle back.

the exhaust should noticeably diminish after this process, although a little blue smoke is still normal.

FINAL CHECKS AFTER BREAK-IN

After your road test repair any problems you noticed while you were driving. If you noticed a problem with the brakes, suspension, or exhaust system mark your observations on the repair order so that you can communicate these to the customer when she picks up her vehicle. If the check engine or MIL light came on during your road test extract the DTC from the PCM. If the code relates to a component such as the throttle position sensor or manifold absolute pressure sensor check the wiring connections and vacuum supply to the part, as applicable. Next, diagnose the fault using the manufacturer's procedure.

Give a final check to all the fluid levels. Take the time to put the vehicle back up on the lift, and make sure there are no leaks from the engine. Repair any faults now before the customer can notice them and return angrily to you. A small oil leak could lead to serious engine damage once the customer is driving the vehicle. Take one last look under the hood at wires, hoses, vacuum lines, fuel lines, and belts.

Clean the steering wheel, seat, and fenders. Check the seat for any spots, and use an appropriate cleaner if needed. Put a fresh floor mat in the vehicle. Set the radio station presets to a variety of channels with good reception **(Figure 10)**. Adjust the clock to the proper time. Set the trip odometer to zero so the customer can keep an eye on the mileage.

FIVE-HUNDRED-MILE SERVICE

After five hundred miles of operation on a new or rebuilt engine you should ask the customer to return for a service. Most shops will bill this into the engine overhaul charges so that it appears to be a "free" service to the customer. They are less likely to return if they have to lay out more money shortly after paying the large bill for major engine work. The service work should include an oil change. The oil becomes contaminated quickly with small metal particles as new engine components wear in. Retorquing the intake manifold is occasionally recommended. Cylinder head shifting can affect manifold bolt tightness. Check the adjustment of solid lifters after the valves have had time to wear into their freshly ground or new seats. Verify that the engine is running and idling properly. Finally, road test the vehicle to be sure that it is performing as it should. Take any measures needed to avoid a customer comeback **(Figure 11)**. Repair any concerns that could be related to your engine work. Make a note of other problems, and discuss them with the customer.

Figure 10. The little touches can help build your reputation as one of the finest techs.

Figure 11. You want to avoid this scene; this is very likely a lost customer.

COMMUNICATION WITH THE CUSTOMER

It is best if you can speak directly with the customer when she comes to pick up her vehicle after major work. Be sure you look presentable and are prepared to explain the work you have completed. Describe the problems you found and how you corrected them to ensure proper engine operation. Let her know that you have road tested the vehicle and that you know the engine is running properly. If you noticed any unrelated concerns describe those to the customer so she is not surprised. If an axle boot is cracked or the muffler is leaking let her know before she drives the vehicle. She will be more aware of noises and faults after a major service. Be prepared with a rough estimate of what the additional work would cost.

Let the customer know that the battery was disconnected. Explain that you set the clock and some radio stations, but those and seats or other memory functions may not be exactly as she had them stored previously. Tell her that you set the trip odometer to zero so that she would be aware of when she should return for the five-hundred-mile service. Explain the importance of her return for this oil change and checkup.

Before handing over the keys explain how she should drive the vehicle for the first few hours and then for the next 2000 miles. In the short term she should avoid maximum engine load or extensive idling. The more she varies the engine speed the better it is for the rings. Ask her to idle the engine for thirty seconds before shutting the engine down after a hard drive. During the first 2000 miles of operation the customer should avoid overrevving the engine or driving at excessive speeds. She can drive it aggressively but not to the point of abuse.

Make sure that she is aware that the rings may not fully seat for a few thousand miles. This can cause increased oil consumption, which is normal. Ask her to check and refill the oil and other fluids at every fuel fill while the engine is still breaking in. If she keeps a note about the amount of oil she has had to add you will be able to gauge whether a problem is indicated **(Figure 12)**.

A final step in excellent customer service is to call the customer after a week and make sure the engine is running properly and the customer does not have any concerns. Remind her about the five-hundred-mile service and its importance to the longevity of the engine. Thank her for her business, and let her know you will be looking forward to checking out her vehicle soon. This follow-up call can be an important one. It is hard to overcome the negative statements an unhappy customer might make to others in the community. A satisfied customer is the best advertisement you can get for your shop; take the time to be sure you have done the excellent work you have been trained to perform.

Figure 12. Good communication and customer relations are part of being an excellent technician.

Summary

- The first twenty minutes of operation are critical to the new or rebuilt engine. Keep the engine at 1500 to 2000 rpm to ensure good lubrication.
- New or freshly machined parts will have added friction that requires excellent lubrication and a fully functional cooling system.
- Prime the oil pump by driving it manually or using an oil priming tool. The oil pressure gauge should register pressure before you start the engine. If priming is not possible monitor the oil pressure gauge immediately after startup to verify proper oil pressure.
- Double-check the oil level and start the engine. Block the throttle open so that the rpm stays between 1500 and 2000 rpm.
- While the engine is running for its first twenty minutes check for leaks, noises, proper thermostat and fan operation, and adequate charging voltage.
- Monitor the oil pressure and coolant temperature; shut down the engine if problems occur.
- Road test the vehicle to help seat the rings. Accelerate moderately to 50 mph, and let the engine decelerate down to 30 mph. Repeat this cycle a dozen times.
- Listen for unusual noises, and monitor the gauges carefully while driving. Make a mental note of any concerns so you can repair them back at the shop.
- After your road test, repair any related problems you noticed. Top off all the fluid levels as needed. Write down other concerns, and prepare a rough estimate for the work needed to resolve them.
- Before returning the vehicle to the customer be sure the interior and exterior are clean. Reset the clock, radio, and any other memory functions that you can.
- When the vehicle returns for the five-hundred-mile service change the oil and filter. You may also need to retorque the intake manifold and adjust the valve clearances. Check for any fluid leaks.
- Good communication with the customer is essential. Explain your work and provide guidance for how to operate the vehicle during the break-in period.
- Follow up with a call to the customer about a week later. Be sure she is satisfied with the engine, and remind her about the importance of the five-hundred-mile service.

Review Questions

1. When first starting a new or rebuilt engine you should hold the rpm up between __________ and __________ rpm in order to ensure __________.
2. Describe the driving procedure typically used to help break in the new rings.
3. What problems can a rattle or knock cause if left unrepaired?
4. The customer should return the vehicle after __________ miles so you can __________.
5. Technician A says that new parts will endure higher friction during break-in. Technician B says that the cooling system will have to control higher than normal engine temperatures. Who is correct?
   A. Technician A only
   B. Technician B only
   C. Both Technician A and Technician B
   D. Neither Technician A nor Technician B
6. Technician A says to start the engine and look for oil pressure; if it is low shut the engine down. Technician B says to increase the engine speed to 3500 rpm immediately. Who is correct?
   A. Technician A only
   B. Technician B only
   C. Both Technician A and Technician B
   D. Neither Technician A nor Technician B
7. Each of the following tasks should be performed during the first fifteen minutes of operation *except:*
   A. Check the engine for fluid leaks.
   B. Monitor the coolant temperature and oil pressure.
   C. Check the charging system voltage.
   D. Adjust the idle speed.
8. Technician A says to remove the oil pressure gauge and install the pressure switch before the road test. Technician B says to set the radio stations while you are checking the drivability on the road test. Who is correct?
   A. Technician A only
   B. Technician B only
   C. Both Technician A and Technician B
   D. Neither Technician A nor Technician B
9. Technician A says the coolant and thermostat should be changed after 2000 miles of break-in. Technician B says the customer should return at five hundred miles for an oil change and inspection. Who is correct?
   A. Technician A only
   B. Technician B only
   C. Both Technician A and Technician B
   D. Neither Technician A nor Technician B
10. Technician A says that it is important to communicate to the customer how she should drive the vehicle for the first few hours and the first few thousand miles. Technician B says to warn the customer that the engine may consume some oil during the break-in process. Who is correct?
    A. Technician A only
    B. Technician B only
    C. Both Technician A and Technician B
    D. Neither Technician A nor Technician B

Appendix A

ASE PRACTICE EXAM FOR ENGINE REPAIR

1. Technician A says that you should know the location of the eye wash station. Technician B says that you should wear work gloves when grinding or cleaning on a wire wheel. Who is correct?
   A. Technician A only
   B. Technician B only
   C. Both Technician A and Technician B
   D. Neither Technician A nor Technician B
2. You can use each of the following tools to evaluate an engine's mechanical condition *except* a:
   A. Brush hone
   B. Compression tester
   C. Vacuum gauge
   D. Stethoscope
3. Technician A says to use a micrometer to measure a component accurately to .0001 in. Technician B says that an error in measurement of .001 in. during an engine overhaul could cause serious problems. Who is correct?
   A. Technician A only
   B. Technician B only
   C. Both Technician A and Technician B
   D. Neither Technician A nor Technician B
4. Technician A says to torque bolts in a circular pattern around the pan or component. Technician B says that she has learned to gauge proper torque with an ordinary ratchet. Who is correct?
   A. Technician A only
   B. Technician B only
   C. Both Technician A and Technician B
   D. Neither Technician A nor Technician B
5. Technician A says that you may find both metric and standard fasteners on the same vehicle. Technician B says that you can use standard wrenches on either type of bolt. Who is correct?
   A. Technician A only
   B. Technician B only
   C. Both Technician A and Technician B
   D. Neither Technician A nor Technician B
6. Technician A says that good communication skills can make the difference between a good technician and an excellent technician. Technician B says that most of the diagnostic and repair skills needed to repair today's cars are learned on the job. Who is correct?
   A. Technician A only
   B. Technician B only
   C. Both Technician A and Technician B
   D. Neither Technician A nor Technician B
7. Technician A says that a bigger engine makes more power than a smaller one. Technician B says that the more air an engine can take in and exhaust, the more power it can make. Who is correct?
   A. Technician A only
   B. Technician B only
   C. Both Technician A and Technician B
   D. Neither Technician A nor Technician B
8. Technician A says that the intake and exhaust valves should be closed when the spark plug fires. Technician B says that both valves are open at the end of the exhaust stroke. Who is correct?
   A. Technician A only
   B. Technician B only
   C. Both Technician A and Technician B
   D. Neither Technician A nor Technician B

9. Technician A says that the camshaft rotation must be perfectly timed with the crankshaft rotation. Technician B says that a band connects the camshaft and the crankshaft. Who is correct?
   A. Technician A only
   B. Technician B only
   C. Both Technician A and Technician B
   D. Neither Technician A nor Technician B
10. A V8 engine has a 3.75-in. bore and a 3.85-in. stroke. The engine displacement is:
    A. 91 c.i.d.
    B. 170 c.i.d.
    C. 340 c.i.d.
    D. 433 c.i.d.
11. Technician A says that increasing an engine's stroke will enable the engine to produce more torque. Technician B says that modifications to increase volumetric efficiency decrease horsepower. Who is correct?
    A. Technician A only
    B. Technician B only
    C. Both Technician A and Technician B
    D. Neither Technician A nor Technician B
12. Technician A says that a higher-octane fuel resists knocking better than a lower-octane fuel. Technician B says that using a higher-octane fuel may improve engine performance on a modern vehicle. Who is correct?
    A. Technician A only
    B. Technician B only
    C. Both Technician A and Technician B
    D. Neither Technician A nor Technician B
13. Each of the following is a toxic emission from an IC engine *except:*
    A. CO
    B. HC
    C. NOx
    D. N_2
14. Technician A says that the oil filter should be changed at every other oil change. Technician B says that when the oil pump pressure relief valve sticks open the engine will lose oil pressure. Who is correct?
    A. Technician A only
    B. Technician B only
    C. Both Technician A and Technician B
    D. Neither Technician A nor Technician B
15. Technician A says that if the engine calls for a 10W-40 oil you should use a 10W-50 oil in the summer to keep the engine cooler. Technician B says that if the oil does not show the API "donut" you should not use the oil. Who is correct?
    A. Technician A only
    B. Technician B only
    C. Both Technician A and Technician B
    D. Neither Technician A nor Technician B
16. Technician A says to check oil pressure with the engine at NOT. Technician B says to put dye in the oil and use a black light to help locate the source of a leak if it is not obvious. Who is correct?
    A. Technician A only
    B. Technician B only
    C. Both Technician A and Technician B
    D. Neither Technician A nor Technician B
17. Technician A says to mix coolant with water in at least a 50/50 concentration. Technician B says that in very cold climates you should use straight coolant. Who is correct?
    A. Technician A only
    B. Technician B only
    C. Both Technician A and Technician B
    D. Neither Technician A nor Technician B
18. An engine overheats only at idle. Technician A says the thermostat should be changed. Technician B says the lower radiator hose could be collapsed. Who is correct?
    A. Technician A only
    B. Technician B only
    C. Both Technician A and Technician B
    D. Neither Technician A nor Technician B
19. Technician A says that when a head gasket is leaking you will probably notice blue smoke coming from the tailpipe. Technician B says to check the cylinder head flatness when replacing a blown head gasket. Who is correct?
    A. Technician A only
    B. Technician B only
    C. Both Technician A and Technician B
    D. Neither Technician A nor Technician B
20. Technician A says that if a variable intake system malfunctions it may cause power loss at higher rpms. Technician B says that a restricted exhaust system will cause a lack of engine power. Who is correct?
    A. Technician A only
    B. Technician B only
    C. Both Technician A and Technician B
    D. Neither Technician A nor Technician B
21. Technician A says that you must change the oil more frequently on turbocharged engines. Technician B says that if a supercharger bypass valve fails detonation may occur. Who is correct?
    A. Technician A only
    B. Technician B only
    C. Both Technician A and Technician B
    D. Neither Technician A nor Technician B
22. A fine-running engine should develop ________ vacuum at idle.
    A. 10 in. Hg
    B. 14 in. Hg
    C. 18 in. Hg
    D. 24 in. Hg

23. An engine has a steady vacuum reading at idle, but at 2500 rpm the needle fluctuates rapidly. Technician A says the valve guides are probably worn. Technician B says the valve springs may be weak. Who is correct?
    A. Technician A only
    B. Technician B only
    C. Both Technician A and Technician B
    D. Neither Technician A nor Technician B
24. Technician A says that platinum plugs will show a worn, collapsed electrode if they have a lot of miles on them. Technician B says that a plug with heavy tan deposits indicates that the engine is burning oil. Who is correct?
    A. Technician A only
    B. Technician B only
    C. Both Technician A and Technician B
    D. Neither Technician A nor Technician B
25. An engine has cranking compression readings of 150 psi, 155 psi, 160 psi, and 75 psi. A wet test is performed on the low cylinder, and the compression increases to 95 psi. The most likely cause is:
    A. A blown head gasket
    B. Worn rings
    C. A bent pushrod
    D. A burned valve
26. An engine shows 45 percent leakage during a cylinder leakage test. Air is heard coming from the oil dipstick tube. Technician A says an exhaust valve is probably burned. Technician B says the rings are likely worn. Who is correct?
    A. Technician A only
    B. Technician B only
    C. Both Technician A and Technician B
    D. Neither Technician A nor Technician B
27. Technician A says that black smoke is caused by a rich fuel mixture. Technician B says that a blown head gasket may produce white smoke from the tailpipe. Who is correct?
    A. Technician A only
    B. Technician B only
    C. Both Technician A and Technician B
    D. Neither Technician A nor Technician B
28. Technician A says that loose flywheel bolts can cause an engine knocking noise. Technician B says that weak lifters will produce valvetrain clatter. Who is correct?
    A. Technician A only
    B. Technician B only
    C. Both Technician A and Technician B
    D. Neither Technician A nor Technician B
29. Technician A says that some engines must be removed with the transaxle attached. Technician B says that to remove an engine on a front-wheel-drive vehicle you may need to remove the front axles. Who is correct?
    A. Technician A only
    B. Technician B only
    C. Both Technician A and Technician B
    D. Neither Technician A nor Technician B
30. Technician A says that it is best to loosen cylinder head bolts when the engine is warm. Technician B says to use a coffee can or similar bin to store all the engine hardware as it comes off. Who is correct?
    A. Technician A only
    B. Technician B only
    C. Both Technician A and Technician B
    D. Neither Technician A nor Technician B
31. Technician A says that you should remove the main caps to remove the pistons from the engine. Technician B says you may have to mark the main and/or rod caps before disassembly. Who is correct?
    A. Technician A only
    B. Technician B only
    C. Both Technician A and Technician B
    D. Neither Technician A nor Technician B
32. Technician A says that pistons are usually removed through the bottom of the engine. Technician B says that you may have to remove the ring ridge after removing the pistons. Who is correct?
    A. Technician A only
    B. Technician B only
    C. Both Technician A and Technician B
    D. Neither Technician A nor Technician B
33. Technician A says that you should replace a cast crankshaft with a forged crankshaft if you add a turbocharger to the engine. Technician B says that you may need to install domed pistons. Who is correct?
    A. Technician A only
    B. Technician B only
    C. Both Technician A and Technician B
    D. Neither Technician A nor Technician B
34. Technician A says that too much cylinder out of round may cause oil consumption. Technician B says that too much cylinder taper can cause the rings to break. Who is correct?
    A. Technician A only
    B. Technician B only
    C. Both Technician A and Technician B
    D. Neither Technician A nor Technician B
35. Technician A says that you should measure the main bore alignment with a telescoping gauge and a micrometer. Technician B says you should replace the main bearings during an overhaul. Who is correct?
    A. Technician A only
    B. Technician B only
    C. Both Technician A and Technician B
    D. Neither Technician A nor Technician B

36. Technician A says that you measure ring end gap with the new rings on the pistons. Technician B says you use a feeler gauge to measure ring end gap and ring side clearance. Who is correct?
    A. Technician A only
    B. Technician B only
    C. Both Technician A and Technician B
    D. Neither Technician A nor Technician B

37. Technician A says that too much crankshaft endplay can cause a knocking on acceleration. Technician B says that crankshaft endplay is usually checked using plastigauge. Who is correct?
    A. Technician A only
    B. Technician B only
    C. Both Technician A and Technician B
    D. Neither Technician A nor Technician B

38. Technician A says that on a modern engine if two cylinders have .010 in. of taper all the cylinders should be bored. Technician B says that a brush hone will take care of up to .005 in. of out of round in the cylinders. Who is correct?
    A. Technician A only
    B. Technician B only
    C. Both Technician A and Technician B
    D. Neither Technician A nor Technician B

39. Technician A says that if the rocker arm is worn the valve is likely to burn. Technician B says the exhaust valve face must be about $^{1}/_{64}$ in. wide to prevent burning. Who is correct?
    A. Technician A only
    B. Technician B only
    C. Both Technician A and Technician B
    D. Neither Technician A nor Technician B

40. Technician A says that a cylinder head should be checked for cracks using magnaflux, dye penetrant, or pressure testing if it was overheated. Technician B says to look closely between the valve seats for cracks. Who is correct?
    A. Technician A only
    B. Technician B only
    C. Both Technician A and Technician B
    D. Neither Technician A nor Technician B

41. Technician A says that most valve seats are cut at a 30° angle. Technician B says that the valve face is typically cut at about 45°. Who is correct?
    A. Technician A only
    B. Technician B only
    C. Both Technician A and Technician B
    D. Neither Technician A nor Technician B

42. Technician A says that a timing belt may require replacement as part of a routine maintenance operation. Technician B says that you should replace timing chains every 60,000 miles. Who is correct?
    A. Technician A only
    B. Technician B only
    C. Both Technician A and Technician B
    D. Neither Technician A nor Technician B

43. The timing belt has jumped two teeth on the camshaft. Technician A says the engine will likely run roughly if at all. Technician B says the engine compression will be lower. Who is correct?
    A. Technician A only
    B. Technician B only
    C. Both Technician A and Technician B
    D. Neither Technician A nor Technician B

44. Technician A says that improperly timed balance shafts will cause lower power. Technician B says the timing marks must be within two teeth of their markings to provide good engine performance. Who is correct?
    A. Technician A only
    B. Technician B only
    C. Both Technician A and Technician B
    D. Neither Technician A nor Technician B

45. Technician A says that a snapped timing belt may cause the valves to knock against the pistons. Technician B says that an engine with a snapped timing belt or chain will crank over faster than normal. Who is correct?
    A. Technician A only
    B. Technician B only
    C. Both Technician A and Technician B
    D. Neither Technician A nor Technician B

46. Technician A says that most hydraulic lifters require adjustment during routine maintenance. Technician B says that to adjust some valves you have to fit the correct size shim into the follower. Who is correct?
    A. Technician A only
    B. Technician B only
    C. Both Technician A and Technician B
    D. Neither Technician A nor Technician B

47. Technician A says that if the valves' adjustment is too loose they are likely to burn. Technician B says that if the valves' adjustment is too tight the engine may run roughly and backfire. Who is correct?
    A. Technician A only
    B. Technician B only
    C. Both Technician A and Technician B
    D. Neither Technician A nor Technician B

48. Technician A says that a new oil pump should be sealed with a $^{1}/_{2}$-in. bead of RTV silicone. Technician B says that the oil pump should be packed with chassis grease to keep it lubricated. Who is correct?
    A. Technician A only
    B. Technician B only
    C. Both Technician A and Technician B
    D. Neither Technician A nor Technician B

49. Technician A says it is a good idea to replace the spark plugs, air filter, and fuel filter after an engine replacement. Technician B says to replace the generator after an engine overhaul. Who is correct?
    A. Technician A only
    B. Technician B only
    C. Both Technician A and Technician B
    D. Neither Technician A nor Technician B

50. Technician A says that you should change the coolant after five hundred miles are put on a freshly rebuilt engine. Technician B says that you should adjust the timing chain at five hundred miles. Who is correct?
    A. Technician A only
    B. Technician B only
    C. Both Technician A and Technician B
    D. Neither Technician A nor Technician B

Appendix B

METRIC CONVERSIONS

	To convert these	to these,	multiply by:
TEMPERATURE	Degrees Celsius	Degrees Fahrenheit	1.8 then + 32
	Degrees Fahrenheit	Degrees Celsius	0.556 after – 32
LENGTH	Millimeters	Inches	0.03937
	Inches	Millimeters	25.4
	Meters	Feet	3.28084
	Feet	Meters	0.3048
	Kilometers	Miles	0.62137
	Miles	Kilometers	1.60935
AREA	Square Centimeters	Square Inches	0.155
	Square Inches	Square Centimeters	6.45159
VOLUME	Cubic Centimeters	Cubic Inches	0.06103
	Cubic Inches	Cubic Centimeters	16.38703
	Cubic Centimeters	Liters	0.001
	Liters	Cubic Centimeters	1000
	Liters	Cubic Inches	61.025
	Cubic Inches	Liters	0.01639
	Liters	Quarts	1.05672
	Quarts	Liters	0.94633
	Liters	Pints	2.11344
	Pints	Liters	0.47317
	Liters	Ounces	33.81497
	Ounces	Liters	0.02957
	Milliliters	Ounces	0.3381497
	Ounces	Milliliters	29.57
WEIGHT	Grams	Ounces	0.03527
	Ounces	Grams	28.34953
	Kilograms	Pounds	2.20462
	Pounds	Kilograms	0.45359
WORK	Centimeter-Kilograms	Inch-Pounds	0.8676
	Inch-Pounds	Centimeter-Kilograms	1.15262
	Meter-Kilograms	Foot-Pounds	7.23301
	Foot-Pounds	Newton-Meters	1.3558

	To convert these	to these,	multiply by:
PRESSURE	Kilograms/Square Centimeter	Pounds/Square Inch	14.22334
	Pounds/Square Inch	Kilograms/Square Centimeter	0.07031
	Bar	Pounds/Square Inch	14.504
	Pounds/Square Inch	Bar	0.0689
	Pounds/Square Inch	Kilopascals	6.895
	Kilopascals	Pounds/Square Inch	0.145

Appendix C

SPECIAL TOOL SUPPLIERS

Ajax Lifting Equipment, Roseville, Mich.
Bear Automotive Service Equipment Co., Milwaukee, Wis.
Blackhawk Automotive Inc., Waukesha, Wis.
Federal Mogul Corporation, St. Louis, Mo.
Fluke Corporation, Everett, Wash.
KD Tools Danaher Tool Group, Lancaster, Pa.
K-Line Industries Inc., Holland, Mich.
Lisle Corp., Clarinda, Iowa
Mac Tools, Washington Courthouse, Ohio
Magnaflux Corp., Chicago, Ill.
Mitchell International, San Diego, Calif.
OTC Division, SPX Corp., Owatonna, Minn.
Sealed Power Corp., Muskegon, Mich.
Snap-on Tools Inc., Kenosha, Wis.
Sioux Tools Inc., Sioux City, Iowa
Sunnen Products Co., St. Louis, Mo.
Sun Test Equipment, Division of Snap-on Tools, Kenosha, Wis.
Winona–Van Norman Machine Co., Winona, Wis.

Bilingual Glossary

Air Bag System (SRS) A safety restraint system using inflatable bags to keep the passenger from hitting the dashboard or the driver from hitting the steering wheel.
Sistema de bolsa de aire (SBA) *Sistema de restricción que utiliza bolsas inflables para prevenir que el pasajero se golpee en el tablero o que el conductor se golpee en el volante.*

Air Conditioner Compressor A pump that cycles refrigerant throughout the air conditioning system.
Compresor del aire acondicionado *Bomba que hace circular el refrigerante por todo el sistema de aire acondicionado.*

American Petroleum Institute (API) The API rates the quality of motor oil and upgrades standards periodically to designate current oils with the highest rating.
Instituto Americano del Petróleo (IPA) *El IPA designa el grado de calidad del aceite para el motor y mejora los estándares periódicamente para designar los aceites actualizados con el máximo rendimiento.*

Anaerobic Sealant A sealant that will only cure in the absence of oxygen and takes up no space.
Sellador anaeróbico *Sellador que sólo cura si hay ausencia de oxígeno y que no ocupa espacio.*

Axle A shaft that connects the drive train to the wheel hub.
Eje *Árbol que conecta el tren de accionamiento al cubo de la rueda.*

Backlash The clearance between two gears.
Contrapresión *Espacio entre dos engranajes.*

Balance Shaft A shaft in the engine that rotates to reduce vibrations in the engine.
Eje de balance *Eje en un motor que da vueltas para reducir las vibraciones de la máquina.*

Ball Joint A flexible pivot to allow for turning a front wheel to the right or to the left and arcing up and down as the wheel encounters bumps.
Cojinete *Pivote flexible que permite dar vueltas a la llanta delantera hacia la derecha o la izquierda y arquearse hacia arriba o hacia abajo cuando la llanta se encuentra con topes.*

Bearing Crush The extension of the bearing half beyond the seat that is crushed into place when the bearing cap is tightened.
Agolpamiento del cojinete *Extensión del balero a la mitad del asiento que se agolpa en su lugar cuando se aprieta la tapa del balero.*

Bearing Inserts An interchangeable type of bearing. The engine bearings are a set of steel backed half rounds with a soft face to protect the shaft that rides in them.
Encartes de cojinete *Un tipo de cojinete intercambiable. Los cojinetes del motor son un juego de media rueda con soporte de acero con cara suave para proteger el árbol montado sobre los mismos.*

Bearing Spread The distance between the outside parting edges is larger than the diameter of the bore.
Extensión del cojinete *La distancia entre las dos orillas de separación es mayor que el diámetro de la barrena.*

Belt Tension Gauge Used to measure drive belt tension. The belt tension gauge is installed over the belt and indicates the amount of belt tension.
Manómetro de tensión de la banda *Se usa para medir la tensión de la banda de la trasmisión. El manómetro de tensión de la banda se instala sobre la banda e indica la cantidad de tensión que hay en ella.*

Bleeder Valve A valve that opens to allow removal of air from a hydraulic system.
Válvula purgadora *Válvula que se abre para permitir la salida del aire del sistema hidráulico.*

Blow-by The unburned fuel and combustion by-products that leak past the piston rings and enter the crankcase.
Soplado *Subproductos de combustión y de carburantes no quemados que escapan por los anillos del pistón y que entran en el cárter.*

Bolt Grade Tensile Strength of a bolt. There is a standard and metric classification of the tensile strength of bolts.
Tensor graduado *De la resistencia de un perno. Existe una clasificación métrica y estándar del tensor de la resistencia de los pernos.*

Bolt Head Top of a bolt that fits a wrench or socket to apply torque to the bolt.
Cabeza del perno *Parte del perno sobre la cual se ajusta la llave o el casquillo para aplicar torsión o apretar el perno.*

Bolt Shank The long shaft of the bolt excluding the head, the threaded portion.
Mango del perno *Parte larga y enroscada del perno sin la cabeza; la porción enroscada.*

Boost Control Solenoid A solenoid used on a supercharging system typically controlled by the PCM to maintain the boost pressure at a safe level.
Solenoide de control de impulso *Solenoide que se usa en un sistema de supercarga típicamente controlado por el MCM para mantener la presión de impulso a un nivel seguro.*

Boost Pressure A positive pressure created by a turbocharger.
Presión de impulso *Presión positiva creada por un turbosobrealimentador.*

Boring The process of enlarging a hole. An engine cylinder is bored oversize when it is too worn to properly fit original pistons.
Perforación o mandrinado *Proceso para agrandar un agujero. Se perfora el cilindro de un motor para agrandarlo cuando está demasiado gastado para que acomode a los pistones originales.*

Bottom Dead Center (BDC) A term used to indicate that the piston is at the very bottom of its stroke.
Punto muerto inferior (PMI) *Término usado para indicar que el pistón está al fondo de su golpe.*

British Thermal Units (BTUs) A standardized measure of the potential energy available in a fuel.
Unidades térmicas inglesas (UTI) *Medida estandarizada de la energía de potencia de un combustible.*

Bypass Valve An oil filter safety feature to prevent engine failure. The valve opens when there is a pressure differential of 5 to 15 psi between the outside and inside of the oil filter element.
Válvula de desvío *Filtro de aceite como característica de seguridad para prevenir que falle el motor. La válvula se abre cuando hay una presión diferencial de 5 a 15 psi entre el elemento exterior e interior del filtro de aceite.*

Corporate Average Fuel Economy (CAFE) CAFE standards dictate the average fuel economy of a manufacturer's product line.
Corporación para promediar la economía del combustible (CPEC) *Los lineamientos de la CPEC dictan la economía promedio de combustible de la línea de productos de un fabricante.*

Cam Bearing Driver A tool used to install camshaft bearings in the block of an engine.
Rueda de cojinete de leva *Herramienta usada para instalar los cojinetes del árbol de levas en un bloque del motor.*

Cam Ground Machined into a slightly oval shape to allow for expansion. As the piston warms up it will become round to fit the cylinder better.
Levas aplastado por máquina *en forma de casi un óvalo para permitir que se expanda. Mientras se va calentando el pistón se pone redondo para quedar mejor en el cilindro.*

Camshaft The shaft, driven by the crankshaft, containing lobes to operate the engine valves.
Árbol de levas *El eje, movido por el cigüeñal, contiene lóbulos para manejar las válvulas del motor.*

Carbon Dioxide (CO_2) Non-flammable, colorless, odorless gas that is produced by an engine during combustion.
Bióxido de carbono (CO_2) *Gas inflamable, sin color y sin olor que produce el motor durante la combustión.*

Carbon Monoxide (CO) An odorless, colorless, and toxic gas that is produced as a result of incomplete combustion.
Monóxido de carbono (CO) *Gas sin olor, sin color y tóxico que se produce como resultado de una combustión no apropiada.*

Catalytic Converter An emission control device in the exhaust system that converts toxic HC, CO and NOx gases into harmless H_2O and CO_2.
Transformador catalítico *Aparato de control de emisión en el sistema de escape que convierte los gases tóxicos de HC, CO y NOx en inofensivos H_2O y CO_2.*

Center Punch A special tool used to center a drill bit, to mark components, or to remove roll pins.
Punzón para marcas *Herramienta especial que se usa para centrar la marca de la broca, para marcar componentes o para quitar horquillas.*

Clean Air Act of 1990 An important piece of legislation that set many restrictions on vehicular emissions and dictated the OBD II diagnostic system.
Decreto del Aire Limpio de 1990 *Importante ley del gobierno para sentar muchas restricciones en emisiones vehiculares y para dictar el sistema de diagnóstico OBD II.*

Clutch A device that connects the engine with the power train. This device also makes engaging and disengaging of the engine and power train possible with the driver's foot to facilitate shifting transmission gears.
Cloche, embrague *Pieza que conecta el motoral manejo del motaro. Esta pieza también permite embragar y desembragar el motor con el pie del conductor para facilitar el cambio de mecanismos de la trasmisión.*

Combustion The process of a controlled burn.
Combustión *Proceso controlado de consumo.*

Combustion Chamber The volume of the cylinder above the piston with the piston at TDC.
Cámara de combustión *Volumen del cilindro encima del pistón con el pistón en su PMS.*

Compression Pressure The reduction in volume of gas. Compression pressure is formed in the cylinder when the piston moves from BDC to TDC with the valves closed.
Presión de compresión *Reducción del volumen de un gas. La presión de compresión se forma en el cilindro cuando el pistón se mueve de su PMI a su PMS con las válvulas cerradas.*

Compression Ratio A comparison between the volume above the piston at BDC and the volume above the piston at TDC.
Relación de compresión *Comparación entre el volumen encima del pistón a su PMI y el volumen encima del pistón en su PMS.*

Compression Tester A tool used to determine the engine's ability to seal the combustion chamber.
Probador de compresión *Herramienta usada para determinar la habilidad del motor para sellar la cámara de combustión.*

Conformability The ability of the bearing material to conform itself to slight irregularities of a rotating shaft.
Conformidad *Habilidad del material de los cojinetes para conformarse a las pequeñas irregularidades de un eje de rotación.*

Connecting Rod The link between the piston and crankshaft.
Varilla de mando *Enlace entre el pistón y el cigüeñal.*

Connecting Rod Side Clearance The clearance between two rods on the same crankshaft journal. Some clearance is needed to prevent binding between the connecting rods.
Juego transversal de la varilla de mando *Espacio libre entre dos varillas en el mismo extremo del cigüeñal. Se necesita algo de juego o espacio para prevenir que se doblen los ejes de conexión.*

Coolant Hydrometer A tool that compares the weight of water with the weight of the coolant mixture. This indicates the coolant strength and antifreeze protection level in the cooling system.
Hidrómetro de enfriamiento *Herramienta que compara el peso del agua con el peso de la mezcla del refrigerante. Esto indica la potencia del enfriador y el nivel de protección del anticongelante en el sistema de refrigeración.*

Cooling System Pressure Tester A tester used to put pressure on the cooling system to make sure that there are no leaks in the system or to help find the source of a leak.
Probador de presión en el sistema de enfriamiento *Probador que se usa para poner presión en el sistema de enfriamiento para asegurarse que no haya goteo en el sistema o para encontrar la fuente de goteo.*

Core Plugs Metal plugs screwed or pressed into the block or cylinder head at locations where drilling was required or sand cores were removed during casting.
Mandriles o espigas de soplado *Mandriles metálicos atornillados o presionados en el bloque o cabeza de cilindro en lugares que se necesitaba barrenar o en donde se quitaron machos de arena durante el moldeo.*

Crankshaft A mechanical shaft that converts the reciprocating motion of the pistons into rotary motion.
Cigüeñal *Eje mecánico que convierte el movimiento recíproco de los pistones en movimiento rotativo.*

Crankshaft Endplay The measure of how far the crankshaft can move lengthwise in the block.
Juego axial del cigüeñal *Medida del movimiento a lo largo del bloque que puede hacer el cigüeñal.*

Crosshatch A criss-cross pattern left on the cylinder walls after a cylinder is honed or deglazed. The angle of the crosshatches is important for oil distribution along the cylinder wall.
Cuadrícula *Patrón cruzado que queda en las paredes del cilindro después de que el cilindro se afila o se desbarniza. El ángulo de los entrecruces es importante para la distribución del aceite en la pared del cilindro.*

Cupping A deformation of the valve head caused by excessive heat and combustion pressures.
Acopación *Deformación de la cabeza de la válvula causada por presiones excesivas de calor y combustión.*

Cylinder Bore A long stem that fits into a drill with multiple flexible arms with abrasive balls on the end. The balls form the circumference of the tool and are fitted into the cylinder.
Barrena del cilindro *Tubo largo que queda dentro de la broca y tiene muchos brazos flexibles con pelotas abrasivas en la punta. Las pelotas forman la circunferencia de la herramienta y se meten en el cilindro.*

Cylinder Brush A drill with abrasive balls held on flexible shafts outward from the drill shaft used to deglaze a cylinder wall.
Cepillo del cilindro *Broca con pelotas abrasivas detenidas por ejes flexibles que salen del eje de la broca y que se usa para desbarnizar la pared de un cilindro.*

Cylinder Head On most engines, the cylinder head contains the valves, valve seats, valve guides, and valve springs, and forms the upper portion of the combustion chamber.
Cabeza del cilindro *En la mayoría de los motores, la cabeza del cilindro contiene válvulas, asientos de válvulas, guías de válvulas, y resortes de válvulas, y forma la parte superior de la cámara de combustión.*

Cylinder Head Gasket A gasket used to prevent compression pressures, gases, and fluids from leaking from cylinder to cylinder or externally. It is located between the cylinder head and the engine block.
Junta de la cabeza del cilindro *La junta se usa para prevenir que las presiones de la compresión, los gases y los líquidos goteen de cilindro a cilindro o al exterior. Se sitúa entre la cabeza del cilindro y el bloque motor.*

Cylinder Hone An abrasive tool used to smooth out worn cylinders with ridges and grooves.
Rectificador del cilindro *Herramienta abrasiva que se usa para darles acabado a los cilindros gastados que tienen estrías y rayas.*

Cylinder Leakage Tester A tool that determines the sealing ability of the piston ring, intake or exhaust valve, and head gasket. It uses a controlled amount of air pressure to determine the amount of leakage past the different components.
Probador de goteo del cilindro *Herramienta que usa una cantidad controlada de presión de aire para determinar la cantidad de goteo que pasa los segmentos del pistón, las válvulas de admisión y las de escape y la junta de la cabeza del cilindro.*

Deployed When an air bag has inflated it has been deployed.
Desplegada *Cuando una bolsa de aire se infla, se ha desplegado.*

Detonation Improper and dangerous combustion that occurs if the air/fuel mixture in the cylinder is burned too fast.
Detonación *Combustión inapropiada y peligrosa que sucede si la mezcla de aire con combustible se quema muy rápido en el cilindro.*

Diagnostic Trouble Codes (DTCs) Codes that an on-board automotive computer will set when it has detected a problem with a system or component. These can be retrieved using a scan tool to assist the technician in locating the problem. When the computer sets a DTC it will also illuminate a malfunction indicator light (MIL) to alert the driver that there is a system malfunction.
Códigos de diagnóstico de problemas (CDP) *Códigos que la computadora incorporada al automóvil señalará cuando detecte un problema con un sistema o componente. Éstos pueden recuperarse cuando se usa una herramienta de exploración que ayuda al técnico a localizar el problema. Cuando la computadora señala un CDP también se prenderá una luz indicadora de funcionamiento defectuoso (LIFD) que alerta al conductor sobre una falla en el sistema.*

Dial Bore Gauge An instrument used to measure the cylinder bore for wear, taper, and out of round.
Manómetro Dial bore gauge *Instrumento que se usa para medir el diámetro interior de la boca del cilindro para desgaste, despunte y descentrado.*

Dial Caliper A precision measuring instrument used to measure inside or outside diameter as well as depth.
Disco de calibre *Instrumento de medición precisa que se usa para medir el diámetro interior y exterior como también lo hondo.*

Dial Indicator An instrument used to measure the travel of a plunger in contact with a moving component. In an engine this is commonly used to check crankshaft run out or endplay.
Comparador mecánico *Herramienta que se usa para medir el desplazamiento de un pistón en contacto con otro componente en movimiento. Se usa comúnmente en un motor para revisar el juego axial del cigüeñal.*

Diesel Fuel A heavy fraction of petroleum used for combustion in diesel engines.
Combustible de Diesel *Fracción pesada de petróleo que se usa en la combustión de motores diesel.*

Distributor The mechanism within the ignition system that directs the secondary voltage to the correct spark plug.
Distribuidor *Mecanismo en el sistema de encendido que dirige el voltaje secundario a la bujía correcta.*

Duration The length of time, expressed in degrees of crankshaft rotation, that the valve is open.
Duración *Cantidad de tiempo, expresada en grados de rotación del cigüeñal, que la válvula está abierta.*

Dynamometer A device used to load an engine in order to test the engine's power output.
Dinamómetro *Mecanismo que se usa para cargar un motor con el propósito de probar la potencia de salida del motor.*

Electrolyte A solution that is capable of conducting electricity. The sulfuric acid used in most automotive batteries.
Electrolito *Solución que es conductor de electricidad. El ácido sulfúrico que se usa en la mayoría de las baterías automotrices.*

Electronic Ignition System An ignition system using electronic components to turn the ignition coil on and off to fire the spark plugs.
Sistema de encendido electrónico *Sistema de encendido que usa componentes electrónicos para prender y apagar la bobina de encendido para prender las bujías.*

Embeddability The ability for a material (the engine bearing) to conform and allow dirt and metal particles to embed themselves into the soft face of the bearing to protect the rotating shaft.
Fijabilidad *Habilidad de un material (del cojinete de eje de motor) de ajustarse y permitir que la mugre y las partículas metálicas se fijen en la superficie suave del cojinete para proteger el eje de rotación.*

Engine Bearings Bearings used inside an engine to prevent excessive wear of the crankshaft and camshaft.
Cojinetes del eje del motor *Cojinetes que se usan adentro del motor para prevenir el desgaste excesivo del cigüeñal y del árbol de levas.*

Engine Block The main structure of the engine that houses the pistons and crankshaft. Most other engine components attach to the engine block.
Bloque motor *Estructura principal del motor que alberga los pistones y el cigüeñal. La mayoría de los demás componentes se adhieren al bloque motor.*

Engine Coolant Temperature Sensor (ECT) A heat sensitive resistor placed in a coolant passage used to send an electrical signal to the PCM about the temperature of the coolant.
Detector de la temperatura del enfriador del motor (STEM) *Resistencia sensitiva al calor que se coloca en el pasaje del enfriador y que se usa para mandar señales eléctricas al STEM sobre la temperatura del enfriador.*

Engine Cradle Engine support frame used in a front-wheel drive vehicle with a transaxle.
Caballete de soporte del motor *Armazón de soporte del motor que se usa en un vehículo de tracción delantera con un transeje.*

Engine Crane A tool used to lift and lower an engine.
Grúa de motor *Herramienta que se usa para subir y bajar un motor.*

Engine Displacement A measure of the volume of the engine that the pistons move through.
Desplazamiento del motor *Medida del volumen del motor en el que se mueven los pistones.*

Engine Stand A special holding fixture that attaches to the back of the engine, supporting it at a comfortable working height. In addition, most stands allow the engine to be rotated for easier disassembly and assembly.
Banco para el motor *Artefacto especial para sujetar que se coloca en la parte trasera del motor, y que lo detiene a una altura conveniente para trabajar. Además la mayoría de los bancos permiten que el motor dé vueltas para desarmarlo y armarlo fácilmente.*

Environmental Protection Agency (EPA) A governmental agency responsible for protecting the environment. It sets many regulations for the automotive industry.
Secretaría de Protección del Ambiente (SPA) *Agencia gubernamental responsable de proteger el ambiente. Establece muchas reglas para la industria automotriz.*

Ethanol A grain alcohol containing oxygen, used as an octane and oxygen enhancer in gasoline.
Etanol *Alcohol de grano que contiene oxígeno, y que se usa para incrementar el octano y oxigeno en la gasolina.*

Ethylene Glycol The base fluid used in most automotive coolants.
Etilenglicol *Líquido base que se usa en la mayoría de los refrigerantes automotrices.*

Exhaust Gas Recirculation System (EGR) The EGR system recirculates a small amount of exhaust gas back into the engine to reduce engine temperatures and production of NOx emissions.
Sistema de recirculación del gas residual de escape (SRGRE) *El sistema de RGRE vuelve a circular una pequeña cantidad de gas residual de escape en el motor para reducir sus temperaturas y la producción de emisiones de NOx.*

Feeler Gauge Set Thin metallic strips of known thickness used for measuring.
Juego de calibrador de espesores *Tiras metálicas delgadas de grosor conocido que se usan para medir.*

Fillet Small, rounded corners machined on the edges of journals to increase strength.
Filetes *Pequeños rincones redondos hechos a máquina en las orillas de las muñequillas para aumentar la resistencia.*

Firing Order The order in which the cylinders in an engine fire.
Orden de encendido *La secuencia en que se encienden los cilindros de un motor.*

Flat Rate A set time that it should take a technician to complete a certain job. The customer is charged this amount no matter if it takes the technician a longer or shorter amount of time to get the job done. The flat rate times vary from vehicle manufacturers.
Tarifa de precio único *Un tiempo establecido que debe tomarle a un técnico para terminar cierto trabajo. Se le cobra al cliente una cantidad sin importar si al técnico le toma más o menos tiempo para terminar el trabajo. Los horarios de precio único varían entre los productores de vehículos.*

Flex Plate A stamped steel coupler bolted to the rear of the crankshaft. The flex plate provides a mount for the torque converter on an automatic transmission and gears around the edge into which the starter motor engages.
Placa flexible *Conectador de acero estampado que está atornillado en la parte trasera del cigüeñal. La placa flexible proporciona la montura para el convertidor de par en la transmisión automática y se engrana alrededor de la orilla en la que se ajusta el motor de encendido.*

Flywheel A heavy circular component located on the rear of the crankshaft that uses inertia to keep the crankshaft rotating smoothly between power strokes.
Volante *Un componente circular pesado que se encuentra en el trasero del cigüeñal y que usa inercia para mantener al cigüeñal dando vueltas suavemente entre los tiempos de combustión.*

Forced Induction A system that forces air into the intake to build more power out of the engine. Supercharging and turbocharging are two kinds of forced induction systems.
Inducción forzada *Un sistema que empuja aire en la admisión para generar más potencia en el motor. La sobrealimentación y la turbosobrealimentación son dos tipos de sistemas de inducción forzada.*

Freewheeling A freewheeling engine has clearance between the valves and pistons even when the timing mechanism breaks.
Marcha por efecto fuerza *Un motor en marcha por efecto fuerza tiene un huelgo entre las válvulas y pistones cuando se rompe el mecanismo cronométrico.*

Fuel Injection A system that injects charges of fuel under high pressure into the intake manifold near the cylinder.
Inyección *Sistema que inyecta el combustible bajo alta presión en el distribuidor de admisión cerca del cilindro.*

Fuel Pressure Regulator (FPR) An FPR regulates how much pressure is delivered to the fuel rail and injectors.
Regulador de la presión del combustible (RPC) *RPC regula cuánta presión se reparte al distribuidor de combustible y a los inyectores.*

Fuel Volatility A measure of how readily a fuel vaporizes.
Volatilidad del combustible *Medida de cuan rápidamente se evapora un combustible.*

Full Floating Wrist Pins Piston pins that are not anchored to the piston or to the connecting rod but are allowed to swivel in both components. They are held in place with snaprings.
Ejes del pistón de flotación completa *Ejes del pistón que no están ancladas al pistón o al eje de unión pero se les permite girar en ambos componentes.*

Gasket A rubber, felt, cork, or metallic material used to seal surfaces of stationary parts.
Junta *Material de hule, fieltro, corcho o metal que se usa para sellar superficies de partes estacionarias.*

Harmonic Balancer A component attached to the front of the crankshaft used to reduce the torsional or twisting vibration that occurs along the length of the crankshaft.
Equilibrador armónico *Componente pegado al frente del cigüeñal que se usa para reducir la torsión o vibración de torsión que sucede a lo largo del cigüeñal.*

Hazard Waste Material that could cause injury or death to a person, or could damage or pollute land, air, or water.
Residuo peligroso *Material que puede causar daño o muerte a una persona, o puede dañar o ensuciar la tierra, el aire o el agua.*

Headers Exhaust manifolds with long individual runners made of tube steel used as a method of improving engine breathing.
Cabezas *Colectores múltiples de escape con largos apoyos deslizantes individuales hechos de tubos de acero que se usan como método de mejorar la respiración del motor.*

Heater Core A small heat exchange unit similar to a radiator that is located inside the vehicle. Coolant runs through the heater core, allowing the vehicle to use this heat to warm the inside of vehicle.
Núcleo del calentador *Pequeña unidad de intercambio de calor similar a un radiador que se encuentra dentro del vehículo. El enfriador corre por el núcleo del calentador permitiendo que el vehículo use este calor para calentar el interior del vehículo.*

Helicoid A spiral thread used to replace damaged threads and retain the original size.
Helicoide *Un tornillo espiral que se usa para reemplazar los tornillos dañados y que retiene el tamaño original.*

Hemispherical Chamber A low turbulence, half circle-shaped combustion chamber used on many older high-performance engines and a few new ones. The valves are on either side of the combustion chamber to assist breathing and the spark plug is in the center to shorten flame travel.
Cámara hemisférica *Cámara de combustión en forma de medio círculo de baja turbulencia que se usa en muchos motores antiguos de alto rendimiento y en algunos nuevos. Las válvulas están en uno de los lados de la cámara de combustión para ayudar a respirar y la bujía está en el centro para acortar el deslizamiento de la flama.*

Horsepower The measure of the rate of work.
Caballo de fuerza *Medida del índice de trabajo.*

Hot Tank Cleaning tank that sprays chemical solutions onto the components along with soaking them.
Tanque caliente *Tanque limpiador que rocía soluciones químicas en los componentes a la vez que los remoja.*

Hub The hub allows a place for a wheel to be mounted to the axle.
Cubo *El cubo proporciona un lugar para montar la rueda en el eje.*

Hydrocarbon (HC) A chemical compound, made up of hydrogen and carbon. One of the toxic gases emitted from an IC engine.
Hidrocarburo (HC) *Composición química hecha de hidrógeno y carbono. Uno de los gases tóxicos que se emiten de un motor de CI.*

Hypereutectic Piston A piston with high silicon content used to increase strength without adding weight. It is also distorted minimally when heated.
Pistón hipereutéctico *Pistón con alto contenido de silicón que se usa para aumentar la resistencia sin añadir peso. Se distorciona mínimamente cuando se calienta.*

Ignition Timing The precise time at which the spark plugs are fired. This is usually determined by the PCM on modern vehicles.
Calado *El tiempo preciso en que se encienden las bujías. Esto lo determina generalmente el MCM de motores modernos.*

Impact Screwdriver A screwdriver that uses a force of a hammer to ratchet and unscrew hard to remove screws.
Destornillador de impacto *Destornillador que usa la fuerza de un martillo para trinquetar y remover tornillos difíciles.*

Intercooler A component that cools the compressed air that comes out of a turbocharger before it enters the intake of the engine.
Intercambiador de calor, refrigerador intermedio o intercooler *Componente que enfría el aire comprimido que sale del turbo cargador antes de que entre a la admisión del motor.*

Interference Angle Valve machining process in which the seat angle is 1/2 to 1 degree different from the valve face angle to provide a more positive seal.
Ángulo de interferencia *Proceso mecanizado de válvula en el que el asiento tiene un ángulo de 1/2 a 1 grado de diferencia del ángulo de la cara de la válvula para proporcionar un sello más positivo.*

Interference Engine An engine design in which, if the timing belt or chain breaks or is out of phase, the valves will contact the pistons.
Motor de interferencia *Diseño de motor en el que, si la banda de tiempo o la cadena se rompen o se salen de fase, la válvula se pone en contacto con los pistones.*

Internal Combustion (IC) Engines An engine that houses the chamber in which combustion occurs. All current gas and diesel motors used in production vehicles are IC engines.
Motores de combustión interna (CI) *Motor que alberga la cámara en donde sucede la combustión. Todos los motores de gas y de diesel usados corrientemente en vehículos de producción, son motores de CI.*

International Lubricant Standardization and Approval Committee (ILSAC) An organization that developed an oil rating that incorporates both the SAE and API oil ratings. If an oil meets the ILSAC standard the container will show a starburst symbol.
Comité Internacional de Aprobación y Estandarización de Lubricantes (CIAEL) *Organización que desarrolló un grado de aceite que incorpora tanto los grados de aceite de SAE que de API. Si un aceite iguala los estándares del CIAEL, el pote mostrará un símbolo de ráfaga de estrellas.*

Jack Stands Mechanical support devices used to hold the vehicle off the floor after it has been raised by the floor jack.
Bancos de trabajo *Artefactos de apoyo mecánico que se usan para detener el vehículo en el aire después de que lo levantó el gato.*

Journal A round, machine surface protected by bearings on the crankshaft or camshaft on which the shaft or connecting rods ride.
Muñequilla *Superficie mecanizada redonda protegida por cojinetes en el cigüeñal o el árbol de levas en los que se montan los ejes.*

Keepers Little metal inserts that hold the valve spring retainer in place.
Armaduras de protección de concentración *Pequeñas tiras metálicas que mantienen en su lugar el resorte retenedor de la válvula.*

Keys A small block insert used on shafts to prevent a gear or hub from spinning on the shaft.
Cuñas *Un pequeño bloque que se usa en ejes para prevenir que la palanca o el cubo giren en el eje o árbol.*

Kick Down Cable A cable that is used to shift down the transmission under hard acceleration.
Cable de cambio de velocidad *Cable que se usa para reducir la velocidad de la trasmisión bajo aceleración fuerte.*

Knock Sensors A piezo-electric sensor that is used to pickup noises (vibrations) from the cylinders due to engine misfire, pinging, or detonation.
Detectores de detonación *Detector piezoeléctrico que se usa para recoger ruidos (vibraciones) de los cilindros causados por el fallo del encendido, zumbido o detonación del motor.*

Lift The maximum distance the valve is lifted from its seat. This distance is determined by multiplying cam lobe lift by the rocker arm ratio.
Desnivel entre tramos *Distancia máxima que se levanta la válvula de su asiento. Esta distancia se determina al multiplicar el desnivel del lóbulo de leva por la proporción de la palanca del basculador.*

Lifters, Mechanical (Solid) or hydraulic connections between the camshaft and the valves. Lifters follow the contour of the camshaft lobes to lift the valve off its seat.
Levantadores mecánicos *(sólidos) o hidráulicos de conexión entre el árbol de levas y las válvulas. Los levantadores siguen el contorno de los lóbulos del árbol de levas para levantar la válvula de su asiento.*

Machinist's Rule A multiple scale ruler used to measure distances or components that do not require precise measurement.
Regla del maquinista *Regla de escala múltiple que se usa para medir distancias o componentes que no necesitan medición precisa.*

Magnafluxing A process used to check ferrous engine components for cracks. The component is magnetized, then sprayed with fluid containing metal particles. If a crack exists the metal particles will line up along the crack.
Aplicación del flujo magna o magnaflujo *Proceso que se usa para encontrar grietas en los componentes de motor ferroso. El componente se imana y luego se rocía con líquido que contiene partículas metálicas. Si hay una grieta, las partículas metálicas se forman a lo largo de la grieta.*

Main Bore The housing that is machined to receive a main bearing and hold the crankshaft in place.
Barreno principal *Albergue que se hace con máquina para recibir al cojinete principal y mantener al cigüeñal en su lugar.*

Main Cap A cap that forms the other half of the main bore holding the crank in place commonly using a two- or four-bolt system.
Funda principal *Funda que forma la otra mitad del interior principal que mantiene el cigüeñal en su lugar usando comúnmente un sistema de dos o cuatro tornillos.*

Major Thrust Side The side of the piston skirt that pushes against the cylinder wall during the power stroke.
Empuje lateral principal *El lado de la falda de un pistón que empuja hacia la pared del cilindro durante el tiempo de combustión.*

Malfunction Indicator Light (MIL) An amber light on the dashboard that is used to alert the driver that the onboard diagnostic system has detected a fault.
Luz indicadora de funcionamiento defectuoso (LIFD) *Luz color ámbar en el tablero usada para indicar al conductor que el sistema de diagnóstico incorporado ha detectado una falla.*

Manifold Absolute Pressure Sensor (MAP) A sensor that measures the pressure in the intake manifold used to inform the PCM about the load on the engine.
Detector de presión absoluta del colector (SPAC) *Detector que mide la presión en el colector de admisión usado para informar al MCM sobre la carga del motor.*

Margin A small surface on a valve between the head and the face of the valve to prevent the valve from burning.
Margen *Pequeña superficie en una válvula entre la cabeza y la cara de la válvula para prevenir que se queme la válvula.*

Mass Air Flow Sensor (MAF) The MAF sensor measures the mass of air going into the engine. The PCM uses this information to determine how much fuel to deliver.
Detector de flujo de aire (SFA) *El detector de FA mide la masa de aire que entra en el motor. El SFA usa esta información para determinar la cantidad de combustible que manda.*

Material Safety Data Sheets (MSDS) Forms that include information about the safety precautions, toxicity, and flammability of a product.
Hojas de datos de la seguridad del material (HDSM) *Formas que incluyen información sobre precauciones de seguridad, toxicidad e inflamabilidad de un producto.*

Methanol A natural gas also known as wood alcohol used to oxygenate fuel and improve octane or mixed in a 15% concentration with gasoline to create M85 fuel.
Metanol *Gas natural también conocido como alcohol de caña que se usa para oxigenar el combustible y para mejorar el grado de octano, o mezclado en una concentración al 15% con gasolina para crear el combustible M85.*

Metric System The international system of measurement.
Sistema métrico *Sistema internacional de medición.*

Micrometer A precision measuring instrument designed to measure outside, inside, or depth measurements.
Micrómetro *Instrumento de medición precisa diseñado para hacer mediciones exteriores, interiores y de profundidad.*

Misfire Circumstance when complete combustion does not occur in a cylinder. A total misfire means that no combustion occurred.
Falsa explosión *Circunstancia en que la combustión completa no sucede en el cilindro. Una falsa explosión total significa que no ocurrió la combustión.*

National Institute for Automotive Service Excellence (ASE) ASE provides the only nationally recognized certification of service technicians through standardized written tests.
Instituto Nacional para la Excelencia del Servicio Automovilístico (INESA) *INESA es la única que proporciona la certificación de técnicos de servicio reconocida nacionalmente mediante exámenes escritos estandarizados.*

Naturally Aspirated An engine without a forced induction system.
Aspiración natural *Motor sin sistema de inducción forzado.*

Normal Operating Temperature (NOT) The temperature at which the engine is designed to run. At NOT the pistons fit best in the cylinders for optimum performance and combustion.
Temperatura de operación normal (TON) *Temperatura en la cual se diseña un motor a correr. En TON los pistones quedan mejor en los cilindros dando rendimiento y combustión máximos.*

Occupational Safety and Health Administration (OSHA) OSHA sets regulations for workplace environments that protect the health and safety of workers.
Administración de Salud y Seguridad Ocupacional (ASSO) *ASSO establece regulaciones para ambientes laborales que protegen la salud y la seguridad de los trabajadores.*

Octane Rating A rating that describes the ability of a gasoline to resist engine knock.
Índice de octano *Índice que describe la habilidad de la gasolina en resistir la detonación del motor.*

Oil Consumption When oil gets by the piston rings and enters the combustion chamber it is consumed as it is burned during combustion. Blue smoke exiting the tailpipe is an indication of excessive oil consumption.
Consumo de aceite *Cuando el aceite llega a los segmentos del pistón y entra a la cámara de combustión, es consumido mientras se quema durante la combustión. El humo azul que sale del escape indica que hay un consumo excesivo de aceite.*

Oil Cooler Located below the radiator and similar to a small radiator, an oil cooler exchanges the heat of the oil to lower oil temperatures.
Radiador del aceite *Se encuentra debajo del radiador y es similar a un radiador pequeño; un radiador del aceite intercambia el calor del aceite para bajar las temperaturas del aceite.*

Oil Galleries The main oil supply lines formed in the engine block.
Galerías de aceite *Líneas principales de suministro de aceite que se forman en el bloque motor.*

Oil Pressure Gauge A gauge that is used to monitor the engine's oil pressure.
Manómetro de aceite *Manómetro que se usa para monitorear la presión del aceite del motor.*

Oil Pressure Relief Valve (PRV) The pressure relief valve in the oil pump assembly opens when a predetermined pressure is reached to control maximum oil pressure.
Válvula descongestionadora de aceite (VDA) *La válvula descongestionadora en el ensamblaje de la bomba del aceite se abre cuando llega a una presión predeterminada. Así hay control máximo de la presión del aceite.*

Oil Pressure Switch or Sending Unit A switch or sending unit is threaded into the side of the block into an oil passageway, used to send an electrical signal to a gauge or light on the dashboard.
Presóstato de seguridad de aceite o envase de envío *Un presóstato o envase de envío se enrosca a un lado del bloque dentro del pasaje del aceite y se usa para enviar una señal eléctrica al manómetro o a la luz en el tablero.*

Oil Pump A rotor- or gear-type positive displacement pump used to take oil from the engine sump and deliver it under pressure to the oil galleries. The oil galleries direct the oil to the high friction areas of the engine.
Bomba de aceite *Bomba de movimiento rectilíneo alternativo de tipo rotor o de palanca que se usa para tomar aceite de la caldera del motor y enviarlo bajo presión a las cámaras de aceite. Las cámaras de aceite lo dirigen a las áreas de alta fricción del motor.*

Organic Additive Technology A type of coolant that is free of silicates and phosphates. The additives extend the working life of the coolant.
Tecnología de preservante orgánico *Tipo de refrigerante que está limpio de silicatos y fósforos. Los preservantes extienden la vida activa del refrigerante.*

Oversquare Engine An engine in which the cylinder bore diameter is larger than its stroke. This generally allows the engine to revolve faster than an undersquare engine and may produce more horsepower.
Motor oversquare *Motor en el que el diámetro de la barrena del cilindro es mayor que su recorrido. Esto generalmente permite que el motor gire más rápido que un motor undersquare y puede producir más caballos de fuerza.*

Oxides of Nitrogen (NOX) NOx are toxic emissions produced in the IC engine particularly when combustion temperatures exceed 2500°F.
Monóxido de nitrógeno (NO_X) *NO_X son emisiones tóxicas producidas en un motor de CI particularmente cuando las temperaturas de combustión exceden los 2500°F.*

Oxygen Sensor (O_2S) A sensor that monitors the amount of oxygen present in the exhaust pipe. The PCM uses this information to carefully trim the rate of fuel delivery for minimum emissions and maximum fuel economy.
Detector de oxígeno (DO_2) *Detector que controla la cantidad de oxígeno presente en el tubo de escape. El MCM usa esta información para cortar cuidadosamente el suministro de combustible produciendo emisiones mínimas y economía máxima del combustible.*

Oxygenated Fuel Fuel that has added oxygen to create a better and leaner burn inside an engine.
Combustible oxigenado *Combustible al que se le ha añadido oxígeno para crear una mejor y más limpia calcinación dentro del motor.*

Pentroof Combustion Chamber A combustion chamber that places the intake and exhaust valves opposite each other for good cross flow of air and centers the spark plug in the chamber. It is a modification of the hemi head, adding corners to edges to improve turbulence and quenching.
Cámara inclinada de combustión *Cámara de combustión que coloca las válvulas de admisión y de escape una frente a la otra para mejor fluidez del aire y centra las bujías en la cámara. Es una modificación de la hemi head que añade rincones a las orillas para mejorar la turbulencia y esfuerzo de temple.*

PH Level A value expressing acidity or basicity in terms of the relative amounts of hydrogen ions (H+) and hydroxide ions (OH–) present in a solution. An appropriate PH level helps determine the condition of engine coolant.
Nivel de acidez *Valor que expresa la acidez o la alcalinidad en términos de las cantidades relativas de iones de hidrógeno (H+) e iones de hidróxido (OH–)presentes en una solución. El nivel de acidez apropiado ayuda a determinar la condición del refrigerante del motor.*

Pin Boss A strengthened bore machined into the piston that accepts the piston or wrist pin to attach the piston to the connecting rod.
Saliente interior del pistón *Perforación reforzada hecha con máquina en el pistón que acepta el pistón o el muñón (pasador de pie de biela, pasador del pistón, muñequilla del cigüeñal) para sujetar el pistón al eje.*

Piston An engine component in the form of a hollow cylinder that is enclosed at the top and open at the bottom. Combustion forces are applied to the top of the piston to force it down during the power stroke. The piston, when assembled to the connecting rod, is designed to transmit the power produced in the combustion chamber to the crankshaft.
Pistón *Componente del motor en forma de cilindro hueco que está cerrado en la parte superior y abierto en el fondo. Se aplican fuerzas de combustión a la parte superior del pistón para forzarlo a bajar durante el tiempo de combustión. El pistón, cuando se ensambla con el eje, está diseñado para trasmitir la fuerza que se produce en la cámara de combustión hacia el cigüeñal.*

Piston Pin Also called a wrist pin. A component that connects the piston to the connecting rod. There are three basic designs used: a piston pin anchored to the piston and floating in the connecting rod, a piston pin anchored to the connecting rod and floating in the piston, and a piston pin full floating in the piston and connecting rod.
Eje del pistón *También conocido como muñón (pasador de pie de biela, pasador del pistón, muñequilla del cigüeñal). Componente que conecta el pistón con el eje de conexión. Hay tres diseños básicos: el eje del pistón anclado al pistón y que flota en el eje de conexión, el eje del pistón anclado al eje de conexión y que flota en el pistón, el eje del pistón que flota completamente en el pistón y en el eje de conexión.*

Piston Rings Components that fit around the piston to seal the combustion pressure in the combustion chamber and prevent oil from entering the combustion chamber.
Segmentos del pistón *Componentes que se ajustan alrededor del pistón para sellar la presión de la combustión en la cámara de combustión e impiden que el aceite entre en la cámara de combustión.*

Piston Slap A sound that results from the piston hitting the side of the cylinder wall.
Ruido del pistón *Ruido que resulta cuando el pistón golpea la pared del cilindro.*

Pitch The pitch of a bolt is a measure of how many threads per inch on an English bolt or the distance between two threads in millimeters on a metric bolt.
Declive *El declive de un tornillo o perno es la medida del número de roscas por pulgada en un perno inglés o la distancia entre dos roscas en milímetros en un perno métrico.*

Plastigauge A string-like plastic of a precise thickness available in different diameters that is used to measure the clearance between a bearing and a journal. The diameter of the plastigauge is exact, thus any crush of the gauge material provides an accurate measurement of oil clearance.
Plastimanómetro *Plástico en forma de cadena de grosor preciso que se encuentra en diferentes diámetros y se usa para medir el espacio entre el cojinete y la muñequilla. El diámetro del plastimanómetro es exacto, por lo que cualquier compresión en el material del manómetro proporciona una medida exacta del deshago de aceite.*

Plenum A common chamber in the intake manifold designed to distribute the intake charge more evenly along the runners of the intake.
Cámara impelente *Cámara común en el colector de toma diseñada para distribuir la carga de entrada más uniformemente en los puntos deslizantes de toma.*

Positive Crankcase Ventilation System (PCV) The PCV system prevents toxic blow-by gases from being discharged into the atmosphere.
Sistema de ventilación del cárter positivo (SVCP) *El sistema de VCP impide que gases tóxicos emitidos se descarguen en la atmósfera.*

Positive Displacement Pump A pump that delivers the same amount of oil with every revolution, regardless of speed.
Bomba de movimiento rectilíneo alternativo *Bomba que reparte la misma cantidad de aceite con cada revolución, sin importar la velocidad.*

Power Train Control Module (PCM) The on-board computer that controls engine management, transmission functions, and self-diagnostics.
Módulo de control del motor (MCM) *Computadora incorporada que controla el manejo del motor, las funciones de la transmisión y los autodiagnósticos.*

Preignition Improper combustion that is the result of spark occurring too soon.
Preencendido *Combustión inapropiada que resulta cuando se enciende la chispa demasiado pronto.*

Pressure Cap The cap located on the top of the radiator or cooling system reservoir that seals the system and maintains pressure in the cooling system up to a designated pressure. The cap is also used to prevent a vacuum from forming in the radiator after the engine begins to cool.
Tapón de presión *Tapón colocado en la parte superior del radiador o sistema de refrigeración que sella el sistema y mantiene la presión en el sistema de refrigeración a una presión designada. El tapón se usa también para prevenir que se forme un vacío en el radiador después de que se empieza a enfriar el motor.*

Pressure Plate The pressure plate compresses the clutch against the flywheel to engage the clutch and allow power to flow from the engine to the transmission.
Placa de presión *La placa de presión comprime el cloche hacia el volante para ajustar el cloche y permitir que la energía fluya del motor a la transmisión.*

Puller A tool used to remove a pressed-on part such as a harmonic balancer to ensure that the part can be removed off the shaft without any damage.
Extractor *Herramienta usada para quitar una parte prensada tal como el equilibrador armónico para asegurarse que la parte se remueva del eje sin causar daños.*

Pushrod A connecting link between the lifter and rocker arm. Engines designed with the camshaft located in the block use pushrods to transfer motion from the lifters to the rocker arms.
Empujador largo *Enlace de conexión entre el levantador y la palanca de basculador. Los motores diseñados con el cigüeñal situado en el bloque usan empujadores largos para transferir movimiento entre los levantadores y las palancas de basculador.*

Quench The cooling of gases in the combustion chamber as a result of having a large cool surface area compared to the volume of gases. The quench area of the combustion chamber is at the small corner of a wedge combustion chamber.
Temple *El enfriamiento de los gases en la cámara de combustión como resultado de tener una gran área de superficie fría comparada con el volumen de los gases. El área de temple de la cámara de combustión está en un rincón pequeño acuñado de la cámara de combustión.*

Radiator A component consisting of a series of tubes and fins that transfer the heat from the coolant to the air.
Radiador *Componente que consiste en una serie de tubos y aletas que transfieren calor del refrigerante al aire.*

Reformulated Gasoline (RFG) RFG reduces the light and heavy compounds in gasoline that contribute to higher hydrocarbon emissions.
Gasolina reformulada (GRF) *GRF reduce los compuestos pesados y livianos en la gasolina que aportan altas emisiones de hidrocarburo.*

Reid Vapor Pressure (RVP) A measurement of heated gasoline pressure used to determine the fuel's volatility.
Presión del aire (método Reid) PAMR *Medida de presión de la gasolina calentada usada para determinar la volatilidad del combustible.*

Relay An electrical device that uses a low current signal to switch a high current circuit on and off.
Relé *Dispositivo eléctrico que usa una señal de baja corriente para prender y apagar un circuito de alta corriente.*

Release Bearing A bearing located in the clutch assembly that moves the pressure plate release levers or diaphragm spring during clutch engagement and disengagement.
Cojinete de tope *Cojinete localizado en el ensamblado del cloche que mueve la palanca de liberación de la placa de presión o el muelle de diafragma durante el embrague y el desembrague.*

Retrusion The sinking of the valve seat into the cylinder head.
Retrusión *Hundimiento del asiento de la válvula en la cabeza del cilindro.*

Ridge Reamer A cutting tool used to remove the ridge formed at the end of ring travel at the top of the cylinder.
Escariador de cresta *Herramienta para cortar usada para quitar las estrías que se forman al final del movimiento de la rosca en la parte superior del cilindro.*

Right-to-Know Laws Laws requiring employers to educate employees about workplace hazards and available protections.
Leyes de derecho de información *Leyes que requieren que los patrones eduquen a sus empleados sobre los riesgos del área de trabajo y las protecciones disponibles.*

Ring Compressor A tool that compresses the ring against the piston for piston installation into the cylinder.
Compresor de anillos *Herramienta que prensa el anillo contra el pistón para instalar el pistón en el cilindro.*

Ring End Gap The distance of the gap between the ends of the ring when installed in the cylinder. This is checked to ensure proper ring fit when installing new rings.
Ranura del segmento del anillo *Distancia de la ranura entre las orillas del anillo cuando se instala en el cilindro. Se revisa para asegurarse que el anillo quede bien cuando se instalan nuevos anillos.*

Ring Expander A tool that is used to spread a ring open to remove it from a piston.
Ensanchador de anillos *Herramienta usada para abrir el anillo cuando se quita del pistón.*

Ring Groove Cleaner A tool used to clean carbon off the groove that a ring sets in.
Limpiaranuras de anillos *Herramienta que se usa para limpiar las ranuras de carbón donde se coloca el anillo.*

Ring Ridge A small ridge left on a cylinder wall after an engine has been worn by the piston rings. The cylinder wall has been worn by the rings except for this spot, leaving a small ridge at the top of the cylinder. This needs to be removed before removing the pistons.
Estría del anillo *Pequeña estría que queda en la pared del cilindro después de que los anillos del pistón han gastado el motor. Los anillos gastan la pared del cilindro excepto este lugar en el que queda una pequeña estría en la parte superior del cilindro. Necesita quitarse antes de quitar los pistones.*

Ring Side Clearance The clearance on the side of the ring when seated in the ring groove. Too little clearance can cause the ring to seize as it expands. Too little clearance will prevent the ring from sealing properly.
Espacio libre del lado del anillo *Espacio lateral libre del anillo cuando sentado en la ranura del anillo. Tener poco espacio libre puede causar que el anillo se agarrote cuando se expanda. Poco espacio libre va a prevenir que se cierre apropiadamente el anillo.*

Rocker Arm A pivot that transfers the motion of the pushrods or followers to the valve stem.
Palanca de basculador *Pivote que transfiere el movimiento de los empujadores largos o los rodillos de leva al vástago de la válvula.*

RTV (Room Temperature Vulcanization) Sealer A gasket making material that is used instead of a gasket. This material dries in air with the help of moisture.
Obturador (vulcanizador a la temperatura ambiente) VTA *Material de junta que se usa en lugar de una junta. Este material se seca al aire con ayuda de la humedad.*

Scan Tool A tool used to interface with the PCM and extract engine sensor data and DTCs. On OBDII vehicles each vehicle uses a standardized connection to hook up the scan tool.
Herramienta exploratoria *Herramienta que se usa para interconectarse con el MCM y sacar datos del detector del motor y los CDP. En los vehículos OBDII cada uno usa una conexión estandarizada para engancharse a la herramienta exploratoria.*

Scuffing Scraping and heavy wear between two surfaces, typically between the piston and cylinder wall.
Desgaste por fricción *Residuo y desgaste severo entre dos superficies, generalmente entre el pistón y la pared del cilindro.*

Seal A seal is used between a stationary part and a moving one to prevent fluid leakage.
Obturación *Se usa entre las partes estacionarias y móviles para prevenir el goteo de líquido.*

Seal Driver A tool used to safely install seals.
Controlador de obturación *Herramienta que se usa para instalar obturaciones de manera segura.*

Set Screws Small screws used to keep a part stationary on a shaft.
Tornillos sin tuerca *Tornillos o pernos pequeños que se usan para mantener una parte estacionaria en el eje.*

Short Block The engine block without cylinder heads installed.
Bloque corto *Bloque motor que no tiene instaladas las cabezas del cilindro.*

Slave Cylinder A hydraulic piston assembly that actuates the clutch release lever.
Cilindro subordinado *Ensamblado de pistón hidráulico que acciona la palanca de embrague.*

Society of Automotive Engineers (SAE) SAE sets many standards within the automotive industry.
Sociedad de Ingenieros Automotrices (SIA) *SIA establece muchos estándares en la industria automotriz.*

Squish The area of the combustion chamber where the piston is very close to the cylinder head. The air/fuel mixture is rapidly pushed out of this area as the piston approaches TDC, causing turbulence and forcing the mixture toward the spark plug. The squish area can also double as the quench area.
Squish *Área en la cámara de combustión en donde el pistón está muy cerca a la cabeza del cilindro. La mezcla de aire y combustible sale rápidamente de esta área cuando el pistón se acerca al PMS, causando turbulencia y forzando la mezcla hacia la bujía. El área squish puede funcionar como el área de temple.*

Standard System United States customary system of measurement.
Sistema estándar o normalizado *Sistema estadounidense común de medidas.*

Static Electricity An electrical charge that forms with friction but produces no electron flow.
Electricidad estática *Carga eléctrica que se forma con fricción pero que no produce flujo eléctrico.*

Stethoscope A tool used to locate the source of engine and other noises.
Estetoscopio *Herramienta que se usa para localizar la fuente del motor y otros ruidos.*

Stoichiometric Ratio The theoretically perfect mixture of air and fuel to support combustion, 14.7 parts of air to one part of fuel.
Relación estequiométrica *Mezcla teóricamente perfecta de aire y combustible para apoyar la combustión, 14.7 partes de aire por una parte de combustible.*

Straightedge A tool machined straight to allow checks for component warpage. A straightedge is placed on a cylinder head and feeler gauges are inserted into any gaps to measure warpage.
Regla rodante *Herramienta exacta hecha a máquina que permite revisar torceduras de los componentes. Se coloca una regla rodante en la cabeza del cilindro y se insertan calibradores de espesores en cualquier abertura para medir las torceduras.*

Straight Time A pay system where technicians get paid for the amount of hours they work rather than by the number of billable hours they produce.
Tiempo registrado *Sistema de pago en el que se le paga al técnico por las horas que trabajó en lugar del número de horas facturables que produce.*

Stroke The distance traveled by the piston from TDC to BDC and vice versa.
Recorrido *Distancia que recorre el pistón de su PMS a su PMI y viceversa.*

Supercharger A belt driven pump that is used to force air into an engine's intake.
Sobrecargador *Bomba movida por banda que se usa para forzar aire en la admisión del motor.*

Surface to Volume (S/V) Ratio The amount of surface area in a combustion chamber compared to the volume of the chamber. A combustion chamber with a lower S/V ratio tends to emit fewer hydrocarbons.
Índice de superficie a volumen (S/V) *La cantidad del área de superficie en una cámara de combustión comparada con el volumen de la cámara. Una cámara de combustión con un índice de S/V bajo tiende a emitir menos hidrocarburos.*

Synthetic Oils Man-made oil or specially blended oil that offers better lubricating qualities at high and low temperatures.
Aceites sintéticos *Aceite fabricado o especialmente mezclado que ofrece mejor calidad de lubricación a altas y bajas temperaturas.*

Telescoping Gauge A precision tool used in conjunction with outside micrometers to measure the inside diameter of a hole. Telescoping gauges are sometimes called snap gauges.
Manómetro o calibre telescópico *Herramienta de precisión que se usa conjuntamente con micrómetros exteriores para medir el diámetro interior de un agujero. A los manómetros o calibres telescópicos algunas veces se les llama calibres de mordaza.*

Thermostat A mechanical device that opens and closes to allow the engine to reach normal operating temperature quickly and help maintain the desired temperature.
Termostato *Aparato o mecanismo que se abre y se cierra para permitir que el motor alcance una temperatura de operación normal rápidamente y ayuda a mantener la temperatura deseada.*

Thread Sealer A compound used to seal threads that may enter a water or oil passage.
Sellador de roscado *Compuesto que se usa para sellar el roscado que puede entrar en un pasaje de agua o de aceite.*

Thrust Bearing A double-flanged bearing used to cushion the movement of the crankshaft forward and backward in the engine.
Cojinete de empuje *Cojinete con enfaldillado doble que se usa para amortiguar el movimiento hacia enfrente y hacia atrás del cigüeñal en el motor.*

Timing Chain Tensioner A device, typically hydraulic/mechanical, that is used to maintain proper tension on the timing chain.
Tensor de la cadena de tiempo *Aparato, típicamente mecánico/hidráulico que se usa para mantener la tensión apropiada en la cadena de tiempo.*

Timing Mechanism The components used to maintain a proper coordination of timing between the crankshaft, camshaft, and balance or auxiliary shafts.
Mecanismo de tiempo *Los componentes que se usan para mantener la coordinación apropiada del tiempo entre el cigüeñal, el árbol de levas y los ejes de balance o auxiliares.*

Top Dead Center (TDC) Term used to indicate that the piston is at the very top of its stroke.
Punto muerto superior (PMS) *Término que se usa para indicar que el pistón está en su golpe más alto.*

Torque A rotating force around a pivot point; for example, the twisting force applied to a bolt or shaft. The twisting force of the crankshaft is the engine's torque.
Torque o par motor *Fuerza de rotación alrededor del punto de giro; por ejemplo, la fuerza de torsión que se aplica a un perno o a un eje. La fuerza de torsión del cigüeñal es la torsión del motor.*

Torque Angle Gauge A gauge used to measure the number of degrees a bolt is turned.
Torsiómetro de ángulo *Manómetro o calibre que se usa para medir el número de grados que se gira un perno.*

Torque Converter A fluid coupling between the engine and the automatic transmission used to multiply engine torque.
Convertidor del par motor *Acoplamiento por líquido entre el motor y la transmisión automática que se usa para multiplicar la torsión del motor.*

Torque Plate A thick metal block bolted to the cylinder block at the cylinder head mating surface to prevent twisting during honing and boring operations.
Placa de par *Bloque metálico grueso atornillado al bloque del cilindro en la superficie de ajuste de la cabeza del cilindro para prevenir la torsión durante las operaciones de bruñido y mandrinado.*

Torque-O-Meter A torque wrench with a dial on it to visually watch the torque being applied.
Torsiómetro de indicador *Llave de torsión con un indicador instalado para ver literalmente cuando se aplica la torsión.*

Torque-to-Yield Bolts Bolts that are torqued to within 2% of their yield strength.
Pernos torque to yield *Pernos que se tuercen hasta el 2% de su límite de elasticidad.*

Torque Wrench A wrench that measures the amount of twisting force (torque) applied to a fastener.
Llave tensiométrica *Llave que mide la cantidad de la fuerza de torsión que se aplica a un remache.*

Transaxle A transmission that incorporates a differential typically used on FWD vehicles.
Transeje *Transmisión que incorpora un diferencial que se usa generalmente en vehículos de tracción delantera.*

Transmission A device that transmits power to the wheels and multiplies torque from the engine.
Transmisión *Aparato que transmite potencia a las ruedas y multiplica la tensión que viene del motor.*

Transmission Bell Housing A housing at the front of the transmission that bolts to the engine. The transmission bell housing gives protection to the clutch or torque converter.
Cárter de caja de cambio *Caja en el frente de la transmisión que se atornilla al motor. El cárter de caja de cambio da protección al cloche o al convertidor del par motor.*

Turbocharger A small pump driven by exhaust gases to force air into an engine's intake.
Turboalimentador *Bomba pequeña impulsada por gases quemados que forzan aire en la admisión del motor.*

Turbulence Rapid movement of the air/fuel mixture.
Turbulencia *Movimiento rápido de la mezcla del aire y combustible.*

Under Square Engine An engine in which the stroke is bigger than the bore. This design potentially delivers higher torque but may not spin as fast as an over square engine.
Motor under square *Motor en el que el golpe es mayor que la superficie interior. Este diseño da un torque o par de mayor potencia pero puede que no gire tan rápido como el motor over square.*

Vacuum A pressure lower than atmospheric pressure. Vacuum in the engine is created when the volume of the cylinder above the piston is increased as the piston moves from TDC to BDC.
Vacío *Presión menor que la presión atmosférica. El vacío en el motor se crea cuando el volumen del cilindro sobre el pistón se incrementa mientras el pistón se mueve de su PMS al PMI.*

Vacuum Gauge A tool that is used to measure the amount of vacuum in the intake manifold of an engine.
Vacuómetro o indicador de vacío *Herramienta que se usa para medir la cantidad de vacío en el colector de admisión de un motor.*

Valve A component that opens and closes to control the flow of gases into and out of the engine cylinder.
Válvula *Componente que abre y cierra el control de paso de gases hacia adentro y hacia fuera del cilindro del motor.*

Valve Face The part of the valve seals on the cylinder head seat that seals the combustion chamber.
Cabeza de la válvula *Esta parte de la válvula se junta herméticamente en el asiento de la cabeza del cilindro para obturar la cámara de combustión.*

Valve Float A condition that allows the valve to remain open longer than it is intended. At high rpm the valve spring may not be able to hold the valve on the profile of the camshaft.
Flotación de la válvula *Condición que permite que la válvula se mantenga abierta más de lo necesario. En un alto índice de rpm el muelle de la válvula puede que no detenga la válvula en el perfil del árbol de levas.*

Valve Follower Followers are used on many overhead camshaft (OHC) engines as a direct link between the camshaft and the valves.
Casquillo de la válvula *Casquillos que se usan en muchos motores de árbol de levas a la cabeza (ALC) como un enlace directo entre el árbol de levas y las válvulas.*

Valve Guide A part of or a sleeve in the cylinder head that supports and guides the valve stem.
Guía de la válvula *Parte o camisa en la cabeza del cilindro que apoya y guía el vástago.*

Valve Overlap The length of time, measured in degrees of crankshaft revolution, that the intake and exhaust valves of the same combustion chamber are open simultaneously.
Solapamiento de la válvula *Período de tiempo, medido en grados de la revolución del cigüeñal, en el que las válvulas de admisión y las de escape de la misma cámara de combustión se abren simultáneamente.*

Valve Seals Valve seals slide over the valve stem to keep oil out of the combustion chamber.
Obturaciones de la válvula *Las obturaciones de la válvula se resbalan en el vástago para mantener el aceite fuera de la cámara de combustión.*

Valve Seat A machined surface of the cylinder head that provides the mating surface for the valve face. The valve seat can be either an integral machined part of the cylinder head or a pressed-in insert.
Asiento de la válvula *Superficie hecha a máquina de la cabeza del cilindro que proporciona una superficie de ajuste a la cabeza de la válvula.*

Valve Spring Compressor A tool used to compress valve springs, assisting installation and removal of the springs.
Compresor del muelle de válvula *Herramienta que se usa para comprimir los muelles de válvula y que ayudan a instalar y quitar los muelles.*

Valve Spring Tester A device used to measure the valve spring pressure.
Probador del resorte de la válvula *Aparato que se usa para medir la presión del resorte de la válvula.*

Valve Timing The timing when the valves will open and close as defined by the camshaft and crankshaft timing.
Reglaje de las válvulas *Reglaje de cuando las válvulas se abren y se cierran como lo define el reglaje del árbol de levas y del cigüeñal.*

Valve Tip The very end of the valve stem. The rocker arm or follower rides on the often hardened valve tip.
Hongo o pie *El final del vástago. El eje o tapa está montado en la cabeza comunmente endurecida de la válvula.*

Valvetrain Clatter A high pitched clattering noise heard from the valve train due to worn valvetrain components or misadjusted valves.
Ruido del tren de válvulas *Un ruido de alto tono que se escucha del tren de válvulas producido por los componentes del tren de válvulas ya gastados o por válvulas mal ajustadas.*

Variable Valve Timing A timing mechanism that changes valve timing during different engine operating conditions to improve performance, emissions, and fuel economy.
Reglaje de válvula variable *Mecanismo de reglaje que cambia el reglaje de la válvula durante las diferentes condiciones de operación del motor para mejorar el funcionamiento, las emisiones y el ahorro del combustible.*

Vee Blocks Machined blocks with a vee notched into them used for spinning shafts to check for run out.
Bloques o cilindros en V *Bloques hechos a máquina que tienen una v muescada y que se usan con los bloques giratorios para revisar el mal funcionamiento.*

Vehicle Identification Number (VIN) An alphanumeric code consisting of seventeen characters used to properly identify the vehicle and its major components.
Número de identificación del vehículo (NIV) *Código alfanumérico que consiste de 17 caracteres que se usan para identificar apropiadamente un vehículo y sus componentes principales.*

Viscosity A rating of an oil's resistance to flow.
Viscosidad *Proporción de la resistencia del aceite para fluir.*

Volumetric Efficiency (VE) How well an engine breathes. The actual amount of air taken into the cylinder compared to the amount of air the cylinder can fit.
Rendimiento volumétrico (RV) *Eficacia de respiración del motor. La cantidad actual de aire que entra en el cilindro comparada con la cantidad de aire que puede caber en el cilindro.*

Warranty Repair A repair that is free of charge to the customer because the manufacturer guaranteed it would work properly or be replaced free of charge for a certain time period or mileage.
Reparación de garantía *Reparación sin costo para el cliente ya que el fabricante garantizó que funcionaría apropiadamente o se reemplazaría sin costo por un período de tiempo o millaje.*

Waste Gate A relief valve that will reduce the build up of boost pressure from a turbocharger.
Válvula de descarga *Válvula de seguridad que reducirá el aumento de la presión de alimentación de un turboalimentador.*

Wedge Combustion Chamber A combustion chamber that is shaped like a wedge, creating turbulent airflow and providing a squish and quench area.
Cámara de combustión en cuña *Cámara de combustión en forma de cuña que crea un flujo de aire turbulento y provee un área squish y un área de temple.*

Wrist Pin Also called the piston pin, the pin that connects the piston to the connecting rod.
Eje del pistón *Eje del pistón; eje que conecta al pistón con la biela.*

Index

D

E

P

Q

R

S

W